Concrete Structures

Repair, Rehabilitation and Retrofitting

Concrete Structures

Repair, Rehabilitation and Retrofitting

J Bhattacharjee
MTech (Str. Engg), MPhil

Professor and Advisor
Amity University
Noida, UP

ex-Joint Director General
Military Engineering Service (MES)
Ministry of Defence
Govt of India

CBS Publishers & Distributors Pvt Ltd

New Delhi • Bengaluru • Chennai • Kochi • Kolkata • Lucknow • Mumbai
Hyderabad • Jharkhand • Nagpur • Patna • Pune • Uttarakhand

Concrete Structures
Repair, Rehabilitation and Retrofitting

ISBN: 978-93-85915-90-1

Copyright © Author and Publisher

First Edition: 2017

Reprint: 2019, 2022, 2023

Published by **Satish Kumar Jain** and produced by **Varun Jain** for

CBS Publishers & Distributors Pvt Ltd

4819/XI Prahlad Street, 24 Ansari Road, Daryaganj, New Delhi 110 002, India.
Ph: 011-23289259, 23266861

Website: www.cbspd.com
e-mail: delhi@cbspd.com

Corporate Office: 204 FIE, Industrial Area, Patparganj, Delhi 110 092
Ph: 011-4934 4934   Fax: 011-4934 4935

e-mail: publishing@cbspd.com; publicity@cbspd.com

Branches

- **Bengaluru:** Seema House 2975, 17th Cross, KR Road, Banasankari 2nd Stage, Bengaluru 560 070, Karnataka, India
  Ph: +91-80-26771678/79   Fax: +91-80-26771680   e-mail: bangalore@cbspd.com
- **Chennai:** 7, Subbaraya Street, Shenoy Nagar, Chennai 600 030, Tamil Nadu, India
  Ph: +91-44-26680620, 26681266   Fax: +91-44-42032115   e-mail: chennai@cbspd.com
- **Kochi:** 42/1325, 1326, Power House Road, Opp KSEB, Power House, Ernakulum Kochi 682 018, Kerala, India
  Ph: +91-484-4059061-65,67   Fax: +91-484-4059065   e-mail: kochi@cbspd.com
- **Kolkata:** 147, Hind Ceramics Compound, 1st Floor, Nilgunj Road, Belghoria, Kolkata-700056, West Bengal, India
  Ph: +033-25633055, 033-25633056   e-mail: kolkata@cbspd.com
- **Lucknow:** Basement, Khushnuma Complex, 7 Meerabai Marg (Behind Jawahar Bhawan),Lucknow-226001, UP, India
  Ph: +0522-4000032   e-mail: tiwari.lucknow@cbspd.com
- **Mumbai:** PWD Shed, Gala no 25/26, Ramchandra Bhatt Marg, Next to JJ Hospital Gate no. 2, Opp. Union Bank of India, Noorbaug, Mumbai-400009, Maharashtra, India
  Ph: 022-66661880/89   e-mail: mumbai@cbspd.com

Representatives

- Hyderabad   0-9885175004
- Patna   0-9334159340
- Jharkhand   0-9811541605
- Pune   0-9923910676
- Nagpur   0-9421945513
- Uttarakhand   0-9716462459

Printed at: Rashtriya Printers, Dilshad Garden, Delhi, India

Foreword

The book *Concrete Structures: Repair, Rehabilitation and Retrofitting* is a well-organized and comprehensive textbook, which will be useful not only for the undergraduate and postgraduate students but also for the young teachers and practising civil engineers. I foresee this book to fill a long-standing gap in the neglected field of civil engineering.

The book is divided into six parts which comprise 28 chapters. The six parts deal with durability and deterioration of concrete structures; damage assessment; repair materials; repair and rehabilitation of concrete structures; retrofitting of structures; and case studies of various types of concrete structures. I appreciate the author's insight into the subject as I discover, on going through the script, that he has very well measured the length and breadth of the subject based on his long practical experience before attempting this book.

The author, Prof J Bhattacharjee, has written a number of papers for various national/international seminars/conferences and journals on various topics of civil engineering. He has more than 44 years of professional experience in various government and private sectors as a practicing engineer including teaching experience. He has worked in defence (MES) and retired as Joint Director General. He worked as a Head (Engineering) in the multinational consultancy firm Gherzi Eastern Ltd for more than four years after retirement. He is Professor and Advisor in Department of Civil Engineering, Amity University, Noida, UP, for the last more than four years, and has worked for sometime as the head of the department.

It gives me great pleasure to write the Foreword to this book written by a veteran professional in the field of civil engineering. I heartily congratulate the author for his path-breaking effort in producing this pioneering scholarly work and wish him all the very best.

KB Rajoria
Former Engineer-in-Chief
PWD, Delhi

Preface

The rapid progress and changes in the field of maintenance, repair, rehabilitation and retrofitting of concrete structures in the recent past have motivated me to write this book, based on my personal experience in the field. The book will serve as a textbook for civil engineering undergraduate and postgraduate students for their study, and practicing engineers working on site for actual execution. Some of the institutions have introduced separate courses as the repair, rehabilitation and retrofitting of the concrete structures. The main purpose of this text is to provide sufficient information on the state-of-art relating to repair, rehabilitation and retrofitting of concrete structures. This field is a major branch of civil and structural engineering and is getting momentum everyday. Thus, to keep up with the developments, I have tried to consolidate the information in a concise form, which will be useful for students preparing for their undergraduate, postgraduate and competitive examinations and also for the practicing engineers.

This book is divided into six parts which comprise 28 chapters. The six parts deal with durability and deterioration of concrete structures, damage assessment, repair materials, repair and rehabilitation of concrete structures, retrofitting of structures, and case studies of various types of concrete structures. At the end of each part, short answer and theory questions have been provided to enhance the knowledge of students. Hopefully this book will be useful to the students, faculty members and practicing engineers.

As most of the information has been derived from many sources, including those which are specially mentioned in the bibliography, I wish to express my indebtedness to these individuals and publications for the contribution. Further thanks are recorded to the Bureau of Indian Standards to quote certain regulations. The efforts have been made to avoid errors to the best of my ability, utilizing my personal experience in the subject. However, I will be grateful if mistakes or errors, if any, are pointed out and also welcome suggestions for improvement.

Special thanks to the Vice Chancellor and Director (ASET), Amity University, Noida, and Head of Department of Civil Engineering, Amity University, for their encouragement and motivation.

Finally, I am grateful to my wife Lakshmi and children Meghna and Vivek for their forbearance and patience in putting up with inconvenience, when I was engaged in the preparation of the text.

I am thankful to CBS Publishers and Distributors, especially I would like to put on record the sincere efforts of Mr YN Arjuna and his team comprising Mrs Ritu Chawla, Mrs Poonam Kapoor Bhatia, Mr Manish Raj and Mr Parmod Kumar for bringing out the book in the present form.

J Bhattacharjee

Contents

Foreword by KB Rajoria .... *v*

Preface .... *vii*

PART 1: DURABILITY AND DETERIORATION OF CONCRETE STRUCTURES

1. Understanding Concrete and RCC Structures — 3–4

1.1 Introduction   3
1.2 What is to be Learnt   3

2. Physical Causes — 5–17

2.1 Introduction   5
2.2 Durability Aspect of Concrete   5
2.3 Major Causes for Distress in Concrete   6
2.4 Shrinkage Problem in Concrete   8
2.5 Freeze and Thaw on Concrete   10
2.6 Weathering Action on Concrete   11
2.7 Crazing on Concrete   12
2.8 Honeycombing on Concrete   12
2.9 Swelling Effect of Concrete   12
2.10 Pop Outs on Concrete   12
2.11 Creep Problem on Concrete   13
2.12 Abrasion, Erosion and Cavitation on Concrete   13
2.13 Temperature Variation   13
2.14 Fire Effect on Concrete   14
2.15 Thermal Movement in Concrete   14
2.16 Subgrade Movement   15
2.17 Formwork Movement   15
2.18 Settlement and Movement   15
2.19 Foundation Settlement   15
2.20 Design Errors   15
2.21 Construction Errors   16
2.22 Cracks due to Construction Overload   16
2.23 Cracks due to Externally Applied Loads   17
2.24 Accidental Loadings Effect   17

3. Chemical Causes — 18–23

3.1 Chemical Attack on Concrete   18
3.2 Carbonation Attack on Concrete   19
3.3 Chloride Attack on Concrete   19
3.4 Sulphate Attack on Concrete   19

3.5   Acid Attack on Concrete   20
3.6   Alkali Reaction on Concrete   20
3.7   Hydrolysis and Leaching on Concrete   20
3.8   Salt Attack on Concrete   20
3.9   Aggressive Water Attack   21
3.10   Crystallization of Salts in Pores   21
3.11   Seawater Attack on Concrete   22
3.12   Biological Attack on Concrete   22
3.13   Mechanism of Miscellaneous Chemical Attack   23

4. Corrosion   24–33

4.1   Basic Principles of Corrosion   24
4.2   Corrosion Mechanism of Embedded Metals   25
4.3   Corrosion Process   26
4.4   Symptoms to Causes of Distress and Deterioration   29
4.5   Damages due to Corrosion   30
4.6   Codal Provisions for Different Exposure Conditions   31
4.7   Corrosion Protection Technique   32

5. Evaluation of Concrete and Reinforced Concrete Structures   34–38

5.1   Introduction   34
5.2   Reviewing the Records   34
5.3   Site Survey   34
5.4   Shoddy Workmanship   35
5.5   Pattern Cracking   35
5.6   Isolation Cracks   35
5.7   Disintegration and Spalling   36
5.8   Erosion   36
5.9   Joint Seals and Seepage   36
5.10   Special Cases of Spalling   37
5.11   Crack Surveys   37
5.12   Surface Mapping   38
5.13   Joint Inspections   38
5.14   Core Drilling   38

6. Causes of Distress and Deterioration of Concrete and Reinforced Concrete Structures   39–47

6.1   Introduction   39
6.2   Chemical Reactions   39
6.3   Causes of Distress and Deterioration of Concrete   40
6.4   Symptoms to Causes of Distress and Deterioration   44
       Short Answer Questions   45
       Theory Questions   46

PART 2: DAMAGE ASSESSMENT

7. Condition Survey and Nondestructive Testing System   51–64

7.1   Condition Survey   51
7.2   Nondestructive Testing Methods   51
7.3   Recent Developments on NDT Instruments   62

8. Semidestructive Testing System   65–93

8.1 Introduction   65
8.2 Chemical Testing of Concrete   83
8.3 Diagnostic Methods of Corrosion Damage   84
8.4 Investigation Strategies   85
8.5 Determination of Concrete Quality and Composition   89
8.6 Systematic Assessment of Fire Affected Structures   89

9. Destructive Testing System   94–105

9.1 Introduction   94
9.2 Investigation of Damage   98
9.3 Testing System of Hardened Concrete   100
9.4 Evaluation of Cracks   100
9.5 Destructive Testing Systems   101
    Short Answer Questions   103
    Theory Questions   104

PART 3: REPAIR MATERIALS

10. Materials for Repair   109–111

10.1 Introduction   109
10.2 Chemical Admixtures: Classification   109
10.3 Essential Parameters of Coating   110
10.4 Repair Materials (Special)   110

11. Function of Repair Materials   112–133

11.1 Patching Materials   112
11.2 Resurfacing Materials   117
11.3 Sealing Materials   124
11.4 Waterproofing Materials   128
11.5 Bonding Materials   131

12. Special Types of Repair Materials   134–158

12.1 Introduction   134
12.2 Chemical Admixtures   135
12.3 Mineral Admixtures   138
12.4 Fly Ash   138
12.5 Ground Granulated Blast Furnace Slag (GGBS)   139
12.6 Condensed Silica Fume (CSF)   139
12.7 Admixtures for Repair/Rehabilitation   139
12.8 Epoxy Resins   140
12.9 Polymeric Materials   141
12.10 Organic Polymers   142
12.11 Polymeric Repair Materials   144
12.12 Fiber Reinforced Concrete   146
12.13 Behavior of Steel Fiber Reinforced Concrete   147
12.14 Application of SFRC to Repair of Distress Structures   149
12.15 Behavior of FRC with Other Fibers   149
12.16 Fibre Reinforced Polymer Composites   152
12.17 Ferrocement   154

12.18 SIFCON and SIMCON Materials   156
12.19 Miscellaneous Materials   157
12.20 Requirements for the Repair Materials   158

13. Selection and Evaluation of Materials

159–171

13.1 Introduction   159
13.2 Selection of Repair Materials   159
13.3 Classification of Repair Materials   161
13.4 Evaluation Tests for Repair Materials   161
       Short Answer Questions   169
       Theory Questions   170

PART 4: REPAIR AND REHABILITATION OF CONCRETE STRUCTURES

14. Planning and Designing of Concrete Repair

175–178

14.1 Introduction   175
14.2 Planning a Repair   176
14.3 Causes and Repair Approaches for Spalling and Disintegration   177
14.4 Repair Methods for Spalling and Disintegration   177

15. Repair of Cracks

179–190

15.1 Introduction   179
15.2 Repair Procedure   180
15.3 Durable Repair Design   181
15.4 Durable Repair Application   181
15.5 Repair Method   182
15.6 Methods of Repair   182

16. Concrete Removal and Preparation for Repair

191–202

16.1 Introduction   191
16.2 General Classification of Concrete Removal Methods   191
16.3 Removal Methods   191

17. Strategies and Techniques for Repair/Rehabilitation/Retrofitting of Structures

203–208

17.1 Introduction   203
17.2 Techniques to Restore Original Strength   204
17.3 Strengthening RC Members   205
       Short Answer Questions   206
       Theory Questions   208

PART 5: RETROFITTING OF STRUCTURES

18. Techniques for Strengthening/Retrofitting

211–226

18.1 Introduction   211
18.2 Structural Concrete Strengthening   213
18.3 Strengthening with External Reinforcement   217
18.4 Short Spanning   218
18.5 External Post-tensioning   218
18.6 Section Enlargement   219

18.7 Strengthening by SIMCON 219
18.8 Dam Safety: Concrete Repair/Retrofitting Techniques 220
18.9 General Guidelines for Seismic Rehabilitation of Existing Buildings 222

19. Repair/Retrofit of Nonengineered Structures 227–231

19.1 Introduction 227
19.2 Repair Materials 227
19.3 Repair Techniques 228
19.4 Strengthening of Walls 229
19.5 Strengthening of Pillars 230
19.6 Techniques for Global Strengthening 230

20. Retrofit of Historical and Heritage Structures 232–235

20.1 Introduction 232
20.2 Basic Principles of Structural Restoration 232
20.3 Condition Assessment 232
20.4 Strengthening of Masonry Walls 232
20.5 Strengthening of Arches, Vaults and Domes 233
20.6 Strengthening of Towers and Spires 233
20.7 Reduction of Seismic Effects on Structures 233
20.8 Strengthening of Soils and Foundations 233
20.9 Archaeological Reconstruction 234

21. Retrofit of Reinforced Concrete Structures 236–256

21.1 Introduction 236
21.2 Building Deficiencies 237
21.3 Retrofit Strategies 243
21.4 General Remarks 255
21.5 Summary 256

22. Retrofit of Foundations 257–271

22.1 Overview 257
22.2 Introduction 257
22.3 Deficiencies of Foundations 258
22.4 Condition Assessment of Foundations 258
22.5 Methods of Analysis 259
22.6 Types of Interventions 259
22.7 Methods of Execution 266
22.8 Implemented Case 270
22.9 Summary 271

23. Retrofit Using FRP Composites and Base Isolation Technique 272–287

23.1 Overview 272
23.2 Introduction 272
23.3 Strengthening of Masonry Walls 276
23.4 Strengthening of Reinforced Concrete Beams 278
23.5 Strengthening of Reinforced Concrete Columns 280
23.6 Strengthening of Beam–Column Joints 282
23.7 Summary of Use of FRP 282
23.8 Base Isolation 282
Theory Questions 287

PART 6: CASE STUDIES

24. Retrofit of Buildings and Other Structures — 291–296

24.1 Retrofit of a Building 291
24.2 A Double Storeyed Load Bearing Residential Building at Mumbai 292
24.3 Rehabilitation of RCC Overhead Reservoir at Siliguri, WB 293

25. Repair/Rehabilitation of Bridges — 297–312

25.1 Assessment and Retrofitting of Masonry Arch Bridge 297
25.2 Retrofitting/Strengthening of Masonry Bridge Structure 300
25.3 Rehabilitation of Concrete Bridges 302
25.4 Rehabilitation of a Road Overbridge at Bankim Setu 309

26. Repair/Rehabilitation of Marine Structures — 313–327

26.1 Repair/Rehabilitation of Jetty in Mumbai 313
26.2 Rehabilitation of Oil Jetty at New Mangalore Port 319
26.3 Repair/Rehabilitation of Marine Structures: Kandla Port 321
26.4 Structural Reassessment of Offshore Platforms 325

27. Repair/Rehabilitation of Monuments — 328–333

27.1 Structural Preservation of Lord Jagannath Temple at Puri 328

28. Repair/Rehabilitation of Irrigation Structures — 334–351

28.1 Hirakud Dam 334
28.2 Lar Dam (Iran) 337
28.3 Willingdon Dam (Tamil Nadu) 339
28.4 Massingir Dam (Mozambique) 341
28.5 Upper Indravati Hydroelectric Project 343
28.6 Remedial Measures for the Gates of Hirakud Dam 345
Theory Questions 351

Bibliography 353–355
Index 357–361

Part 1

Durability and Deterioration of Concrete Structures

1. Understanding Concrete and RCC Structures
2. Physical Causes
3. Chemical Causes
4. Corrosion
5. Evaluation of Concrete and Reinforced Concrete Structures
6. Causes of Distress and Deterioration of Concrete and Reinforced Concrete Structures

1
Understanding Concrete and RCC Structures

Time takes its toll on concrete structures, which creates a real problem for the use of concrete in a country's infrastructure. Knowing the right principles and procedures for the repair, rehabilitation and retrofitting of concrete structures is a critical element to financial success. Tearing down existing structures and rebuilding them up from the ground can be cost prohibitive and also against green building concept where we want to save energy for the future generation and even sometime for our own generation also. Learning and perfecting the ways to make the most of existing structures are key elements, when it comes to sustainable living and safe living conditions.

Many people look at concrete and see nothing but well concrete. But the knowledgeable mind sees much more. Are there stress cracks in the surface? Were expansion joints provided properly? Does the color of concrete indicate a proper curing time? Is the surface a slick, glass-like finish or brushed finish? Is the material flaking away? Can existing flaws be repaired in such a way to guarantee structural integrity?

Many people take concrete for granted. Yet, it is one of the strongest building materials of many bridges, highways, and other significant infrastructures. Working with new structures of concrete is very different from repairing, rehabilitating and retrofitting of existing concrete structures. Both types of work have their rules of thumb and their engineering elements. It often requires more experience to repair concrete than it does to install it as a new construction. This is what will be learned here.

Are we capable of doing a site inspection and evaluation of existing concrete structures? If not, would we like to do? This is our chance to learn a great deal about existing concrete structures. A few experts may be needed for complicated compromises in concrete construction, but many situations can be assessed personally. How much do we know about our options, when it comes to fixing problems with concrete repairs? Maintenance is a great place to start, and thus procedure will be discussed in detail. If we are looking for specifics on concrete work and codes, we have to come to the right place.

1.2 WHAT IS TO BE LEARNT

The following main topics will be covered/learnt about concrete/RCC structures:

- Determining physical causes of deterioration
- Determining chemical causes of deterioration

- Determining reasons and effect of corrosion
- Evaluation of concrete
- Determining quality of concrete
- High performance concrete
- In-place strength of concrete
- Durability testing
- Performance specifications and data
- The effects of fly ash
- The effects of various admixtures
- Certifications
- Quality control
- Compressive strength of concrete
- Aggregates
- Thermal characteristics
- Crack resistance
- Mechanical behavior
- Corrosion inhibitors
- Post-tension concrete
- Prestressed concrete
- Reinforcing steel in concrete structures
- Symptoms of deterioration
- Testing of concrete
- Repair materials
- Function of repair materials
- Special types of repair materials
- Evaluation and selection of materials
- Planning/designing of repair/rehabilitation/retrofitting
- Strategies for rehabilitation
- Techniques for retrofitting
- Retrofit of various types of structures
- Various case studies

2

Physical Causes

2.1 INTRODUCTION

Concrete is the most extensively used material for construction of various types of structures such as buildings, bridges, dams, jetty, docks, roads, etc. Normally much emphasis is given on concrete compressive strength rather than on environmental factors, which are known to affect the concrete durability. This is one of the main reasons for serious deterioration of concrete structures mainly reinforced concrete structures. Maintenance and repair/rehabilitation of constructed facilities is presently the most significant challenge facing the concrete industry, the issue of durability has replaced concerns about strength as the most pressing problem of the day.

2.2 DURABILITY ASPECT OF CONCRETE

Concrete is said to be durable, if it withstands the conditions for which it has been planned/designed, without deterioration, over a period of intended service life. The term durability of concrete is used to characterize, in broad terms, the resistance of concrete to a variety of physical or chemical attacks due to internal or external causes. The internal causes include alkali aggregate reaction, volume change due to differences in thermal properties of aggregate and cement paste and above all the permeability of concrete. The external causes may be due to weathering, occurrence of extreme temperature variations, abrasion, and electrolytic action and attack by natural or industrial liquids and gases The environmental penetration by water, chlorides, sulphates, carbon dioxide and oxygen into reinforcement affect the durability of concrete. In general, the following factors influence the durability of the concrete:

- Water–cement ratio and water content
- Curing of concrete
- Cover to reinforcement
- Cement content and its properties
- Aggregates
- Mix design
- Workability
- Use of admixtures
- Thermal incompatibility of concrete constituents
- Transition zone between aggregate and cement matrix
- Environmental interaction
- Condition of reinforcing bars used.

Each of these factors, jointly or in isolation, controls the susceptibility to deterioration mechanism in concrete. All other factors mentioned above except cover to reinforcement and condition of reinforcing bars, control the pore structure of the concrete, which are directly related to the permeation properties. Therefore, measurement of permeation properties of concrete would result in assessment of durability of concrete. Permeability determines the following:

- Saturation of concrete with water
- Ingress of moisture and air
- Corrosion of steel
- Cracking and spelling of cover concrete
- Water tightness of the structure
- Thermal insulation properties.

2.2.1 Changes in the IS:456-2000 in Terms of Durability

Some significant following changes were made in IS:456-2000 version to improve durability of concrete structures:

- All three grades of OPC, 33 grade, 43 grade, 53 grade and sulphate resisting Portland cement have been included.
- Minimum grade of concrete for RCC work is M20.
- The permissible limits for solids in water have been modified keeping in view the durability requirements.
- The clause of admixtures has been modified and quality control aspect, while use of admixtures is emphasized.
- Durability clause has been enlarged to include detailed guidance concerning the factors affecting durability.
- The table of exposure conditions has been modified to include severe and extreme exposure conditions.
- This clause also covers requirements for shape and size of member, depth of concrete cover (minimum cover to reinforcement), concrete quality, requirement against exposure to aggressive chemicals and sulphate attack, minimum cement requirement and maximum w/c ratio, limits of chloride content, alkali silica reaction, and importance of compaction, finishing and curing.

- Clause on quality assurance measures has been incorporated.
- Cover to reinforcement has been specified, based on durability for different exposure conditions. This is one of the most important stipulations made basically on durability considerations.

Now, again some more changes will be required in provisions of IS:2000 code, based on experience gained in intervening period.

2.2.2 Protective Measures for Durable Design

The following measures need to be taken for durable design of any structure:

- Use of selected structural formwork
- Proper concrete composition including special additions of admixtures
- Proper reinforcement detailing including minimum concrete cover
- A special skin quality concrete including skin reinforcement
- Limiting or avoiding crack development and limiting crack widths
- Provision of coatings as additional protective measures
- Inspection and maintenance procedures including monitoring procedures to be specified
- Special active protective measures such as cathodic protection or monitoring by way of sensors.

2.3 MAJOR CAUSES FOR DISTRESS IN CONCRETE

Once the evaluation phase is over for a structure, the next step is to establish the cause or causes for the damage that has been detected. Since many of the systems might have been caused by more than one mechanism acting upon the concrete, it is necessary to have an understanding of the basic underlying causes of damage and deterioration. Distress and deterioration in concrete structures may arise from various reasons as brought out below:

- Faulty planning and designing: Wrong mix proportion, lack of protection, inadequate joint provision and under-designed structures.
- Use of inferior materials: Partially hydrated cement, contaminated aggregates, and contaminated water.
- Poor construction practices: Faulty formwork, improper placing, improper compacting, inadequate curing, lack of proper supervision and quality control at site.
- Environmental effects: Thermal movement, moisture movement, freezing and thawing and surface erosion.
- Chemical effects: Carbonation, chloride intrusion and acid attack.
- Internal stresses: Fire, sulphate attack, alkali-aggregate reaction and volume change.
- Abuse of structures: Lack of maintenance, vandalism.
- Mechanical causes: Overloading, abrasion or wear.

The following are the major causes of distress in concrete structures:

- Structural deficiency arising out of faulty designing and detailing as well as wrong assumption in the loading criteria
- Structural deficiency due to defects in construction, use of inferior and substandard materials
- Damages caused due to fire, floods, tsunami and earthquakes, etc.
- Physical deterioration and creep
- Chemical deterioration, and marine environments
- Damages caused due to abrasion, wear and tear
- Damages due to impact, vibration, fatigue
- Settlement of foundation, thermal expansion.

2.3.1 Types of Failures/Damages in Concrete Structures

Deterioration of concrete is an extremely complex subject. It would be simplistic to suggest that it will be possible to identify exactly a specific, single cause of deterioration for every symptom detected during an evaluation of a structure. In most cases, the damage detected will be the result of more than one mechanism. For example, corrosion of reinforcing steel may open cracks that allow moisture greater access to the interior of the concrete. The moisture could lead to additional damage by freezing and thawing. Inspite of the complexity of several causes working simultaneously, given a basic understanding of the various damage causing mechanisms, it should be possible, in most cases, to determine the primary cause or causes of the damage seen on a particular structure and to make intelligent choices concerning selection of repair materials and methods. A structure can be considered to have failed or damaged, not merely when it collapses, but also in cases when it fails to perform the functions for which it was designed. Damage to structures can be broadly classified as under:

- Total or partial collapse
- Dissolution of materials
- Cracking (including pattern cracks)
- Spalling of materials resulting in reduction in size of members
- Deformations such as deflection, buckling, twisting, and distortion.

2.3.2 Causes for Deterioration of Concrete

The common causes are brought out below:
- Accidental loadings
- Chemical reactions:
    i. Acid attack
    ii. Aggressive water attack
    iii. Alkali-carbonate rock reaction
    iv. Alkali-silica reaction
    v. Miscellaneous chemical attack
    vi. Sulphate attack
- Construction errors
- Corrosion of embedded metals
- Design errors:
    i. Inadequate structural design
    ii. Poor design details

- Erosion:
  i. Abrasion
  ii. Cavitation
- Freezing and thawing
- Settlement and movement
- Shrinkage:
  i. Plastic
  ii. Drying
- Temperature changes:
  i. Internally generated
  ii. Externally generated
  iii. Fire
- Weathering

2.4 SHRINKAGE PROBLEM IN CONCRETE

Shrinkage is caused by the loss of moisture from concrete. It may be divided into two general categories:

i. Plastic shrinkage which occurs before setting

ii. Drying shrinkage which occurs after setting.

Shrinkage in concrete means moisture movement in concrete. It may be defined as the volume changes in concrete due to loss of water or moisture caused by evaporation or hydration of cement or carbonation. In practice, shrinkage is simply measured as linear strain. Its units are thus mm/mm. Shrinkage can be further classified into the following categories:

- Plastic shrinkage
- Drying shrinkage
- Autogenous shrinkage
- Carbonation shrinkage

2.4.1 Plastic Shrinkage

Mechanism

It is a fact that during the period between placing and setting, most concrete exhibits bleeding to some degree. Bleeding is the appearance of moisture on the surface of concrete; it is caused by the settling of the heavier components of the mixture. Usually, the bleed water evaporates slowly from the concrete surface. When environmental conditions are such that evaporation is occurring faster than water is being supplied to the surface by the bleeding, high tensile stresses may develop. These stresses can lead to the development of cracks on the concrete surface. Typically, the cracks are isolated rather than patterned. These cracks are generally wide and shallow. The reasons and characteristics are brought out below:

- The primary cause of plastic shrinkage crack is the rapid evaporation of water from the surface of concrete.
- Plastic shrinkage cracks occur within few hours after placing concrete, while it is still plastic and before it attained any significant strength.
- These occur almost entirely on horizontal surfaces exposed to the atmosphere. It is observed that the cracks developed are parallel to one another and normally spaced 0.3 m to 1.0 m apart.
- They can be deep and the width varies from 0.1 mm to 3 mm.
- The magnitude of plastic shrinkage and plastic shrinkage cracks are depending upon ambient temperature, relative humidity and wind velocity.
- It depends on the rate of evaporation of water (bleeding) from the surface of concrete (1 kg/m^2/hr).

The following measures could be taken to reduce or eliminate plastic shrinkage cracks:

- Moistening the subgrade and formworks
- Erecting temporary wind-breakers to reduce the wind velocity over concrete
- Erection of temporary roof to protect green concrete from the hot sun
- Reduction of the time between placing and finishing of concrete. If there is delay, the concrete surface should be covered with polythene sheets.
- Evaporation to be minimized by covering concrete with burlap, fog spray and curing compound.

In addition, it will be beneficial to minimize the loss of moisture from the concrete surface between placing and finishing. Finally, curing should be started as soon as practicable. If cracking caused by plastic shrinkage does occur and if it is detected early enough, revibration and refinishing of the cracked area will resolve the immediate problem of the cracks.

Plastic Settlement Cracks and Plastic Cracks

When there is any obstruction to uniform settlement by way of reinforcement or large pieces of aggregate, then it creates some voids or cracks in concrete. This is called *plastic settlement cracks*. The characteristics and reasons are explained below:

- This happens generally in a deep beam. These settlement cracks and voids are so severe, it needs grouting to seal them off.
- The air-entraining admixtures or water-reducing admixtures can be used for preventing such cracks.
- Normally plastic cracks develop before the concrete has hardened, i.e. between 1 and 8 hours of placement of concrete.
- As the evaporation of water takes place from the surface of concrete, it contracts resulting inducement of tensile stress in concrete causing cracks.
- These cracks can be prevented by restricting the rate of evaporation of water from the surface to be less than 0.5 kg/hr/m^2.

2.4.2 Drying Shrinkage

Mechanism

Drying shrinkage is a long term change in the volume of concrete caused by the loss of moisture. If the shrinkage could take place without restraint, there would be no damage to the concrete. However, the concrete in a structure is always subject to some degree of restraint by foundation, or by another part of the structure, or the difference in shrinkage between the concrete at the surface and that in the interior of a member. This restraint may also be attributed to purely physical conditions such as the placement of footing on a raft foundation or chemical bonding of new concrete to earlier placements or both. The combination of shrinkage and restraints cause tensile stresses that can ultimately lead to cracking.

Symptoms

Visual examination shows typical cracks that are characterized by their fineness and absence of any indication of movement. They are usually shallow, a few centimeters in depth. The crack pattern is typically orthogonal or blocky. This type of surface cracking should not be confused with thermally induced deep cracking, which occurs when dimensional change is restrained in newly placed concrete by rigid foundations or old lifts of concrete. The characteristics and reasons of this crack are listed below:

- It is caused by physical loss (evaporation) and chemical loss (hydration) of water during the hardening process and exposure to unsaturated air.
- The resulting reduction in volume can cause cracks, if the concrete is restrained and tensile strength exceeded.
- These cracks normally appear at about 7–10 days after concreting and around 80% of drying shrinkage takes place by about a year.
- It is influenced by a number of factors, such as cement content, water content, aggregates, curing, humidity and temperature.
- Drying shrinkage cracks are generally confined to nonstructural members like floor toppings, screeds and rendering, and parapet walls.
- The total drying shrinkage is made up of irreversible and reversible shrinkage. On initial drying out, an appreciable amount of total shrinkage is irreversible, but after several cycles of wetting and drying, the shrinkage becomes entirely reversible.

Some of the preventive measures are as follows:

- Use of minimum water content
- Use of highest possible aggregate content
- Providing adequate and early curing
- Eliminating external restraints as much as possible
- Providing sufficiently close-spaced reinforcement.

Additionally, placing the concrete at as low a temperature as practical; dampening the subgrade and the forms; dampening aggregates, if they are dry and absorptive; and providing an adequate amount of reinforcement to distribute and reduce size of cracks that do occur. Restraint can be reduced by providing adequate contraction joints.

2.4.3 Autogenous Shrinkage

The reasons and characteristics of autogenous shrinkage are as follows:

- If no movement of water to or from the set paste of concrete is allowed, then the shrinkage developed is known as autogenous shrinkage.
- The shrinkage is caused by the loss of water consumed or used up in the hydration of cement.
- The magnitude of this shrinkage is very small and not of much significance.

2.4.4 Carbonation Shrinkage

The characteristics are as follows:

- Carbonation is the reaction of carbon dioxide present in the atmosphere, with the hydrated cement materials in the presence of moisture.
- The simultaneous reaction of CO_2 with hydrated cement minerals in concrete induces contraction of concrete, which is known as carbonation shrinkage.

2.5 FREEZE AND THAW ON CONCRETE

The disintegration or deterioration takes place when the following conditions are present:

- Freezing and thawing temperature cycles within the concrete
- Porous concrete that absorbs water (water filled pores and capillaries).

2.5.1 Mechanism of Deterioration

As the temperature of a critically saturated concrete is lowered during cold weather, the freezable water held in the capillary pores of cement paste and aggregates expands upon freezing. If subsequent thawing is followed by refreezing, the concrete is further expanded, so that repeated cycles of freezing and thawing have a cumulative effect. By their very nature, concrete hydraulic structures are vulnerable to freezing and thawing simply because there is ample opportunity for portions of this structure to become critically saturated. Concrete is especially vulnerable in area of fluctuating water levels or under spraying conditions. Exposure in such areas as the tops of walls, piers, parapets, and slabs enhances the vulnerability of concrete to the harmful effects of repeated cycles of freezing and thawing and may lead to pitting and scaling. It involves the development of osmotic and hydraulic pressure during freezing, principally in the paste, similar to ordinary frost action. The freezing and thawing mechanism action is brought out below:

- Freeze and thaw deterioration generally occurs on horizontal surfaces that are exposed to water, or on vertical surfaces that are at the water line in submerged portions of structures.
- The freezing water contained in the pore structure expands, as it is converted into ice. The expansion causes localized tension forces that fracture the surrounding concrete matrix.
- The first stage is the development of fine closely spaced cracks parallel to the edge of the exposed concrete.
- The concrete soon becomes filled with a dark deposit of calcium carbonate and are commonly called D-cracks.

- As the deterioration continues, small pieces of concrete between the cracks separate from the body of concrete.
- The deterioration is reduced as the water–cement ratio is reduced, but the only positive way to prevent the problem is to protect concrete by the adequate air-void system.
- The superiority of air entrained with respect to freezing and thawing action is evident from standard graph available.
- Visual examination of concrete damaged by freezing and thawing may reveal symptoms ranging from surface scaling to extensive disintegration.

2.5.2 Preventive Measures

The following remedial measures may be taken against freezing damage:

- Water–cement ratio and total water content used should be as lowest as practicable.
- Adequate air entraining has found to be effective in controlling the freezing damage.
- Use of durable aggregate is also useful to check the freezing effect.
- Adequate curing of concrete prior to exposure to freezing conditions is also important.
- Designing the structure to minimize the exposure to moisture, as for example, providing positive drainage rather than flat surfaces, wherever possible.

2.5.3 Rate of Freeze–Thaw Deterioration

This depends on the following factors:

- Increased porosity (increases rate)
- Increased moisture saturation (increases rate)
- Increased number of freeze-thaw cycles (increases rate)
- Air entrainment (reduces rate)
- Horizontal surfaces that trap standing water (increases rate)
- Aggregate with small capillary structure and high absorption (increases rate).

D-cracking is normally linked to aggregate. It consists of development of fine cracks near free edges of slabs, but the initial cracking starts lower in the slab where moisture accumulates and the course aggregate becomes saturated to the critical level. With cyclic freezing and thawing, the mortar surrounding the aggregate fails leading to the D-cracking. Aggregate of sedimentary, calcareous or siliceous origin (gravel, crushed rock) leads to D-cracking due to frost action.

2.6 WEATHERING ACTION ON CONCRETE

Weathering is normally referred to as a cause of deterioration of concrete. It defines a change in color, texture, strength, chemical composition, or other properties of a natural or artificial material due to the action of weather. The environmental factors that can cause cracking include the following:

- Freezing and thawing
- Wetting and drying
- Cooling and heating.

The salient features of weathering action on concrete are as follows:

- The damage from freezing and thawing is the common weather related physical deterioration.
- Other weathering processes that may cause cracking in concrete are alternate wetting and drying, and heating and cooling.
- If the volume changes due to these processes are excessive, cracks may develop and give the impression that they are on the verge of disintegration.
- The damage due to these factors may appear in the form of general flacking and spalling of concrete from the surface of concrete.
- Concrete generally loses strength with increase in temperature about 300 degree centigrade, damage being greater with aggregate having high coefficient of thermal expansion.

2.7 CRAZING ON CONCRETE

The characteristics of crazing on concrete are as follows:

- Crazing is the development of a network of fine random cracks on the surface of concrete or mortar caused by shrinkage and usually related to finishing and curing procedures.
- Excessive floating and troweling bring water, cement and dust from the aggregates to the surface to produce a surface skin, which has a higher drying shrinkage than the underlying concrete.
- Spreading dry cement on a surface that is too wet to trowel, and sprinkling water on concrete that is too dry to trowel, both produce a skin likely to suffer from crazing.
- Use of highly absorptive aggregates batched in the dry state may contribute to crazing.
- The use of overwet concrete contributes to all the problems associated with drying shrinkage, since shrinkage is almost directly proportional to the amount of mixing water in concrete.
- Improper curing can cause crazing, even if the concrete is properly batched, mixed, placed and finished.
- Use of unvented heater during the curing period is also a cause of crazing.

2.7.1 Preventive Measures

Some of the preventive measures include the following:

- Damping the subgrade before placing the concrete
- Avoiding overfinishing of the surface
- Delaying troweling until the surface moisture has disappeared
- Avoiding sprinkling of dry cement or water on the surface during finishing operation
- Starting curing as soon as possible
- Avoiding use of unvented heaters

- Batching of absorptive aggregates in a moist condition
- Avoiding higher temperature differentials between concrete and curing water.

2.8 HONEYCOMBING ON CONCRETE

Honeycomb consists of exposed pockets of coarse aggregates not covered by a surface layer of mortar. This may be caused due to inadequate consolidation, presence of excess water in concrete or use of leaky formworks, which allow the mortar to escape.

2.8.1 Preventive Measures

Some of the preventive measures are brought out below:

- Good construction practices need to be followed strictly
- Workable concrete to be used
- Good forms (tight and leak-proof) to be used
- Proper vibration (compaction) to be ensured
- Placing of concrete need to be planned in a sequence.

2.9 SWELLING EFFECT OF CONCRETE

Concrete used continuously in water from the time of casting exhibits a net increase in its volume and weight. This increase in volume due to continued hydration of cement is known as swelling. The swelling is due to the absorption of water by the cement gel.

2.10 POP OUTS ON CONCRETE

A pop out is a conical-shaped hole in the surface with a portion of coarse aggregate particle exposed. These occur outdoors on the horizontal and vertical surfaces. They are caused by freezing of water in aggregate particles that have an internal pore structure, which causes them to expand unduly upon freezing. Normally pop outs do not appear during construction, but they start appearing during the first winter following construction and may continue to form for several years.

These do not harm the concrete, but they are unsightly. The rocks that have produced pop outs include chart, shale, clay stone, mudstone, argillaceous limestone, and siltstones. Pop outs can be prevented only by avoiding aggregates which cause them.

2.11 CREEP PROBLEM ON CONCRETE

Concrete, brickwork and timber when subjected to sustained loads not only undergo instantaneous elastic deformation, but also exhibit a time-dependent deformation known as creep. In concrete, creep results in a sustained increase in elastic deformation, which sometimes leads to formation of cracks in brick masonry of framed and load bearing structures. Deferring the removal of centering and application of external load can reduce it. However, when sustained load is removed, the strain decreases immediately by an amount equal to the elastic strain at the given age (instantaneous recovery). This instantaneous recovery is then followed by a gradual decrease in strain, called creep recovery, which is a part of total creep strain suffered by the concrete. If a loaded concrete specimen is viewed as being subjected to a constant strain, the creep decreases the stress progressively with time. This is called relaxation. While 80–85% shrinkage strains occur in six months, only about 75% of creep strains occur in a year.

2.11.1 Various Influencing Factors

All the factors which influence shrinkage also influence creep, which includes the following:
- Types of aggregate
- Type of cement and cement content
- Admixtures used
- Entrained air
- Mix proportions
- Mixing time
- Age of concrete
- Level of sustained stress
- Ambient humidity
- Temperature
- Size of the specimen.

2.12 ABRASION, EROSION AND CAVITATION ON CONCRETE

The terms abrasion, erosion and cavitation are explained below:
- Abrasion refers to wearing away of the surface by friction.
- Erosion refers to wearing away of the surface by fluids.
- Cavitation refers to the damage due to nonlinear flow of water at velocities greater than 12 m/sec.
- The concrete used in roads, floors, pavements and also the one used in the hydraulic structures should exhibit resistance against abrasion, erosion, and cavitation.
- The more the compressive strength, the higher is the resistance to abrasion, erosion and cavitation.
- Due to abrasion, dusting problem arises, which may be harmful in many situations, e.g. industrial floors having material handling system or robots.
- If the aggregate in concrete is wear resistant, it is the property of cement matrix that controls the abrasion resistance.
- Use of steel fibers in cement matrix improves abrasion resistance of concrete and also polymer based systems when applied to concrete improve the abrasion resistance.
- The shape and surface texture of aggregate play an important role in abrasion resistance of the concrete.
- Epoxy screeding and polymer application to the surface is said to be effective against cavitation.

2.13 TEMPERATURE VARIATION

Change in temperature causes a corresponding change in volume of concrete. As was true for moisture-induced volume change, temperature-induced volume change must be combined with restraint before damage can occur.

Basically, there are three temperature change phenomena that may cause damage to concrete. Firstly, there are temperature changes that are generated internally by the heat of hydration of cement in large placements. Secondly, there are temperature changes generated by variations in climatic conditions. Finally, there is a special case of externally generated temperature change-fire damage. For internally and externally generated temperature changes, different mechanisms, symptoms and prevention methods are there.

2.14 FIRE EFFECT ON CONCRETE

Concrete structure gets damaged due to fire, the extent of which depends upon the intensity and duration of the fire. The principal types of damage are:

- Reduction in strength of concrete
- Cracking and spalling of concrete
- Deflection and deformation of members
- Discoloration
- Other miscellaneous functional failures.

Though concrete is not a refractory material, but it has good fire resistance properties. Fire resistance of concrete is determined by three main factors:

1. The capacity of concrete to withstand heat.
2. The conductivity of concrete to heat.
3. The coefficient of thermal expansion of concrete.

In case of reinforced concrete, the fire resistance is not only dependent upon the type of concrete, but also on the thickness of cover to reinforcement. The fire introduces high temperature gradients and as a result of it, the surface layers tend to separate and spill off from the cooler interior. Due to the heating of reinforcement and aggregates, the expansion takes place both laterally and longitudinally of the reinforcement bars resulting in loss of bond and strength of reinforcement. The effect of increase in temperature on the strength of concrete is not much up to a temperature of about 300 degree centigrade, but above 300 degree centigrade, definite loss of strength takes place.

In mortar and concrete, the aggregate undergoes a progressive expansion on heating, while the hydrated products of the set cement, beyond the point of maximum expansion shrinks. These two opposite actions progressively weaken and crack the concrete. Quartz, the principal mineral in sand, granite and gravel expands steadily up to about 573 degree centigrade. At this temperature it undergoes a sudden expansion of 0.85%. This expansion has a disruptive action on the stability of concrete. By the use of fire resistance aggregates, the fire resistance capacity of concrete improves. The best fire resistant aggregates, amongst the igneous rocks are basalt and dolerite.

Limestone expands steadily until temperature reaches to about 900 degree centigrade and then begins to decompose; hence it has been found that dense limestone is considered as a good fire resistant aggregate. Blast furnace slag aggregate and broken bricks aggregate are good resistant to fire. The large series of tests indicated that even the best fire resistant concrete have been found to fail, if the concrete is exposed for a considerable period to a temperature exceeding 900 degree centigrade, while serious reduction in strength occurs at a temperature of about 600 degree centigrade.

2.15 THERMAL MOVEMENT IN CONCRETE

Thermal movement in concrete occurs due to considerable amount of heat generated by heat of hydration, atmospheric temperature and external fire. Due to thermal movement, changes in shape and volume of concrete cause cracks on the concrete structures. The extent of temperature rise depends on the properties of cement used and the shape and size of the components. The heat of hydration may not be significant, but in mass concrete works, it is an important factor to be contended with. The control of heat and avoidance of cracks to maintain the integrity of concrete structures is a subject in itself.

2.15.1 Preventive Measures

Some of the preventive measures include the following:

- Use of Pozzolona cement
- Use of low heat cement
- Precooling of aggregates and mixing with water
- Post cooling of concrete by circulating refrigerated water through pipes embedded in the body of concrete
- Providing joints to relieve the restrains in the structure
- Providing adequate reinforcement to distribute the stresses
- Providing suitable insulation covers.

2.16 SUBGRADE MOVEMENT

If there are local soft pockets in subgrade on which concrete is placed or if there are any pockets or hollows under the building paper, there will be localized settlement of concrete due to the weight of the plastic mass. If this settlement occurs after finishing of the concrete surface, cracks will appear. The occurrence can be prevented by proper attention to compacting and draining the subgrade.

2.17 FORMWORK MOVEMENT

Movement of formwork during the period while the concrete is going from a fluid to rigid material may induce cracking and separation with concrete. A crack open to the surface will allow excess of water to the interior of the concrete. All internal voids may give rise to freezing or corrosion problems, if the voids become saturated. Any movement of formwork, which occurs between the time when concrete begins to lose its fluidity and the time when it has fully set, causes internal cracks to appear in the structure. These cracks are potentially dangerous in the sense that they form a water pocket in the concrete mass, which upon freezing will spall the concrete surface. Rebar corrosion can also result from such water pockets. Such cracks can be prevented by a proper design of the forms, with respect to details and deflections.

2.18 SETTLEMENT AND MOVEMENT

There are two types of mechanism of settlement and movement in concrete structural members. They are:

1. Differential movement
2. Subsidence

Their mechanism is different and symptoms and prevention or corrective measures are also different.

2.19 FOUNDATION SETTLEMENT

Shear cracks occur when there is a differential settlement of foundation. Shear failure is predominant in made-up ground. Some of the preventive measures are brought out below:

- Providing an impermeable apron all around the building
- Prevention of water stagnation around the building
- Providing adequate drainage system
- Avoiding the plantation of fast growing trees in the immediate vicinity of the building.

2.20 DESIGN ERRORS

Design errors may be divided into two general types as brought out below:

1. The first type are those resulting from inadequate structural design. This type has got special mechanism, symptoms and prevention methods. In fact, inadequate design is best prevented by thorough and careful review of all the design calculations, preferably by a third party proof checking and corrective action taken accordingly.
2. The second type are those resulting from poor design details. In the existing structures, problems from poor detailing should be handled by correcting the detailing and not by simply responding to the symptoms.

Poor design details may be due to the following factors:

- Abrupt changes in section
- Insufficient reinforcement at re-entrant corners and openings
- Inadequate provision for deflection
- Inadequate provision for drainage
- Insufficient spacing in expansion joints/ inadequately designed expansion joints
- Incompatibility of materials
- Neglect of creep effect
- Rigid joints between precast units
- Unanticipated shear stresses in piers, columns, or abutments
- Inadequate joint spacing in slabs.

The design and detailing errors that may cause an unacceptable cracking are as follows:

- Improper selection and/or detailing of reinforcement
- Use of poor detailed re-entrant corners in walls, precast members and slabs
- Restraint of members subjected to volume changes due to variations in temperature and moisture
- Lack of adequate contraction joints
- Improper design of foundations results in differential settlement within the structure
- Re-entrant corners provide a location for stress concentration and these are main location of initial cracks, as in the case of window and door openings in concrete walls and beams.

The structures, in which cracking may cause major problem of serviceability need special care in the design and detailing. These structures also need continuous inspection during all the phases of construction.

2.21 CONSTRUCTION ERRORS

Failure to follow specified procedures and good engineering practices or outright carelessness may lead to a number of conditions, which may be grouped together as *construction errors.* Typically, most of these errors may not always lead directly to failure or deterioration of concrete. Instead, they definitely enhance the adverse impact of other mechanisms. The errors normally encountered are as below:

- Adding more water to concrete
- Improper alignment of formwork
- Improper consolidation
- Improper curing
- Improper location of reinforcing bar
- Premature removal of shores or re-shores
- Settling of the concrete
- Settling of the subgrade
- Improper vibration of freshly placed concrete
- Improper finishing of flat work.

2.22 CRACKS DUE TO CONSTRUCTION OVERLOAD

The load induced during construction can be far more severe than those experienced in service. Unfortunately, these conditions may occur at the early ages, when concrete is most susceptible to damage and often result in permanent damage in terms of cracks. A common error occurs when the precast members are not properly supported during transportation and erection. The use of arbitrary or convenient lifting points may cause severe damage/complete collapse of structure in certain cases. A big element lowered too fast, and stopped suddenly carries significant momentum, which is translated into an impact load that may be several times the dead weight of the element. Storage of materials and equipment can easily result in loading conditions during construction for more severe than any load for which the structure is designed. Damage from un-intentional construction overloads can be prevented only if the designers provide information on load limitations for the structure and if the construction personnel/ supervisor heed to these limitations. The broad reasons for cracking, etc. are brought out below:

- Inadequate provision of main steel reinforcement, or inadequate provision of temperature reinforcement or wrong spacing of bars, or absence of corner reinforcement may cause unacceptable cracks in concrete.
- One of the most common occurrences is the displacement of top bars in cantilever thin chajjas porches by the movement of concreting gang, causes cracks at the junction point of cantilever chajjas porches.
- There are number of cases where congestion of reinforcement and difficulties in proper compacting of concrete, particularly at junctions of columns and beams, deep beams, the negative reinforcement over T- and L-beams, needs to be taken care of. In the absence of such care, concrete is sure to crack.

2.23 CRACKS DUE TO EXTERNALLY APPLIED LOADS

Load induced tensile stress may result in cracks in concrete elements. A design procedure specifying the use of reinforcing steel, not only to carry tensile forces, but also obtaining both an adequate crack distribution and reasonable limit on crack width is recommended. Flexural and tensile crack widths can be expected to increase with time for members subjected to either sustained or repetitive loading. A well-distributed reinforcing arrangement offers the best protection against undesirable cracking.

2.24 ACCIDENTAL LOADINGS EFFECT

2.24.1 Mechanism

Accidental loadings may be characterized as short duration, and onetime events such as the impact of barge against a lock wall or an earthquake. These loadings can generate stresses higher than the strength of concrete, resulting in localized or general failure. Determination of whether accidental loading caused damage to the concrete requires knowledge of the events preceding discovery of the damage. Usually, the damage caused by accidental loading is easy to diagnose.

2.24.2 Symptoms

Visual inspection shows spalling or cracking of concrete, which has been subjected to accidental loadings.

2.24.3 Prevention

Accidental loadings by their very nature cannot be prevented. Minimizing the effects of some occurrences by following proper design procedures (an example is the design of earthquake resistant structure) or by proper attention to detailing of structure reduces the impacts of accidental loadings.

3

Chemical Causes

3.1 CHEMICAL ATTACK ON CONCRETE

This category includes specific causes of deterioration that exhibit a wide variety of symptoms. In general, deleterious chemical reactions may be classified as those which occur as the result of external chemical attacking the concrete or those which occur as a result of internal chemical reactions between the constituents of the concrete. The chemical attack on concrete can be classified as follows:

- Acid attack
- Alkali attack
- Carbonation
- Chloride attack
- Leaching
- Salt attack
- Sulphate attack.

Ingress of dissolved substances from the external environment may cause various forms of chemically induced deterioration by reaction with the cement paste or aggregate constituents. Although the resistance of concrete to chemical attack is directly influenced by the following factors:

- Its porosity
- The cement composition used in the concrete
- Condition under which the cement paste hardened

- All determine properties of concrete
- Ability to resist various effects of fluids in the environment.

Most of the problems due to chemical attack arise from the aggressive fluid penetrating into the interior pore space in the concrete. Therefore, damage is in many cases dependent on the permeability of the surface layers and not on the body of the concrete.

The penetration may be influenced by the effects of the following:

- Temperature gradient
- Freezing
- Loading
- Electric current
- Other factors

In general, the main types of aggressive fluid to which concrete gets exposed are one or more of the following:

- Mineral acids
- Some organic acids
- Solutions of sulphates, chlorides, sugars, nitrates, phenols, and ammonium compounds.

3.1.1 Types of Chemical Reaction

The most aggressive solutions are those where hot solutions of above fluids are split on the surface of concrete and absorbed forming a reservoir of chemicals within the concrete.

Essentially the following five main types of chemical reaction can be recognized:

- **Type 1:** Simple leaching of free calcium hydroxide (hydrated lime)
- **Type 2:** Reaction between attacking solutions and cement compounds resulting in the formation of secondary compounds, which are either leached from the concrete, or remain in a nonbound form, resulting in gradual strength loss.
- **Type 3:** Reaction is similar to type 2, but resulting in the crystallization of the reaction products giving rise to expansive forces, which disrupt the concrete.
- **Type 4:** Crystallization of salts directly from the attacking solution causing disruption of the concrete.
- **Type 5:** Corrosion of the embedded steel reinforcement resulting from breakdown of the passivation zone by aggressive solutions.

3.2 CARBONATION ATTACK ON CONCRETE

Carbonation can be defined as the reaction of carbon dioxide (CO_2) with hydrated cement. It is the effect of carbon dioxide in the air on cement products, mainly the hydroxides in the presence of moisture. Calcium hydroxide is converted to calcium carbonate by absorption of carbon dioxide. Calcium carbonate is slightly soluble in water and therefore, when it is formed, it tends to seal the surface pores of the concrete destroying its impermeability system. As a result of the reaction, the layer of concrete, close to the surface of the concrete becomes carbonated and this carbonated layer is not sufficiently alkaline (pH below 9) to protect reinforcing steel.

Apart from the reaction between CO_2 and $Ca(OH)_2$ the former also dissolves any moisture present to form carbonic acid. This is a weak electrolyte, which allows the steel to be oxidized by the atmospheric oxygen, and then reacts with water to form hydrated ferric oxide. The corrosion product occupies a much greater volume than the original metal from which it was formed and thus sets up bursting forces in the surrounding concrete.

3.3 CHLORIDE ATTACK ON CONCRETE

Chlorides can be introduced into concrete by coming into contact with environment containing chlorides, such as seawater or deicing salts. Penetration of chlorides starts at the surface and then moves inward. Chlorides may enter into the concrete from the following sources:

- Cement of the concrete
- Water mixed in concrete
- Aggregates of the concrete
- Admixtures added to the concrete
- By diffusion from atmosphere.

Penetration takes time, depending upon the following factors:

- The amount of chlorides coming into contact with the concrete
- The permeability of the concrete
- The amount of moisture present.

3.4 SULPHATE ATTACK ON CONCRETE

Mechanism

Sulphates are found in most of the soils as calcium, potassium, sodium and magnesium sulphate. Solid salts do not attack concrete, but when present in solution they can react with hardened cement paste. Sulphates are present in seawater, industrial effluents and some ground water. Sulphate attack occurs when pore system in concrete is penetrated by solution of sulphates. Sulphate reaction is dependent on the following parameters:

- Carbonation of sulphate ions
- Cations present in the sulphate solution
- C_3A content of cement
- Density and permeability of the concrete.

3.5 ACID ATTACK ON CONCRETE

Portland cement is a highly alkaline material and is not very resistant to attack by acids. The deterioration of concrete by acids is primarily the reaction between the acid and the products of hydration of cement. Calcium silicate hydrate may be attacked, if highly concentrated acid exists in the environment of the concrete structures. In most cases, the chemical reaction results in the formation of water-soluble calcium compounds that are then leached away. In case of sulphuric acid attack, additional or accelerated deterioration results, because calcium sulphate formed may affect the concrete by the sulphate attack mechanism. If the acid is able to reach the reinforcing steel through cracks or pores in the concrete, corrosion of the reinforcing steel results and causes further deterioration of the concrete.

3.6 ALKALI REACTION ON CONCRETE

The reaction of some forms of silica and carbonates in aggregates with alkalis in cement produces a gel, which causes expansion and cracks. Portland cement concrete made with nonalkali reactive aggregates is highly resistant to strong solutions of most bases (sodium or potassium hydroxides). However, if sodium hydroxide penetrates concrete and becomes concentrated at an evaporating face, physical damage would result from crystallization of sodium carbonate formed by the reaction between NaOH and carbon dioxide from the air.

3.7 HYDROLYSIS AND LEACHING ON CONCRETE

Leaching of lime compounds leads to the formation of salt deposits on the surface of the concrete, known as *efflorescence*. When water percolates through poorly compacted concrete or through cracks or along badly made joints and when evaporation takes place at the surface of the concrete, leaching occurs.

Calcium carbonate formed by the reaction of $Ca(OH)_2$ with CO_2 is left behind in the form of a white deposit.

Efflorescence is more likely to occur in concrete, which is porous near the surface. Thus, the type of formwork plays a vital role in addition to the compaction and w/c ratio. The occurrence of efflorescence is greater when cool wet weather is followed by a dry and hot spell. In this sequence, there is little initial carbonation, the surface moisture dissolves lime and $Ca(OH)_2$ is finally drawn to the surface, where it reacts with CO_2 to form calcium carbonate. Efflorescence can also be caused by the use of unwashed seashore aggregate. The salt coating on the surface of the aggregate particles in the course of lime leads to a white deposit on the surface of the concrete.

Early efflorescence can be removed with a brush and water. Heavy deposits may require acid treatment of the surface of the concrete. The acid used is diluted HCL (1:5 to 1:10). Typically, the quantity of a 1:10 solution of acid used would be 200 g/m^2 (applied by a sponge) the depth of concrete removed by 0.01 mm. There is no danger of continued action of the acid, because it is used up by the reaction with lime, but the concrete should be washed in order to remove the salts, which would be formed. Since the acid removes lime, the surface of the concrete becomes darker. Hence, acid must be applied uniformly in terms of concentration and duration of action. Thus, the influence of leaching is to increase the permeation properties of concrete due to the removal of calcium hydroxide from the pore solution.

3.8 SALT ATTACK ON CONCRETE

Solid salts do not attack concrete, but when present in solution they can react with hardened concrete. It is a more general problem in masonry structures. Efflorescence is a whitish crystalline deposit on the surface. It is the formation of calcium carbonate precipitate on the concrete surface owing to carbonation.

3.8.1 Preventive Measures

Some of the preventive measures include the following:

- Using unsound materials free from salts
- Proper proportioning of concrete
- Proper consolidation and curing
- Preventing the access of moisture to the structure.

3.9 AGGRESSIVE WATER ATTACK

Some waters have low concentrations of dissolved minerals. These soft or aggressive water leaches calcium from cement paste or aggregates. There are indications that this attack takes place very slowly. For an aggressive water attack to have a serious effect on hydraulic structures, the attack must occur in flowing water. This keeps a constant supply of aggressive water in contact with the concrete and washes away aggregate particles that become loosened as a result of leaching of the paste. Normally, visual examination shows concrete surfaces that are very rough in areas, where the paste has been leached. Sand grains may be present on the surface of concrete, making it resemble coarse sand paper. If the aggregate is susceptible to leaching, holes where the coarse aggregate has been dissolved will be evident. Water samples from structures where aggressive water attack is suspected may be analyzed to calculate the Langlier index, which is a measure of the aggressiveness of the water.

3.9.1 Prevention

The aggressive nature of water at the site of a structure can be determined before construction or during a major rehabilitation. Additionally, the water-quality evaluation at many structures can be expanded to monitor the aggressiveness of water at the structure. If there are indications that water is aggressive or becoming aggressive, area susceptible to high flows may be coated with a non-Portland-cement-based coating.

3.10 CRYSTALLIZATION OF SALTS IN PORES

Most of the problems due to salt arise from the aggressive fluid penetrating into the interior pore space in the concrete. Damage is, therefore, in many cases caused depending on the permeability of the surface layers and not on the body on the concrete. In the most aggressive situations, crystallization of slots in pores, directly from the attacking solution, causes disruption of the concrete. Due to chemical attack, in most of the cases, the chemical reaction results in formation of water soluble calcium compounds (salt compounds), which are then leached away by the aqueous solutions. This results in an increase in the porosity and permeability of the system.

In case of sulphuric attack, the formation of calcium sulphate will affect concrete resulting in accelerated deterioration; calcium sulphate formed by the initial reaction can proceed to react with calcium aluminates to form calcium sulpho aluminates, which on crystallization can cause expansive disruption of the concrete. If acids or salt solutions are able to reach the reinforcing steel through cracks or pores in the concrete, corrosion of steel can occur, which will in turn cause cracking and spalling of the concrete. Concrete with increased cement content and a reduced w/c ratio is required to prevent the salt attack. Further, a substantial sacrificial thickness of the same quality of concrete at the surface (cover concrete) has to be provided or protective surface coating has to be given. Portland cement concrete with nonalkali reactive aggregate is highly resistant to strong solutions of most bases (sodium or potassium hydroxide solution). However, if sodium hydroxide penetrates concrete and becomes concentrated at an evaporating face, physical damage would result from crystallization of sodium carbonate formed by the reaction between $NaOH$ and CO_2 from the air.

3.11 SEAWATER ATTACK ON CONCRETE

The marine environment is characterized by wave action, which imposes shock loads and causes erosion of concrete surfaces by abrasion and cavitations. In addition, the concrete is exposed to the aggressive constituents of seawater and subjected to repeated freeze-thaw and wet-dry cycles. Thus, the deterioration of concrete structures in such an environment is both chemical and physical in nature and the type of the attack may be demarcated into three zones depending on the tidal lines.

The upper part above the high tide line is not directly exposed to seawater. Consequently, cracking due to reinforcement corrosion and/or freeze-thaw cycles is the main deterioration phenomenon in this zone. In the tidal zone, the structure is subjected to alternate wet and dry cycles, freeze and thaw cycles, impact of waves, and floating ice, abrasion by sand and gravel, and reinforcement corrosion. The lower zone, submerged in water, is a relatively stable environment, where freeze-thaw action and reinforcement corrosion does not occur. Here the predominant deterioration action is chemical attack, sulphate attack, which causes strength retrogression. In-place concrete density, cement type, and cement content play a vital role in the resistance of concrete to seawater. Concrete made with calcium aluminate, super-sulphated cements, and also those containing supplementary cementing materials, resists seawater fairly well. Such improved resistance as compared to OPC stems from reduced free lime cement in such concrete. In case of reinforced concrete, the absorption of salt establishes anodic and cathodic areas; the resulting electrolytic action leads to an accumulation of corrosion products in the steel with a consequent rupture of surrounding concrete. Thereby it is essential to provide sufficient cover to the reinforcement. So, adopting low w/c and high cement content are essential. A well compacted concrete, good workmanship, especially the construction joints, and types of cements is of vital importance.

3.12 BIOLOGICAL ATTACK ON CONCRETE

The existence of vegetation, such as fast growing trees in the vicinity of compound walls can sometime cause cracks in walls due to expansive action of roots growing under the foundation. Roots of tree generally spread horizontally on all sides to the extent of height of tree above the ground and when trees are located close to the walls; these always should be viewed with suspicion. Sometimes plants take root and begin to grow in fissures of wall, because of seeds contained in bird droppings. If these plants are not removed well in time, these may in course of time develop and cause severe cracking of the wall in question.

When soil under the foundation of building happens to be shrinkable clay, cracking in floors of the building can occur. This is normally due to dehydrating action of growing roots on the soil, which may shrink and cause settlement of foundation. Alternatively it may be due to upward thrust on a portion of the building, when old trees are cut off and the soil that had been dehydrated earlier by roots, swells up on getting moisture from some source, such as rain. Even small plants such as lichen, algae and mass growing on concrete surface attract moisture and create physical and chemical process to deteriorate the concrete. Besides, humic acid produced by microgrowth reacts with cement. Marine borers and marine plants also contribute to the deterioration of concrete. Sometimes in tropical conditions algae, fungi and bacteria use atmospheric nitrogen to form nitric acid, which attack concrete.

3.12.1 Preventive Measures

The following are some of the general measures for avoidance of cracks due to vegetation:

- The trees should not be allowed to grow too close to the buildings, compound walls, garden walls, etc. and taking extra care if soil under the foundation happens to be shrinkable soil/clay. If any saplings of trees start growing in fissures of walls, this need to be removed at the earliest opportunity.

- If some large trees exist close to a building and these are not causing any problem, as far as possible these trees need not be disturbed, if the soil under the foundation happens to be shrinkable clay.

- For any site intended for new construction, vegetation including trees is removed and the soil is shrinkable clay, the construction activity should not be commenced until it has undergone expansion after absorbing moisture and has stabilized.

3.13 MECHANISM OF MISCELLANEOUS CHEMICAL ATTACK

Concrete normally resists chemical attack to a varying degree, depending upon the exact nature of the chemical. This includes an extensive listing of the resistance of concrete to various chemicals. To produce significant attack on concrete, most chemicals must be in solution that is above some minimum concentration. Concrete is seldom attacked by solid dry chemicals. Further, for minimum effect, the chemical solutions need to be circulated in contact with the concrete. Concrete subjected to aggressive solutions under positive differential pressure is particularly vulnerable. The pressure gradients tend to force the aggressive solutions into the matrix. If the low pressure face of the concrete is exposed to evaporation, a concentration of salts tends to accumulate at that face, resulting in increased attack. In addition to the specific nature of the chemical involved, the degree to which concrete resists attack depends upon the temperature of the aggressive solution, the w/c ratio of the concrete, the type of cement used, the degree of consolidation of the concrete, the permeability of the concrete, the degree of wetting and drying of the chemical on the concrete, and the extent of chemically induced corrosion of the reinforcing steel.

3.13.1 Symptoms

Visual inspection of concrete, which has been subjected to chemical attack, usually shows surface disintegration and spalling and the opening of joints and cracks. There also is swelling and general disruption of the concrete mass. Coarse aggregate particles are generally more inert than the cement paste matrix. Therefore, aggregate particles may be seen as protruding from the matrix.

3.13.2 Prevention

Typically, dense concrete with low w/c ratio (max 0.40) provide the greatest resistance. The best-known method of providing long term resistance is to provide a suitable coating.

4

Corrosion

4.1 BASIC PRINCIPLES OF CORROSION

Concrete is a high alkaline material. The pH of newly produced concrete is usually between 12 and 13. In this range of alkalinity, embedded steel is protected from corrosion by a passivating film bonded to the reinforcing bar surface. However, when the passivating film is disrupted, corrosion may take place. The damage is caused by **rebar's corrosion** in concrete. Reinforcement corrosion of bare steel exposed to atmosphere and the steel embedded in concrete. The nature of reinforcement corrosion mechanism can be attributed to three predominant processes, namely **chemical, electrochemical and physical.**

- It is **chemical in the sense** that the alkalinity of concrete can be reduced to a pH level of less than 10.0 by the ingress of carbon dioxide or the passivity of steel can be destroyed by the ingress of chloride thereby initiating corrosion in both cases.

- It is **electrochemical in the sense** that the galvanic cells get established by forming locally or generally cathodic and anodic sites resulting in a flow of current with moist concrete serving as the electrolyte. In this process, the rate of corrosion is influenced by the oxygen supply.

- It is **physical in the sense** that as the corrosion process progresses, the corrosion product (rust) experiences a volume growth as high as six to seven times the original corroding metal. This volume growth exerts physical expansive forces to the concrete surrounding the steel. These forces exceed the tensile strength of concrete, cracking of concrete occurs and as further corrosion takes place, spalling of concrete occurs resulting in exposure of reinforcement.

4.1.1 Influencing Factors

The parameters, which influence the corrosion process, are:

- The cover thickness
- The quality of concrete in the cover region, especially in terms of permeability and diffusivity
- Environment conditions
- pH value in concrete
- Chloride level in concrete
- Presence of cracks, etc.

The problem of corrosion of reinforcement in concrete structures needs serious consideration by the designers and contractors. It is emphasized that a good construction practice can minimize the corrosion problem to a large extent.

4.2 CORROSION MECHANISM OF EMBEDDED METALS

Steel reinforcement is deliberately and almost invariably placed within a few cm of a concrete surface. Under most circumstances, Portland cement concrete provides good protection to the embedded reinforcing steel. This protection is generally attributed to the high alkalinity of the concrete adjacent to the steel and to the relatively high electrical resistance of the concrete. Still, corrosion of the reinforcing steel is among the most frequent causes of damage to concrete.

4.2.1 High Alkalinity and Electrical Resistivity of the Concrete

The high alkalinity of the concrete pore solution can be reduced over a long period of time by carbonation. The electrical resistivity can be decreased by the presence of chemicals in the concrete by carbonation. The chemical most commonly applied to concrete is chloride salt in the form of deicers. As the chloride ions penetrate the concrete, the capability of the concrete to carry an electrical current is increased significantly. If there are differences within the concrete such as moisture content, chloride content, oxygen content, or if dissimilar metals are in contact, electrical potential differences will occur and a corrosion cell may be established. The anodes will experience corrosion while the cathodes will be undamaged. On an individual reinforcing bar there may be many anodes and cathodes, some adjacent, and some widely spaced.

4.2.2 Enhanced Reduction in Load-carrying Capacity of Concrete

As the corrosion progresses, two things occur. First, the cross-sectional area of the reinforcement is reduced, which in turn reduces the load carrying capacity of the steel. Second, the products of the corrosion, iron oxide (rust), expand since they occupy about eight times the volume of the original material. This increase in volume leads to cracking and ultimately spalling of the concrete. For mild steel reinforcing, the damage to the concrete will become evident long before the capacity of the steel is reduced enough to affect its load carrying capacity. However, for prestressing steel, slight reductions in section can lead to catastrophic failure.

4.2.3 Other Mechanisms of Corrosion

In addition to the development of an electrolytic cell, corrosion may be developed under several other situations. The first of these is corrosion produced by the presence of a stray electrical current. In this case, the current necessary for the corrosion reaction is provided from an outside source. A second additional source of corrosion is that produced by chemicals that may be able to act directly on the reinforcing steel.

4.2.4 Symptoms of Metal Corrosion

Visual examination will typically reveal rust staining of the concrete. This staining will be followed by cracking. Cracks produced by corrosion generally run in straight, parallel lines at uniform intervals corresponding to the spacing of the reinforcement. As deterioration continues, spalling of the concrete over the reinforcing steel will occur with the reinforcing bars becoming visible.

4.2.5 Prevention of Metal Corrosion

- Use of concrete with low permeability
- Use of properly proportioned concrete having a low w/c
- Use of as low a concrete slump as practical
- Use of good workmanship in placing the concrete
- Curing the concrete properly
- Providing adequate concrete cover over the reinforcing steel
- Providing good drainage to prevent water from standing on the concrete
- Limiting chlorides in the concrete mixture
- Paying careful attention to protruding items such as bolts or other anchors.

4.3 CORROSION PROCESS

The corrosion process of reinforcement embedded in concrete has two distinct periods namely, initiation period and propagation period. Figure 4.1 shows these periods schematically as service life of a structure. The initiation period, during which the metal, have been embedded in concrete remains passive whilst, with the concrete, environmental changes are taking place that may ultimately terminate passively. The propagation period, during which begins at the moment of de-passivation and involves the propagation of corrosion at a significant rate, until a final state is reached, when the structure is no longer considered acceptable on grounds of structural integrity, serviceability or appearance.

4.3.1 Initiation Period

It is the time taken to initiate corrosion, which can be caused either by ingress of carbon dioxide or chloride ions. The reinforcing steel embedded in concrete is well protected by providing adequate cover thickness and good quality concrete with certain minimum cement content.

Influencing Factors

The factors which influence the corrosion initiation are:

- The environment to which the structure is exposed
- The cover thickness
- Quality of cover concrete in terms of its alkalinity, permeability and diffusion characteristics
- The type of steel
- Critical chloride in concrete
- Presence of cracks.

4.3.2 Propagation Period

After the initiation of corrosion, the propagation begins and this period has two distinct processes. One is that corrosion follows **an electrochemical process** and the other is the **physical process** due to which damage to concrete occurs. During the propagation period, the corrosion progresses at a rate depending on the availability of oxygen and moisture.

Electrochemical Process

The electrochemical process corrosion can be considered as the metallurgy in reverse. Steel is produced from the basic iron ore which is oxide in nature. Energy is added to make the ore into steel and during the electrochemical process by corrosion, electrons get liberated dissipating the energy added and thereby the steel goes back to its oxidized form. This is illustrated in Fig. 4.2.

In respect of reinforcing steel, **the electrochemical process can occur in two situations.** Immediately after production in the factory, the rods come out in light blue colour. During transportation and storage, a thin oxide film gets formed and this acts as a passive layer. However, during handling, it is likely that the passive layer may get mechanically destroyed creating locally depassivated spots.

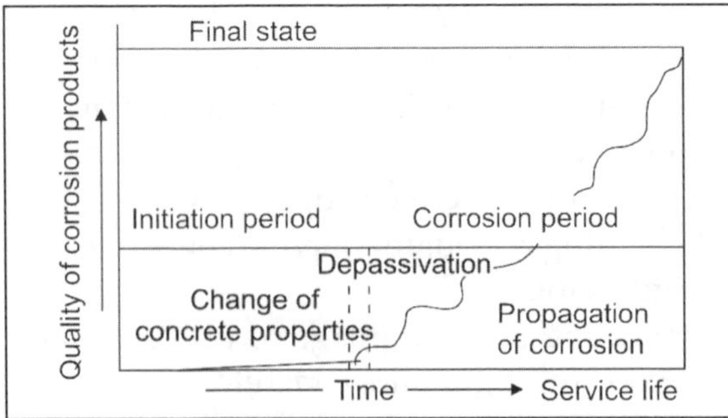

Fig 4.1: Service life model for corrosion affected structures

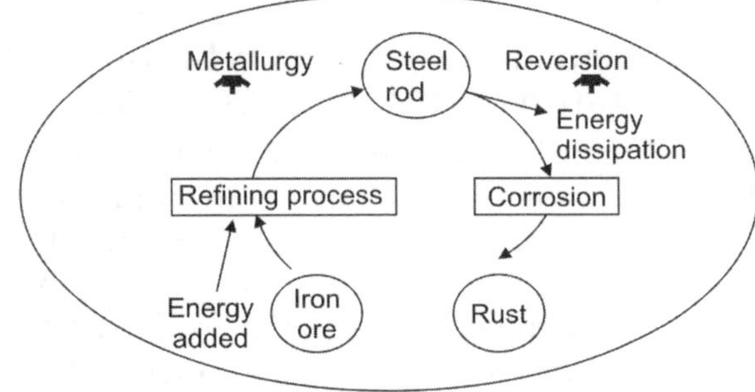

Fig. 4.2: Metallurgy in reverse is corrosion

Such spots in the presence of water and oxygen create galvanic cells, forming anodic and cathodic sites and highly localized corrosion in embedded steel is schematically illustrated in Fig. 4.3.

The process follows an electrochemical phenomenon creating a potential gradient and current flow between the anodic and cathodic location. It is necessary that the reinforcing rods are well protected during storage and this can be achieved by keeping the rods under covered sheds, placing them on wooden supports and providing a cement slurry coating. Another situation of electrochemical process is when the rod is embedded in concrete. Corrosion process of rebar in concrete is illustrated in Fig. 4.4. In this situation, the electrochemical process progress by forming the anodic and cathodic sites, involving chemical reactions.

A moist concrete matrix forms an acceptable electrolyte and the steel reinforcement provides the anode and cathode. Electrical current flows between the cathodes. Electrical current flows between the cathode and anode, and the reaction results in an increase in metal volume as the Fe (iron) is oxidized into $Fe(OH)_2$ and $Fe(OH)_3$ and precipitates as FeOH (rust).

The electrochemical equation is given below:

$$Fe \rightleftharpoons 2e^- + Fe^{++} \quad \text{(anode)}$$
$$Fe(OH)_2$$
$$1/2O_2 + H_2O + 2e^- \rightleftharpoons 2(OH^-) \quad \text{(cathode)}$$
$$4Fe(OH)_2 + 2H_2O + O_2 \longrightarrow 4Fe(OH)_3$$
$$\text{red rust}$$
$$3Fe + 8OH^- \longrightarrow Fe_3OH + 8e^- + 4H_2O$$
$$\text{(black rust)}$$

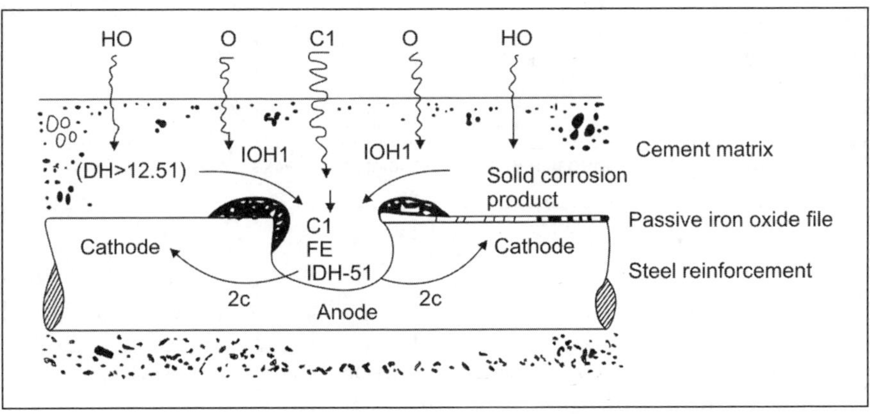

Fig. 4.3: Occurrence of pitting corrosion

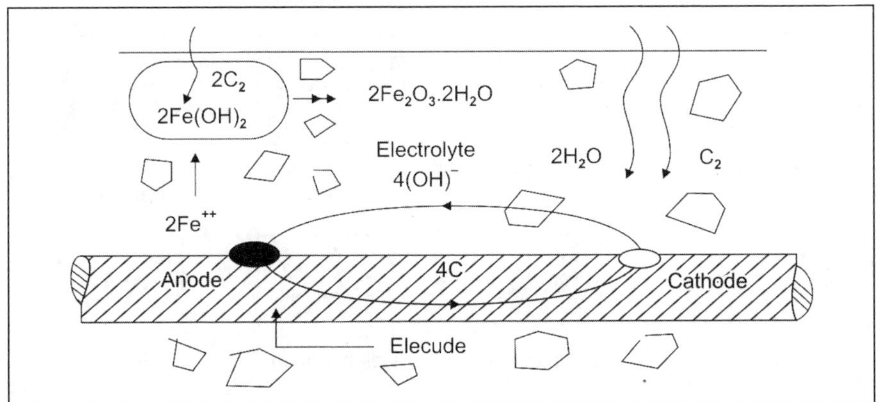

Fig. 4.4: Corrosion process of rebar in concrete

The electrochemical process is greatly influenced by the pH value of concrete and chloride. Water and oxygen must be present for the reaction to take place.

Influencing Factors

The factors, which influence the electrochemical process, can be summarized as follows:

- pH values
- Chloride content
- Moisture within the concrete
- Oxygen supply

In addition to the above factors, electrical resistivity of concrete also influences the electrochemical process. Very dry concrete can have a high resistivity of more than 100 kiloohm · cm. The moisture and other chemicals can reduce the electrical resistivity, thereby increasing the conductivity. It is established that when the resistivity of concrete falls below 5000 ohm·cm, the conductivity of concrete will become high and under such internal environment, the rebar becomes susceptible to corrosion.

Physical Process

The physical process mainly consists of the expansive forces caused by the volume growth of the corrosion product and once the stress induced by this force exceeds the tensile strength of concrete, cracking occurs. As further corrosion takes place, spalling occur. The following Figs 4.5 and 4.6 illustrate the physical symptoms of cracking and spalling mechanism in a slab and a beam respectively.

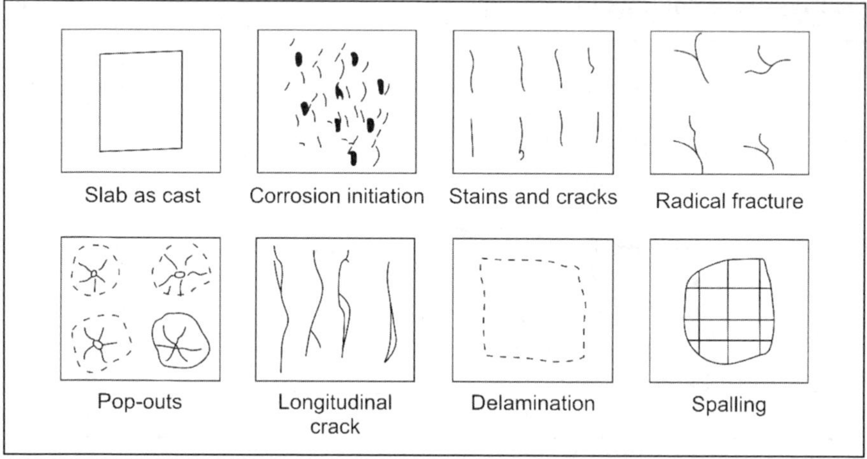

Fig. 4.5: Typical symptoms of corrosion in RC slab

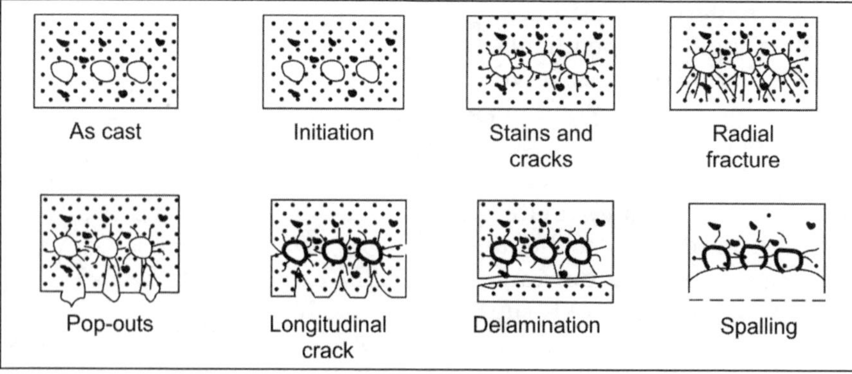

Fig. 4.6: Typical symptoms of corrosion in RC beam

4.4 SYMPTOMS TO CAUSES OF DISTRESS AND DETERIORATION

Given a detailed report of the condition of the concrete in a structure and a basic understanding of the various mechanisms that can cause concrete deterioration, the problem becomes one of relating the observations or symptoms to the underlying causes. When many of the different causes of deterioration produce the same symptoms, the task of relating symptoms to causes is more difficult than it first appears. Although there will usually be a combination of causes responsible for the damage detected on a structure, this procedure should provide a starting point for an analysis.

4.4.1 Evaluate Structural Design to Determine Adequacy

First one has to consider what types of stress could have caused the observed symptoms. For example, tension will cause cracking, while compression will cause spalling. Torsion or shear will usually result in both cracking and spalling. If the basic symptom is disintegration, then overstress may be eliminated as a cause. Second, attempt to relate the probable types of stress causing the damage noted to the locations of the damage. For example, if cracking result from excessive tensile stress is suspected, it would not be consistent to find that type of damage in an area that is under compression. Next, if the damage seems appropriate for the location, attempt to be made to relate the specific orientation of the damage to the stress pattern. If tension cracks are suspected to be roughly perpendicular to the line of externally induced stress, it would not be consistent to find that type of damage in an area that is under compression. Next, if the damage seems appropriate for the location, attempt to be made to relate the specific orientation of the damage to the stress pattern. Tension cracks should be roughly perpendicular to the line of externally induced stress. Shear usually causes failure by diagonal tension, in which the cracks will run diagonally in the concrete section. Visualizing the basic stress patterns in the structure will aid in this phase of the evaluation. If no inconsistency is encountered during this evaluation, then overstress may be the cause of the observed damage. A thorough stress analysis is warranted to confirm this finding. If an inconsistency has been detected, such as cracking in a compression zone, the next step in the procedure should be followed.

4.4.2 Relate the Symptoms to Potential Causes

Depending upon the symptom, it may be possible to eliminate several possible causes. For example, if the symptom is disintegration or erosion, several potential causes may be eliminated by this procedure.

4.4.3 Eliminate the Readily Identifiable Causes

From the list of possible causes remaining after symptoms have been related to potential causes, it may be possible to eliminate two causes very quickly, since they are relatively easy to identify. The first of these is corrosion of embedded metals. It will be easy to verify whether the cracking and spalling noted are a result of corrosion. The second cause that is readily identified is accidental loading, since personnel at the structure should be able to relate the observed symptoms to a specific incident.

4.4.4 Analyze the Available Clues

If no solution has been reached at this stage, all of the evidences generated by field and laboratory investigations should be carefully reviewed. Attention should be paid to the following points:

- If the basic symptom is that of disintegration of the concrete surface, then essentially three possible causes remain: Chemical attack, erosion, and freezing and thawing. Attempts should be made to relate the nature and type of the damage to the location in the structure and to the environment of the concrete in determining which of the three possibilities is the most likely to be the cause of the damage.

- If there is evidence of swelling of the concrete, then there are two possibilities, viz. chemical reactions and temperature changes. Destructive chemical reactions such as alkali-silica or alkali-carbonate attack that causes swelling can be identified during the laboratory investigation. Temperature-induced swelling should be ruled out unless there is additional evidence such as spalling at joints.
- If the evidence is spalling and corrosion and accidental loadings have been eliminated earlier, the major causes of spalling remaining are construction errors, poor detailing, freezing and thawing, and externally generated temperature changes. Examination of the structure should have provided evidence as to the location and general nature of the spalling that will allow identification of the exact cause.
- If the evidence is cracking, then construction errors, shrinkage, temperature changes, settlement and movement, chemical reactions, and poor design details remain as possible causes of distress and deterioration of concrete. Each of these possibilities will have to be reviewed in light of the available laboratory and field observations to establish which is responsible.
- If the evidence is seepage and it has not been related to a detrimental internal chemical reaction by this time, then it is probably the result of design errors or construction errors, such as improper location or installation of a water stop.

4.4.5 Determination of Why the Deterioration has Occurred

Once the basic cause or causes of the damage have been established, there remains one final requirement to understand how the external agent acted upon the concrete. For example, if the symptoms were cracking and spalling and the cause was corrosion of the reinforcing steel, what facilitated the corrosion? Was there chloride in the concrete? Was there inadequate cover over the reinforcing steel? Another example to consider is concrete damage caused by freezing and thawing. Did the damage occur because the concrete did not contain an adequate air-void system, or did the damage occur because the concrete used was not expected to be saturated but, for whatever reason, was saturated? Only when the cause and its mode of action are completely understood should the next step of selecting a repair material be attempted.

4.5 DAMAGES DUE TO CORROSION

In RC structures, the corrosion of reinforcement is unique in the sense that the corrosion process causes extensive damage to the concrete. The corrosion product has a volume growth as high as the original. This volume growth exerts physical expansive forces to the concrete surrounding the steel. Once the stresses induced by the these forces exceed the tensile strength of concrete, cracking of concrete occurs, and as further corrosion takes place, spalling of concrete occurs resulting in exposure of reinforcement. Generally presence of active corrosion process in a reinforcement of a concrete member becomes known only when the symptom, namely, corrosion stain and or cracking is manifested. In good quality concrete the corrosion rate will be very slow. Accelerated corrosion will take place if the pH (alkalinity) is lowered (carbonation) or if aggressive chemicals or dissimilar metals are introduced into the concrete.

Other causes include stray electrical currents and concentration cells caused by an uneven chemical environment. Cracking and spalling of concrete induced by steel corrosion is a function of the following variables:

- Concrete tensile strength
- Quality of concrete over the reinforcing bar
- Bond or condition of the interface between the rebar and surrounding concrete
- Diameter of the reinforcing bar
- Percentage of corrosion by weight of the reinforcing bar.

The structural capacity of a concrete member is also affected by bar corrosion and cracking of surrounding concrete. In flexural and compressive member, cracking and spalling of concrete reduces the effective cross section of the concrete, thereby reducing the ultimate load carrying capacity.

4.6 CODAL PROVISIONS FOR DIFFERENT EXPOSURE CONDITIONS

Steel embedded in concrete is generally well protected by adequate cover thickness, concrete with certain minimum cement content, control on water cement ratio, and good compaction to produce impermeable concrete. However, when discrepancies on above factors exist, the protecting ability of concrete may get destroyed.

4.6.1 Environment

The environment can be broadly classified as marine and marine industrial. It is now recognized that with ever increasing pollution, one of the above environments exists in any given geographical location. This environment, for the purpose of construction of reinforced concrete structures, can further be categorized as exposure conditions. The categorized conditions as per code IS:456-2000 are given in Table 4.1.

4.6.2 Cover Thickness

The cover thickness is the primary physical protection to steel reinforcement. The importance of providing the stipulated cover thickness should be given due consideration during construction. Larger cover thickness obviously will delay the initiation since the time taken for the aggressive chemical such as CO_2 and chloride to penetrate will be longer. Optimum cover thickness is to be chosen taking into account the structural design aspects. However, certain minimum nominal cover thicknesses are to be provided depending on the exposure conditions (environment) and values stipulated in the recent draft code of IS:456-2000 are given in Table 4.2.

Table 4.1: Exposure conditions

Environment	Exposure condition
Mild	• Concrete surfaces protected against weather or aggressive conditions except those situated in coastal area
Moderate	• Concrete surfaces sheltered from severe rain or freezing • Concrete exposed to condensation and rain • Concrete continuously under water • Concrete completely immersed in seawater
Severe	• Concrete surfaces exposed to severe rain, alternate wetting and drying of occasional freezing or severe condensation • Concrete completely immersed in seawater
Very severe	• Concrete surfaces exposed to seawater spray corrosive fumes or severe freezing • Concrete exposed to aggressive subsoil ground water or coastal environment.
Extreme	• Surface of members in tidal zone • Members in direct contact with liquid/solid aggressive chemicals.

Table 4.2: Nominal cover to meet durability requirement

S.No.	Exposure conditions	Nominal concrete cover in mm not less than
1.	Mild	20
2.	Moderate	30
3.	Severe	45
4.	Very severe	50
5.	Extreme	75

4.6.3 Quality of Concrete

Corrosion initiation is largely influenced by the quality of concrete especially in the cover region. Both chemical and physical parameters of concrete need to be considered. The primary chemical parameter is its alkalinity expressed in terms of its pH value. The alkalinity is provided by the internal pore solution of concrete, which is basically calcium hydroxides. This will give the pH value of more than 12.0 and this can be best achieved by using proper cement and certain minimum quantity. Realizing the importance, the minimum requirement of cement content for concrete for different exposure conditions is stipulated in the IS:456-2000 and these are given in Table 4.3.

4.7 CORROSION PROTECTION TECHNIQUE

The detailed analysis of the factors that influence corrosion mechanism and process indicates that corrosion protection requires a multiple approach. There are many possible approaches as listed below:

- Coating to reinforcement
- Galvanized reinforcement
- Improving metallurgically by addition of certain elements
- Using stainless steel
- Using nonferrous reinforcement
- Using corrosion inhibitors
- Coating to concrete
- Cathodic protection, either by means of impressed unit or by sacrificial anodes.
- Electrochemical chloride removal
- Improving the cover concrete.

4.7.1 Coating to Reinforcement

The objective of coating to steel rebar is to provide a sufficiently durable barrier to aggressive materials such as chlorides. Initially, the bar is shot blasted to remove mill-scale. This ensures an adequate bond between the epoxy and the steel. The bar is then heated to a controlled temperature before passing through a spray booth. Hence, electrostatically charged epoxy powder particles are deposited evenly on the surface of the bar. Shortly after spraying, the epoxy starts to act and hardens sufficiently for the bars to be handled. The coating thickness typically varies from 130 micron to 300 micron.

4.7.2 Galvanized Reinforcement

Galvanized reinforcement consists of standard black bar, hot dipped in molten zinc. This process forms a coating which is metallurgically bonded to the surface of the metal. The surface of the zinc reacts with calcium hydroxide in the concrete to form a passive layer, preventing corrosion. However, it is more difficult process to control as uneven coatings can form and also it is more expensive.

4.7.3 Improving Metallurgically by Adding Certain Elements

The bars can be improved for its corrosion resistance by adding certain elements such as chromium and copper during the formation/process itself.

4.7.4 Stainless Steel Reinforcement

Stainless steel is the name given to a family of corrosion resistant steels containing a minimum of 12% chromium. On contact with air, the chromium forms a thin oxide layer on the surface of steel. This is passive and resists corrosion. The austenic group of steels has superior corrosion resistance and is the most suitable for use in reinforced concrete.

4.7.5 Nonferrous Reinforcement

Currently a number of manufacturers who are developing nonferrous reinforcement as an alternative to the conventional steel in traditional structures. A range of man-made fibers is used, the most common being glass, carbon and aramid. The fibers are used either in the form of ropes or combined with suitable resins to form rods. It may be seen that aramid, glass and carbon fibers have ultimate tensile stresses well in excess of that of reinforcing steel and well resistance to corrosion.

Table 4.3: Minimum cement content for different exposure conditions

S.No.	Exposure	Plin concrete in kg/m^3	Reinforcement concrete in kg/m^3
1.	Mild	220	300
2.	Moderate	250	300
3.	Severe	260	350
4.	Very severe	280	375
5.	Extreme	300	375

4.7.6 Corrosion Inhibitors

Certain admixtures can be made to inhibit corrosion of the reinforcement in the presence of chlorides. Examples are:

- The addition of calcium nitrate extends the time to corrosion initiation.
- Total corrosion of samples with calcium nitrate is substantially less
- The corrosion rate, once corrosion is initiated is less with calcium nitrate.

4.7.7 Concrete Coatings

A concrete surface coating of silane-siloxane type was evaluated for its corrosion performance. By several test conducted and the results show the chloride penetration depth is minimized in coated specimen compared with uncoated specimen.

4.7.8 Cathodic Protection

Cathodic protection is a technique by which the electrical potential of the steel is increased to a level at which corrosion cannot take place. It is widely used for both steel and concrete offshore structures, while on land it has been used for the protection of pipelines and similar structures. Two different methods are employed, an impressed current and the use of sacrificial anodes. In the first case, the structure is connected to the negative terminal of a DC power source, ideally using an anode, which does not corrode. In the second case, the reinforcement is connected to anodes with a more negative corrosion potential than steel, such as zinc or aluminum. In both cases electrical continuity of the reinforcement is required. The principle of cathodic protection have been used to remove chlorides from contaminated concrete.

4.7.9 Electrochemical Chloride Removal

This is another emerging area with lot of potentialities. This technique needs only a temporary installation lasting few days. 20 to 50% of the chloride present in concrete can be removed. An electrical current in the range of 1 to 5 A/m^2 is needed for this purpose.

4.7.10 Improving the Concrete

Codes and standards aim to achieve good durability of reinforced and prestressed structures in aggressive environments by specifying the following:

- High cement content
- Low water/cement ratio
- Suitable minimum thickness of cover to the reinforcement
- Careful adequate curing.

5

Evaluation of Concrete and Reinforced Concrete Structures

Proper evaluation of concrete/reinforced concrete requires several steps. The requirements for repair or rehabilitation are generally established in one of three ways. In the worst case, there is a failure of structural integrity that prompts the need for an evaluation. Visual inspection often reveals a need for further evaluation and testing. If cracks, flaking, or other visual defects are encountered, a full investigation and evaluation are usually warranted. The third way to establish the need for an evaluation is periodic inspection and testing. This is the safest way to check the strength and dependability of concrete structures.

5.2 REVIEWING THE RECORDS

Construction documents are normally filed, while getting the permission for construction from various authorities (for example, Municipal Authority). However, normally as built documents are supposed to be made and kept on records. In the case of concrete/RCC construction, reviewing original documents and a paper trail over the history of the structure is one step in a thorough evaluation. Here is an example of existing records that should be available for consideration:

- Design documents
- Plans, elevations and sections
- Specifications
- History reports
- Inspection reports
- Site inspection results
- Laboratory test records
- Concrete records on the materials used in construction and the batching plant
- Instrumentation documents
- Operation reports
- Maintenance reports
- Monument survey data, etc.

5.3 SITE SURVEY

A site survey consists of a visual exam of all exposed concrete/RCC. The surveyor/engineer looks for any implication of distress that may require repair or rehabilitation. The standard procedure is to create a map of potential defects that are encountered. These could include cracks, disintegration, spalling, or joint deterioration. The mapping process is typically done with foldout sketches of the monolith surfaces. Mapping must include inspection delineating of pipe and electrical galleries, filling, and emptying, culverts when possible. Core drilling is a common procedure when evaluating concrete. The material recovered from the drilling can be taken to laboratory for a scientific testing.

5.4 SHODDY WORKMANSHIP

Shoddy workmanship can lead to significant problems in concrete structures. Site inspection can bring such problems to light and result in further testing. Some of the normal defects that are looked for in a visual inspection include the following:

- Cold joints
- Bug holes
- Reinforcing steel that has become exposed
- Honeycombing
- Surface defects that can indicate serious problems.

Workmanship should be monitored closely during the construction process. In a perfect world, corners are cut, and problems can result from the lack of professionalism. Since safety is a paramount concern, routine inspections are needed to confirm that the workmanship in a structure is as it should be.

Cracking

Cracking is a common problem in concrete construction. Home owners see it in basement floors, garage floors, etc. Cracks occur in sidewalks, dams, and retaining walls, etc. Any crack appears is a reason for concern and warrants a thorough inspection and investigation.

5.5 PATTERN CRACKING

Pattern cracks are common. The cracks tend to be short and uniformly distributed throughout a concrete surface. Pattern cracking can have two causes: It can indicate restrained of contraction on the surface layer by the backing or inner concrete, or it can be due to an increase in the volume in the interior concrete. Pattern cracking is referred to as map cracks, crazing, checking, or D-cracking. D-cracking is often found in the lower part of a concrete slab, usually near a joint in the concrete. In case of moisture accumulation, one could find D-cracking.

5.6 ISOLATION CRACKS

Isolation cracks appear as individual cracks. This type of cracking indicates tension on the concrete. This tension is usually perpendicular to the cracks. An individual crack can run in a diagonal, longitudinal, transverse, vertical, or horizontal direction.

5.6.1 Crack Width

Crack width ranges from fine to medium to wide. Fine cracks are typically less than 1 mm wide. A medium crack would be between 1 mm and 2 mm; wide cracks exceed 2 mm.

5.6.2 Crack Activity

Crack activity has to do with the presence of a particular factor causing a crack. Determining crack activity is necessary to determine a mode of repair. If the cause of crack is causing more cracking, then the crack is active. If a specific cause for a crack cannot be determined, the crack must be considered active.

Dormant cracks do not have current movement. Some cracks are considered dormant when any movement of the crack is nominal enough to not interfere with a repair plan.

5.6.3 All Crack Occurrences

Cracks can occur before or after concrete cures. The cracking can be structural or non-structural. It may be hard to determine whether a crack is structural or non-structural by visual inspection only. A full analysis by a structural engineer is normally required to make a full determination of the type of cracking that is encountered.

Structural cracks tend to be wide. Their openings can increase as a result of continuous loading and creep of the concrete. As a rule of thumb, any crack that could be structural in nature should be treated as structural defect and receive a full evaluation from appropriate experts. Table 5.1 shows age of concrete at time of vibration and peak particle velocity of ground vibration.

Table 5.1: Age of concrete at time of vibration (hr) and peak particle velocity of ground vibrations

Upto 3	102 mm/sec
3–11	38 mm/sec
11–24	51 mm/sec
24–48	102 mm/sec
Over 48	178 mm/sec

5.7 DISINTEGRATION AND SPALLING

Disintegration is the deterioration by any means of concrete. The mass of particle being removed from the main body of concrete distinguishes disintegration from spalling. When concrete is poured, the final deposit method must not separate or lose materials in the process.

5.7.1 Scaling

Scaling is a form of disintegration. A common cause of scaling is freezing and thawing conditions. Localized flaking or peeling is normally a form of scaling. Scaling, which can also be referred to as spalling, is rated based on the depth of defect. The rating system is as follows:

- Light spalling is when the loss of mortar does not expose any coarse aggregate.
- Medium spalling occurs when the damage has a depth of up to 1cm.
- Severe spalling is determined when the depth of the damage ranges from 0.5 cm to 1 cm. There is also some loss of mortar surrounding aggregate particles with a depth range of 1 cm to 2 cm.
- Very severe spalling is the loss of coarse aggregate particles, as well as surface mortar and surrounding aggregate, generally to a depth greater than 2 cm.

5.7.2 Dusting

Dusting occurs when there is a development of a powdered material at the surface of hardened concrete. Horizontal concrete surfaces that receive heavy traffic are the most likely surfaces to find dusting occurring. Poor workmanship practices are usually the cause of the problem. Anyone who has worked in construction for any length of time has seen finishers spray water on a concrete surface during the finishing process. This often causes dusting. Concrete that is being poured should be installed as closely as practicable to the final resting place of concrete.

5.7.3 Distortion

Distortion of concrete is also known as movement. In simple way, this is a change in alignment of the components of a structure. This could be wall movement or support movement. When evaluating distortion, historic data is likely to be helpful. Assuming that good records have been maintained over the years, one can find out a history of continuing failure due to distortion.

5.8 EROSION

Erosion is a common concern in concrete construction. Experts generally divide erosion into two categories: Abrasion and cavitations. Abrasion is recognized by the smooth surface it leaves behind on concrete. This is due to repeated rubbing and grinding of debris, equipment, gravel, or other items against concrete. Repeated impact forces that are caused by a collapse of a vapour bubble in rapidly flowing water cause cavitations. Erosion caused by water does not generally leave a smooth surface on concrete. A rough, pitted concrete surface is a sign of cavitations. In severe cases, cavitations can result in structural damage. Statistics indicate that the cavitations generally require a water velocity of at least 100 cm per second.

5.9 JOINT SEALS AND SEEPAGE

Joint seal can fail. The seals are intended to repel water. If a seal fails and water invades a concrete joint, buckling, cracking, erosion, and other problems may occur. Another purpose of joint seals is to prevent debris from entering an expansion joint. When debris embeds in a joint, it can result in a failure with the expansion joint.

Seepage consists of water or other fluids moving through pores or interstices. When conducting a visual inspection, an inspector can check for seepage by looking for the following:

- Water
- Dampness
- Moisture
- Corrosion
- Discoloration
- Staining
- Exudations
- Efflorescence
- Incrustations

Seepage is a common problem around hydraulic structures. Any seepage found should be reported. These data are necessary to maintain historical facts for future review. Cracks that are associated with hardening concrete fall into two categories: those that occur while concrete are hardening and those that occur after the concrete has hardened. Knowing when a crack occurs and what caused the cracking is always of interest to those who evaluate concrete construction. Concrete must remain plastic and flow rapidly when it is being installed.

5.10 SPECIAL CASES OF SPALLING

Two special cases are involved in spalling. The first type is known as pop out. These defects are shallow and are usually conical depressions in the surface of the concrete. When concrete is poured in freezing conditions, especially if some unsatisfactory aggregate particles are present, one can find pop outs.

A pop out is created when water finds its way into coarse aggregate and then freezing begins. The ice pushes off the top of aggregate particle and the adjacent layer of mortar. This leaves shallow pitting. Certain materials are more susceptible to pop outs than others. The second phase of spalling involves the corrosion of reinforcement material. A visual inspection for this type of defect is fairly simple. Evidence will be exposed reinforcement materials that are protruding through the concrete. Rust staining on the reinforcement material is also present in many cases.

5.10.1 Delamination

Delamination can occur when reinforcing steel is placed too close to the surface of a concrete structure. Chloride ions or air that create rust or iron oxide corrodes steel reinforcement material. A corresponding increase in volume up to eight times the original volume amount is possible. The increase volume can crack concrete.

A simple, inexpensive method is available to test for delamination. All one need is a pair of safety glasses and a hammer. One has to tap the concrete with the hammer to check for defects. A sharp "ping" sound is a good sign that the delamination is not present. However, a hollow, echo sound means more tests are needed. This type of test is common when working with small surface areas. Larger surface areas require extensive time to check with a hammer. Horizontal surfaces can be tested by dragging a chain over the area. One has to listen for that "ping" sound. This is a simple way to test more concrete in less time.

Infrared thermography is a more advanced method for inspecting concrete for delamination. The thermal gradients within concrete that is exposed to sunlight can be measured with thermography equipment. Delamination interrupts the heat transfer through concrete. Higher surface temperatures will be present if delamination exists. Infrared thermography is capable of identifying and recording areas that are affected by delamination.

5.11 CRACK SURVEYS

Crack surveys are needed for historical records and to expose potential problems. Obviously, cracks are not supposed to exist in concrete structures. When they are found, the cracks must be identified. Typically, cracks are marked and researched. Once the type and cause of cracks are known, they can be assessed and recorded for future review.

5.11.1 Sizing Cracks

Cracks can be sized with several methods. A simple card that contains lines of various widths can be used to estimate crack size. Crack monitors, which are small, handled microscopes, are used for more specific measurements. There are also transducers that can be used for crack measuring. Once the size of a crack is determined, it should be recorded in historical data of the concrete structure. By doing this, the cracks can be monitored and measured to determine, if they are growing in size.

Obtaining the depth of cracks is often simple, but in some cases, it can be extremely difficult. A small measuring device, such as a feeler gauge, can be used to establish the depth of some cracks. It is not uncommon for simple methods to fail in determining the depth of crack. When this is the case, we must turn sophisticated methods, such as drilling or pulse-velocity measurements.

5.12 SURFACE MAPPING

Surface mapping is important when establishing the history of a concrete surface over time. These types of defects that are sought in surface mapping may include the following:
- Cracking
- Spalling
- Scaling
- Pop outs
- Honeycombing
- Exudation
- Distortion
- Unusual dislocation
- Erosion
- Cavitations
- Seepage
- Joint condition
- Joint materials
- Corrosion of reinforcement materials

For performing surface mapping by hand will need certain tools which are:
- Structural drawings
- Historical data
- Documentation tools, such as a notebook computer or a notepad and pen.
- Tape measure
- Ruler

- Feeler gauge
- Hand microscope
- Knife
- Hammer
- Fine wire
- String
- Flashlight
- Camera outfit
- Tape recorder

5.13 JOINT INSPECTIONS

Joint inspections can be done with a visual tour of all joints. Expansion, contraction, and construction joints should all be inspected. The condition of joints, both good and bad, should be noted for inclusion in the historical record. Some potential defects to look for include spalling, D-cracking, chemical attacks, seepage, and any emission of solids.

5.14 CORE DRILLING

It is expensive. If the quality of concrete in a structure is suspected to be weakened with general inspections, core drilling may be resorted to. Scaling, leaching, or pattern cracking can be the signs of the need for core drilling.

How deep does core drilling go? It depends on the structure. For example, a massive structure may require core sampling to be done at a depth up to 60 cm. The diameter of a core sample should be at least three times the nominal minimum size of aggregate. When there is little mortar bonding the concrete across the diameter of the core, one is likely to see only rubble rather than a solid sample. Core samples must be properly leveled, oriented, and stored for future observation. Written records are also required to maintain consistency in the historical data.

There are times when a bore-hole camera is helpful. The use of such an instrument can reveal facts about the inner condition of a concrete structure. For example, if core samples come up as rubble, the bore-hole camera may be the best alternative.

Causes of Distress and Deterioration of Concrete and Reinforced Concrete Structures

6.1 INTRODUCTION

The list of potential causes of distress and deterioration of concrete is a long one. A few examples include chemical reactions, shrinkage, weathering and erosion. Many other potential causes exist, and we will explore them individually. Understanding the factors that can damage concrete structures is an important element in the business of rehabilitation and repair work.

6.1.1 Accidental Loadings

Accidental loadings are not common. This is why they are accidental. When an earthquake occurs and affects concrete structures, that action is considered an accidental loading. This type of damage is generally short in duration and few and far between in occurrences.

A visual inspection is likely to reveal spalling or cracking when accidental loadings occur. Unfortunately, this type of damage cannot be prevented because the causes are not expected and are difficult to prepare for. For example, an engineer does not expect a ship to hit a piling on a bridge, but it happens. The only defense is to build with as much caution and anticipation as possible.

6.2 CHEMICAL REACTIONS

Concrete damage can occur when chemical reactions are present. It can be surprising how small an amount of chemicals can do serious structural damage to concrete. To expand on this, let us take a look at some examples of chemical reactions and how they affect concrete.

Acid

Most people know that acid can have serious reactions with a number of materials. Concrete can also be affected by acid exposure. When acid attacks concrete, it concentrates on the products of hydration of cement. For example, calcium silicate hydrate can be adversely affected by exposure to acid. Sulfuric acid works to weaken concrete, and if it is able to reach the steel reinforcing members, the steel can be compromised. All of this contributes to a failing concrete structure. Visual inspections may reveal a loss of cement paste and aggregate from the matrix. Cracking, spalling and discoloration can be expected when acid deteriorates steel reinforcements. Laboratory analysis may be needed to identify the type of chemical causing the damage. It is a code violation to embed aluminum conducts and pipes in concrete, unless the aluminum is coated or covered to prevent aluminum concrete reaction or electrolytic action between aluminum and steel.

6.3 CAUSES OF DISTRESS AND DETERIORATION OF CONCRETE

This can happen due to the following:

- Accidental loadings
- Chemical reactions
- Acid attack
- Aggressive water attack
- Alkali–carbonate rock reaction
- Alkali–silica reaction
- Miscellaneous chemical attack
- Sulfate attack
- Construction errors
- Corrosion of embedded metals
- Design errors
- Inadequate structural design
- Poor design details
- Erosion
- Abrasion
- Cavitation
- Freezing and thawing
- Settlement and movement
- Shrinkage
  - Plastic
  - Drying
- Temperature change
  - Internally generated
  - Externally generated
- Fire
- Weathering

How can one create a more defensive concrete, an approved coating or treatment is about the best one can do. Using a dense concrete with a low water–cement (w/c) ratio can provide acceptable protection against mild acid exposure. An aggressive water attack has to do with water that has low concentrations of dissolved minerals. Soft water is aggressive water, and it will leach calcium from cement paste or aggregate. When this type of attack occurs, it is a slow process. The danger is greater in flowing waters. This is due to a fresh supply of aggressive water coming into contact with the concrete.

If we conduct a visual inspection and find rough concrete where the paste has been leached away, it could be an aggressive water defect. Water can be tested to determine if the water quality is responsible for the damage. When testing indicates that water may create problems prior to construction, a cement based coating can be applied to the exposed concrete structures. When conduits are installed the diameter of the conduit should not be more than one-third of the overall thickness of the concrete slab where it is being installed. Alkali-carbonate rock reaction can result in damage to concrete, but it can also be beneficial. Our focus is on the destructive side of this action. This occurs when impure dolomitic aggregates exist. When this type of damage occurs, we are likely to find map or pattern cracking and the concrete will look as if is it swelling.

Alkali-carbonate rock reaction differs from alkali-silica reaction; in that there is a lack of silica gel exudations at cracks. Petrographic examination can be used to confirm the presence of alkali-carbonate rock reaction. To prevent this type of problem, contractors should avoid using aggregates that are or suspected to be reactive. Conduits embedded in concrete must be spaced at a minimum distance that would be equal to not less than three times the diameter of the conduit being installed.

6.3.1 Alkali–Silica Reaction

An alkali–silica reaction can occur when aggregates containing silica that is soluble in highly alkaline solutions may react to form a solid, nonexpansive, calcium-alkali-silica complex or an alkali-silica complex that can imbibe considerable amounts of water and expand. This can be disruptive to concrete.

6.3.2 Various Chemical Attacks

Concrete is fairly resistant to chemical attack. For a substantial chemical attack to have degrading effects of a measurable nature, a high concentration of chemical is required. Solid, dry chemicals are rarely a risk to concrete. Chemicals that are circulated in contact with concrete do the most damage. When concrete is subjected to aggressive solutions under positive differential pressure, the concrete is particularly vulnerable. The pressure can force aggressive solutions into the matrix. Any concentration of salt can create problems for concrete structures. Temperature plays a role in concrete destruction with some chemical attacks. Dense concrete that has a low w/c provides the greatest resistance. The application of an approved coating is another potential option for avoiding various chemical attacks. When concrete joints are created, they must be located in a manner that will not have an adverse effect on the strength of the concrete in which they are installed.

6.3.3 Sulfate Attack

A sulfate attack on concrete can occur from naturally occurring sulfate of sodium, potassium, calcium or magnesium. These elements can be found in soil or groundwater. Sulfate ions in solution will attack concrete. Free calcium hydroxide reacts with sulfate to form calcium sulfoaluminate, and reaction can result in an increase in volume. Additionally, a purely physical phenomenon in which the growth of crystals of sulfate salts disrupts the concrete can occur. Map and pattern cracking and general disintegration of the concrete are signs of a sulfate attack. Preventing sulfate attacks can usually be done with the use of a dense, high quality concrete that has a low water-cement ratio. If pozzolana is used, a laboratory evaluation should be done to establish the expected improvement in performance.

6.3.4 Corrosion

Corrosion of steel reinforcing members is a common cause of damage to concrete. Rust staining will often be present during a visual inspection if corrosion is at work. Cracks in concrete can tell a story. If they are running in straight lines, as parallel lines at uniform intervals that correspond with the spacing of steel reinforcement materials, one can suspect corrosion is at the root of the problem. At time, spalling also occur. Eventually, the reinforcing material will become exposed to a visual inspection. Unless otherwise authorized, it is a code violation to weld crossing bars that will reinforce concrete. Techniques for stopping or controlling corrosion include the use of concrete with low permeability. In addition, good workmanship is needed. Here are some tips to follow:

- Use as low a concrete slump as practical
- Cure the concrete properly
- Provide adequate concrete cover over reinforcing material
- Provide suitable drainage
- Limit chlorides in the concrete mixture
- Pay special attention to any protrusions, such as bolts and anchors.

6.3.5 Design Mistakes

Design mistakes are divided into two categories those that are a result of inadequate structural design and those that are a result of a lack of attention to relatively minor design details. In the case of structural design errors, the result can be anticipated. It will generally result in a structural failure.

Identifying structural design mistakes involves two types of symptoms. Spalling indicates excessively high compressive stress. Cracking and spalling can also indicate light torsion or shear stresses. High tensile stresses will cause cracks. Petrographic analysis and strength testing of concrete is required if any of the concrete elements are to be reused after such failures. The best prevention requires careful attention to detail. Design calculations should be checked thoroughly. Flaws in design details account for most of these types of problems. The following are some examples of design factors to consider.

- Poor design details
- Abrupt changes in section

- Insufficient reinforcement at re-entrant corners and openings
- Inadequate provision for deflection
- Inadequate provision for drainage
- Insufficient travel in expansion joints
- Incompatibility of materials
- Neglect of creep effect
- Rigid joints between precast units
- Unanticipated shear stresses in piers, columns, or abutments
- Inadequate joint spacing in slabs.

6.3.6 Abrasion

Abrasion damage can occur from waterborne debris. The debris typically rolls and grinds against concrete when it is in the water and in contact with concrete structures. Spillway aprons, stilling basin slabs, and lock culverts and laterals are the most likely types of structures to be affected by abrasion. This is usually a result of poor hydraulic design. Another cause for abrasion can be a boat hull hitting a concrete structure.

When groups of reinforcing bars are bundled together as a reinforcement device for concrete, the bundle must not contain more than four bars.

When abrasion is in play, concrete structures tend to wind up with a smooth surface. Long, shallow grooves in a concrete surface and spalling along monolith joints indicate abrasion. The three major factors in avoiding abrasion damage are design, operations, and materials. Here are some tips to keep in mind.

- Use hydraulic model studies to test designs.
- A 45 degree fillet installed on the upstream side of the end sill has resulted in a self-cleaning stilling basin.
- Recessing monolith joints in lock walls and guide walls will minimize stilling basin spalling caused by large impact and abrasion.
- Balanced flows should be maintained into basins by using all gates to avoid discharge conditions where eddy action is prevalent.

- Periodic inspections are needed to locate the presence of debris.
- Basins should be cleaned periodically.
- All materials used must be tested and evaluated.
- Install abrasion-resistant concrete.
- Fiber reinforced concrete should not be used for repairing stilling basins or other hydraulic structures that are subject to abrasion.
- Coatings that produce good results against abrasion include polyurethanes, epoxy-resin mortar, furan-resin mortar, acrylic mortar, and iron aggregate toppings.

6.3.7 Cavitation

Cavitation–erosion is a result of complex flow characteristics of water over concrete surfaces. For damage to occur, the rate of water flow normally has to exceed 12 metre per second. Fast water and irregular surface areas of concrete can result in cavitation. Now we get to the interesting part. The surface irregularity and water speed create bubbles. The bubbles are carried downstream and have a lowered vapor pressure. Once the bubbles reach a stretch of water that has normal pressure, the bubbles collapse. The collapse is an implosion that creates a shock wave. Once the shock wave reaches a concrete surface, the wave causes a very high stress over a small area. When this process is repeated, pitting can occur. This type of cavitation has affected concrete spillways and outlet works of many high dams. Prevention has to do with design, materials, and construction practices. The following list highlights some of the key considerations.

- Include aeration in a hydraulic design
- Use concrete designed with low w/c
- Use hard, dense aggregate particles
- Steel-fiber concrete and polymer concrete can aid in the fight against cavitation
- Neoprene and polyurethane coatings can assist in the fight against cavitation. However, coatings are rarely used because they might prevent the best adhesion to

concrete. Any rip or tear in the coating can cause a complete stripping of the coating over time.

- Maintain approved construction practices
- Bundle reinforcing bars must be enclosed within either stirrups or ties.

6.3.8 Freezing and Thawing

A pattern of freezing and thawing during the curing of concrete is a serious concern. Each time the concrete freezes, it expands. Hydraulic structures are especially vulnerable to this type of damage. Fluctuating water levels and under spraying conditions increases the risk. Using deicing chemicals can accelerate damage to concrete. It will cause pitting and scaling. Core samples are likely to be needed to assess the damage. Prevention is the best cure. Provide adequate drainage, where possible. Work with low w/c concrete. Use adequate entrained air to provide suitable air-void systems in the concrete. Select aggregates best suited for the application. Make sure that the concrete cures properly.

6.3.9 Settlement and Movement

Settlement and movement can be the result of differential movement or subsidence. Concrete is rigid and cannot stand much differential movement. When it occurs, stress cracks and spall are likely to occur. Subsidence causes entire structures, or single elements of entire structures, to move. If subsidence occurs, the concern is not cracking or spalling. The big risk is stability against overturning or sliding. A failure via subsidence is generally related to a faulty foundation. Long-term consolidations, new loading conditions and related faults are contributors to subsidence. Geotechnical investigations are often needed when subsidence is evident. Things to look for when structure movement is suspected included cracking, spalling, misaligned members and water leakage. Specialists are normally needed for these types of investigations. Spiral reinforcement for cast-in place concrete must not be less than 2.5 cm in diameter.

6.3.10 Shrinkage

Shrinkage occurs when concrete is deficient in its moisture content. The shrinkage can occur while the concrete is setting or after it is set. When the condition occurs during setting, it is called plastic shrinkage. Drying shrinkage happens after the concrete is set. Plastic shrinkage is associated with bleeding which is the appearance of moisture on the surface of concrete. This is usually caused by the setting of heavier components in a mixture. Bleed water typically evaporates slowly from the surface of concrete. When evaporation is occurring faster than water is being supplied to the surface by bleeding, high-tensile stresses can develop. The stress can lead to cracks on the concrete surface.

Cracks caused by plastic shrinkage usually occur within a few hours of concrete placement. The cracks are normally isolated. They also tend to be wide and shallow. Patterns cracks are not generally caused by plastic shrinkage. Spacing requirements for shrinkage and temperature reinforcement must be spaced not farther apart than five times a slab's thickness and not farther apart than 20 cm. Weather conditions contribute to plastic shrinkage. If the conditions are expected to be conducive to plastic shrinkage, protect the pour site. This can be done with windbreaks, tarps, and similar arrangements to prevent excessive evaporation. In the event the early cracks are discovered, revibration and refinishing can solve the immediate problem.

Drying shrinkage is a long-term change in volume of concrete caused by the loss of moisture. A combination of this shrinkage and restraints will cause tensile stresses and lead to cracking. The cracks will be fine, and there will no indication of movement. The cracks are typically shallow and only a few cms apart. Look for a blocky pattern to the cracks. The identification can be confused with thermally induced deep cracking that occurs when dimensional change is restrained in newly placed concrete by rigid foundations or by old lifts of concrete.

To reduce drying shrinkage, following precautions need to be taken:

- Use less water in concrete
- Use larger aggregate to minimize paste content.
- Use a low temperature to cure concrete.
- Dampen the subgrade and the concrete forms
- Dampen aggregate, if it is dry and absorbent.
- Provide adequate reinforcement
- Provide adequate contraction joints.

6.3.11 Temperature Changes

Temperature changes can affect shrinkage. The heat of hydration of cement in large placements can create problems. Climatic conditions involving heat also have the capability to affect concrete. Fire damage, while rate, can also contribute to problems associated with excessive heat. The code allows one to assume that the ends of columns built integrally with a structure will remain fixed. This assumption comes into play when one is computing gravity load moments on columns.

One should know that hydration of concrete can raise the temperature of freshly placed concrete by up to 100°F. Rarely the temperature increase is consistent in all of the concrete, and this can generate problems, cracks can occur. The cracks should be shallow and isolated. To avoid this, following measures should be taken:

- Use low-heat cement
- Pour concrete at the lowest reasonable temperature
- Select aggregates with low moduli of elasticity and low coefficients of thermal expansion.

External temperature changes can result in cracking that will appear as regularly spaced cracks. There may be spalling at expansion joints. Using contraction and expansion joints can help prevent this damage.

There are many potential causes for concrete failure. Extended education, experience, and scientific testing are often required to identify clearly the causes of failure. There is always more to learn. Keeping an open mind and immersing one self in the components of concrete are the best ways to achieve success.

6.3.12 Poor Workmanship

Poor workmanship accounts for a number of concrete issues. It is simple enough to follow proper procedures, but there are always times when good practices are not used. The best solution to poor workmanship is to prevent it in the first place. Unfortunately, this is much easier to say than to do. All sorts of problems can occur when quality workmanship is not ensured. The following are some of the key causes for such problems.

- Adding too much water to concrete mixtures
- Poor alignment of formwork
- Improper consolidation
- Improper curing
- Improper location and installation of reinforcing steel members
- Movement of formwork
- Premature removal of shores or reshores
- Settling of concrete
- Settling of subgrade
- Vibration of freshly placed concrete
- Adding water to the surface of fresh concrete
- Miscalculating the timing for finishing concrete
- Adding a layer of concrete to an existing surface
- Use of a tamper
- Jointing

6.4 SYMPTOMS TO CAUSES OF DISTRESS AND DETERIORATION

This is shown in Table 6.1 below:

Table 6.1: Relating symptoms to causes of distress and deterioration of concrete

Causes	Construction faults	Cracking	Disintegration	Distortion/ movement	Erosion	Join failure
Accidental loadings		X				
Chemical reactions		X	X			
Construction errors		X				X
Corrosion		X				X
Design errors		X				
Erosion		X	X		X	
Freezing and thawing		X	X			
Settlement				X		X

SHORT ANSWER QUESTIONS

1. What is the use of cement concrete?
2. What is meant by durable structure?
3. Define the term durability.
4. What are the factors that influence the durability of the concrete?
5. Define permeable and impermeable concrete.
6. Relate permeability and durability.
7. Is permeability a single function?
8. What are the determinations of permeability?
9. What are the changes made in IS: 456-2000 in the view of durability?
10. What is the measure of durability?
11. What is the effect of external causes on durability of concrete?
12. What is the effect of internal causes on durability of concrete?
13. Name the reasons for deterioration of concrete.
14. What are the distresses in concrete structures?
15. Name the physical causes on concrete.
16. Name the chemical causes on concrete.
17. Define shrinkage in concrete.
18. What are the classifications of shrinkage?
19. What are the factors influencing the drying shrinkage?
20. Define physical loss and chemical loss in concrete structures.
21. What is differential shrinkage?
22. Explain the term autogenous shrinkage.
23. Define the term carbonation shrinkage.
24. What are the preventive measures of drying shrinkage?
25. Name the defects normally observed in concrete structures.
26. What is the rate of plastic shrinkage?
27. What is meant by plastic settlement cracks?
28. Define plastic cracks.
29. What is the result of poor construction practices?
30. What are the preventive measures used in thermal movement of concrete?
31. What is the effect of fire on concrete?
32. How will you minimize frost damage?
33. Define D-cracking.
34. What is the effect of weathering on concrete?
35. What is the effect of creep on concrete structures?
36. Classify the chemical attack on concrete.
37. What is the result of crystallization of salts in pores?
38. How will you prevent the salt attack?
39. Expand AAR and ASR.
40. What is the result of water permeates by the aggregates?
41. What is meant by swelling?
42. How will you classify the damages in structure?

43. What is known as abrasion and erosion in concrete?
44. Define the defect pop outs.
45. How to prevent pop outs?
46. What is meant by carbonation and what are the factors influenced?
47. What is the rate of carbonation?
48. What is cavitations in concrete?
49. What is meant by crazing?
50. What are the preventive measures used to reduce crazing?
51. Define honeycombing.
52. How will you prevent honeycombing?
53. What is the effect of hydrolysis and leaching?
54. How will you remove early and heavy efflorescence?
55. How sulfate attack occurs and what are the parameters considered?
56. How will you reduce the sulfate attack on concrete?
57. What is the effect of chloride attack and how is it prevented?
58. How can we reduce the acid attack in concrete?
59. What is the effect of corrosion?
60. What are the factors that influence the corrosion process?
61. What is ultimate result of rebar corrosion?
62. What are the variables considered in steel corrosion?
63. Explain chemical process of corrosion.
64. Explain electrochemical process of corrosion.
65. Explain physical process of corrosion.
66. Name the periods of corrosion process.
67. What are the factors that influence the corrosion initiation?
68. In what factors changes in IS: 456-2000 on durability of concrete?
69. Name the process in propagation period.
70. How is bare steel prevented from corrosion?
71. What is meant by pitting corrosion?
72. How the embedded steel suffers corrosion?
73. Name the physical symptoms of corrosion on beams and slabs.
74. Name the corrosion protection techniques.
75. What are the protective measures considered by the durable design?

THEORY QUESTIONS

1. What are the main topics to be covered/learnt about concrete/RCC structures?
2. Discuss the factors affecting permeability of concrete structures.
3. Define and explain defect in concrete.
4. What are the defects normally observed in concrete structures? Discuss briefly their remedies.
5. What are the causes for distress in RC members? Discuss about the elements that catalyze the distress in RC structural elements.
6. What are the agencies causing deterioration in RC structures? Explain the type of distress caused by the different agencies.
7. Explain clearly about the mechanism of freezing and thawing.
8. Explain various preventive measures in controlling distress RC structures.
9. Explain in detail plastic shrinkage and drying shrinkage.
10. Discuss briefly the strength and behavior of concrete subjected to high temperatures.
11. Write short notes on the following:
    a. Subgrade movement
    b. Formwork movement
    c. Foundation movement
12. Explain about the poor construction practices and errors in design of concrete structures.
13. Discuss briefly the effect of fire on the properties of concrete.
14. Write briefly about chemical attack on the concrete.

15. What is sulfate attack? Explain about the causes and effects of sulfate attack.

16. Explain about hydrolysis and leaching of concrete.

17. Explain the causes and effects of chloride attack and sulfate attack.

18. What is carbonation? Explain the various factors which affect the rate of carbonation.

19. Explain about the causes and effects of alkali reaction in concrete.

20. Explain seawater attack on concrete.

21. Clearly explain about the chemical reactions involved in the corrosion reaction of steel in concrete.

22. Define corrosion. Tabulate the IS codal provisions for cover of various exposure conditions.

23. Explain clearly the various stages of corrosion.

24. Explain the factors affecting corrosion of steel in concrete.

25. What are the factors influencing corrosion of reinforcement? Explain the damages in RC structures due to corrosion of reinforcement.

26. Explain the basic principle of corrosion. Explain the various corrosion protection techniques.

Part 2

Damage Assessment

7. Condition Survey and Nondestructive Testing System
8. Semidestructive Testing System
9. Destructive Testing System

7

Condition Survey and Nondestructive Testing System

7.1 CONDITION SURVEY

Nondestructive evaluation is widely employed for inspecting/assessing the condition of structures. Nondestructive techniques, which are less time consuming and relatively inexpensive, can be used for the following purposes:

- Test on actual structures
- Test at several locations
- Test at various stages
- Access the quality control of actual structures
- Assess the uniformity of the concrete
- Assess the materials used and workmanship
- Assess the construction practices followed
- Assessment of the extent of defects like cracks, voids and honeycombing, etc.
- Assessment of the suspected damage due to poor planning/designing of structures
- Assessment of the partial durability
- Integrity testing of piles
- Monitoring of progressive changes of the defect in structure

The results of Nondestructive tests are most useful, when supplemented by a limited number of destructive/semi-destructive tests. There are more testing techniques with different principles and applications available to evaluate the properties of concrete. The concrete material is so complicated that the efficiency and quality is difficult to be established by one test method only. Most of the tests, which are used, for estimating parameters of concrete provide an excellent means of establishing and evaluating the uniformity of concrete.

7.2 NONDESTRUCTIVE TESTING METHODS

Potential and limitations of various nondestructive techniques cited below are briefly discussed in the following paragraphs to apprise users of their relevance in field application.

- Surface hardness method
- Ultrasonic pulse velocity method
- Resonant frequency method
- Dynamic or vibration method
- Pulse attenuation method
- Pulse echo method
- Radioactive method
- Nuclear method
- Magnetic method
- Electromagnetic method
- Electrical method

- Acoustic emission technique
- Radar technique
- Radiography method

7.2.1 Surface Hardness Test

The surface hardness method consists of impacting the concrete surface in a standard manner. Activating a mass by a given energy and measuring the indentation or rebound achieves this. The most commonly and widely used instrument is a **rebound hammer**. There are several types of hammer having varying impact energy from 0.07 kg·m to 3 kg·m; the high impact energy is used for mass concrete, road pavements and airport runways. The low impact energy hammers (0.07 to 0.09 kg·m) are used for small and low strength materials. A typical rebound hammer is shown in Fig. 7.1.

Test Procedure

The test procedure consists of applying the hammer on the concrete surface and observing the rebound reading indicated by a rider over a scale. Before applying the hammer, the surface of the concrete is cleaned and smoothened. A minimum of 10 readings are compared and each reading should not differ by more than 7 units. The average of remaining readings is determined for evaluating the strength. If more than two readings differ from the average by 7 units, then the entire set of readings is taken afresh. The testing procedure of hammer is shown in Fig. 7.2.

The procedure for determining the rebound values has been specified in ASTMC 805-85, BIS-13311 Part 2 and also in the latest ASTM

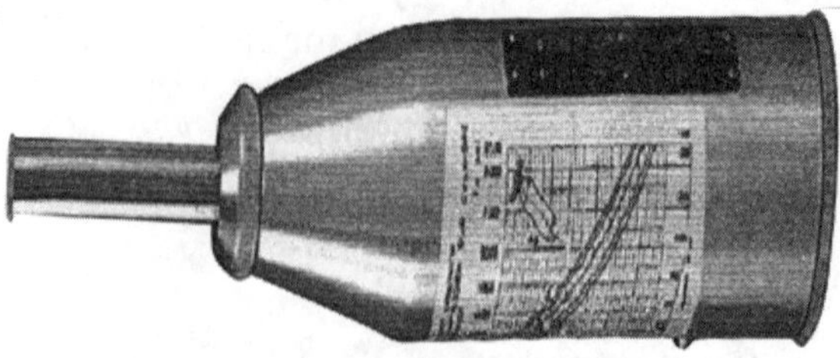

Fig 7.1: A typical rebound hammer

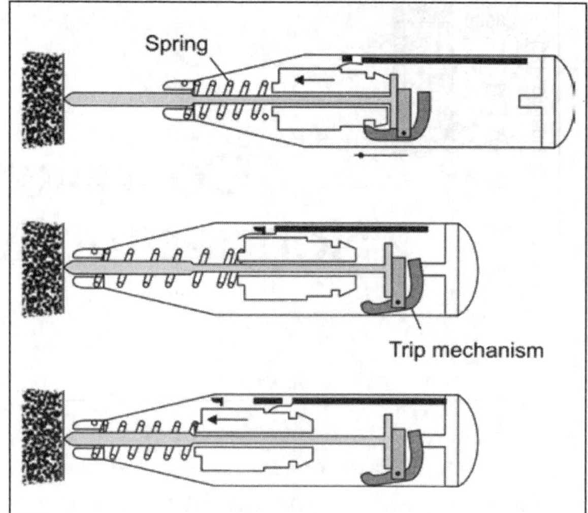

Fig. 7.2: Testing procedure of rebound hammer

specification. Estimation of concrete compressive strength from rebound number is determined from standard calibration curve based on the laboratory results. The calibration curve should be established for each type of concrete. A typical calibration curve is shown in Fig. 7.3.

The main factors that affect the reading are:

- Size and age of concrete
- Surface texture
- Concrete mix characteristics
- Temperature and stress state
- Carbonation level in concrete
- Moisture content

It should be noted that the rebound values reflect the concrete quality up to a depth of 50 mm in the member. In practical situation, the strength prediction can be made to an accuracy of 25%. The surface hardness measurements can be used for the following:

- Checking the uniformity of concrete
- Comparing a given concrete with a specified requirement
- Approximate estimation of strength by using laboratory calibrated graphs
- Abrasion resistance classification

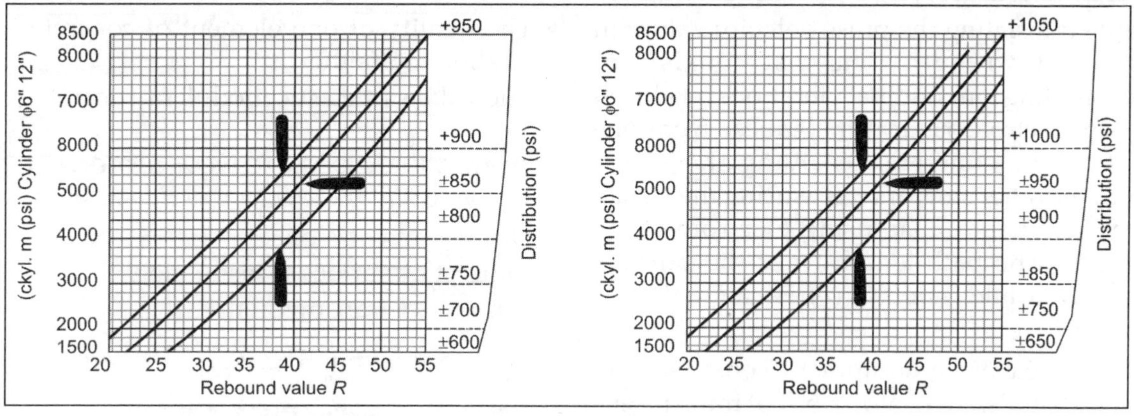

Fig. 7.3: A typical calibration curve

Table 7.1 shows the quality of concrete cover from rebound number.

7.2.2 Ultrasonic Pulse Velocity (UPV) Test

The objective of this method is basically to measure the velocity of the pulses of longitudinal vibrations passing through concrete. The first report of the measurement of the velocity appeared in USA in the mid 1940s. It was found that the velocity depends upon the elastic property and geometry of the material. With the development of reliable measurement techniques, the ultrasonic method has become widely accepted around the world and commercially produced lightweight equipment. If properly used by an experienced operator, a considerable amount of information about the interior of the concrete member can be obtained. Recommendations for the use of this method are given in BS-4408 Part 5, ASTM C597-71, and BIS-13311 Part 1.

Table 7.1: Quality of concrete cover from rebound number

Average rebound number	Quality of concrete
Greater than 40	Very good hard layer
30–40	Good layer
20–30	Fair
Less than 20	Poor concrete
0	Delaminated

Theory of Pulse Propagation through Concrete

Three types of waves are generated by an impulse applied to a solid mass. They are longitudinal (compression), shear (transverse) and surface (Rayleigh) waves. Longitudinal waves are fastest and provide more useful information than others. For an infinite, homogenous, isotropic elastic medium, the compression wave velocity is given by the formula

$$v = \sqrt{k}\, E_d/\rho$$

where,

v = compression wave velocity (km/sec)
$k = (1 - \gamma)/[(1 + \gamma)(1 - 2\gamma)]$
E_d = dynamic modulus of elasticity
ρ = mass density
γ = dynamic Poisson's ratio.

Test Procedure

The velocity of an ultrasonic pulse is influenced by properties of concrete, which determine its elastic stiffness and mechanical strength. Thus, the variations in the pulse velocity values reflect a corresponding variation in the state of concrete under test. In order to evaluate the strength of concrete, calibration charts should be established based on laboratory tests. The calibration chart is influenced by a number of factors such as type of cement, cement content, admixtures, type and size of aggregate, curing conditions and age of concrete. Hence, adequate care and caution should be exercised while translating the pulse velocity values in terms of strength. The test consists of transmitting longitudinal vibrations produced by an electroacoustical transducer from one side of the concrete, receiving the signal from the other side, and measuring the transit time (t) of the pulse using electronic time circuits. The path length (l) of the pulse is measured and velocity (v) is calculated as $v = l/t$.

The measuring equipment consists of an electrical pulse generator, a pair of transducers, an amplifier, and an electronic timing device for measuring the time interval elapsing between the onset of a pulse generated at the transmitting transducer and the onset of its arrival at the receiving transducer. Transducers of 50 to 60 kHz are found to be useful for most of the applications in concrete testing. A typical measuring instrument is shown in Fig. 7.4.

The pulse velocity measurements may be used to establish the following:

- The homogeneity of the concrete
- The presence of cracks, voids and other imperfections
- Changes in the structure of the concrete which occurs with time
- The quality of the concrete in relation to standard requirements
- The quality of one element of concrete in relation to another
- The values of elastic moduli of concrete.

There are three possible ways of measuring pulse velocity as brought out below:

- Direct transmission
- Indirect or surface transmission
- Semidirect transmission

These are shown in Fig. 7.5.

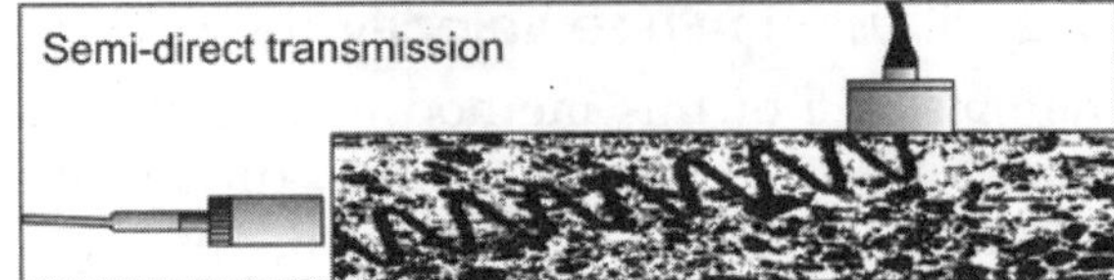
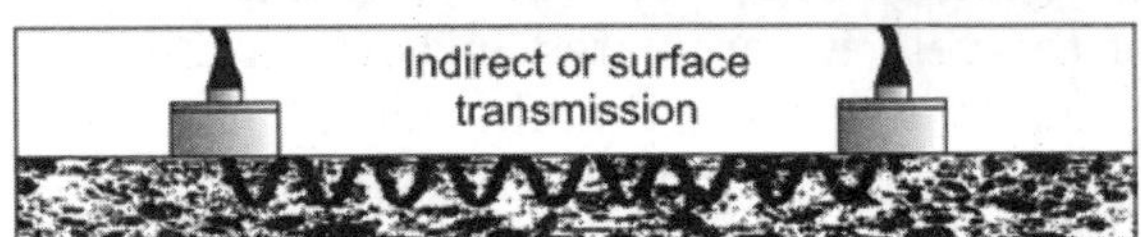

Fig. 7.5: Methods of transmission system

The direct transmission method is generally preferred, since the maximum energy of the pulse is being directed at the receiving transducer and this gives maximum sensitivity. The indirect transmission arrangement is the least sensitive and for a given path length, produces a signal which has only about 2% or 3% of that produced by direct transmission. This arrangement is used when only one face of the concrete is accessible, or when the depth of the surface crack is to be determined or when the quality of the surface concrete relative to the overall quality is of interest. The semidirect transmission arrangement has a sensitivity intermediate between those of the other two arrangements. In this method, there is uncertainty regarding path length. It is

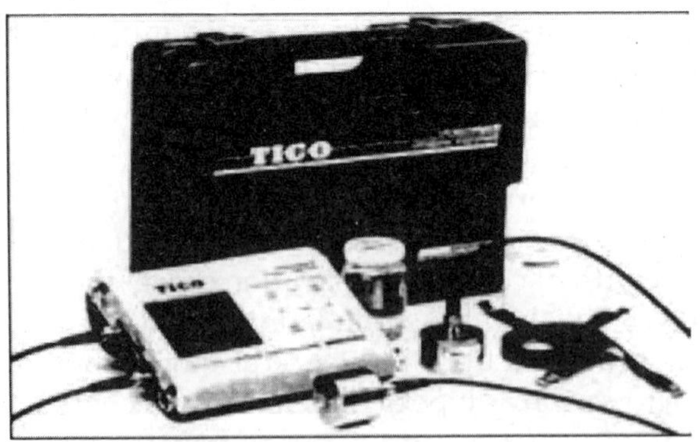

Fig. 7.4: A typical measuring instrument of UPV

generally found to be sufficiently accurate, if the length is measured from center of transducer faces. Some of the factors that influence pulse velocity measurements are:

- Surface condition
- Moisture content
- Temperature of concrete
- Path length
- Shape and size of specimen
- Reinforcement.

Figure 7.6 shows the behavior of ultrasonic pulses in concrete medium under different conditions.

Table 7.2 gives the criteria that are adopted for qualitative certification of concrete under UPV results.

Table 7.2: Quality of concrete from UPV	
UPV (v) km/sec	*Concrete quality*
Greater than 4	Very good
Between 3.5 and 4	Good
Between 3 and 3.5	Poor
Between 2.5 and 3	Very poor
Between 2 and 2.5	Very poor and low integrity
Less than 2	Large voids suspected

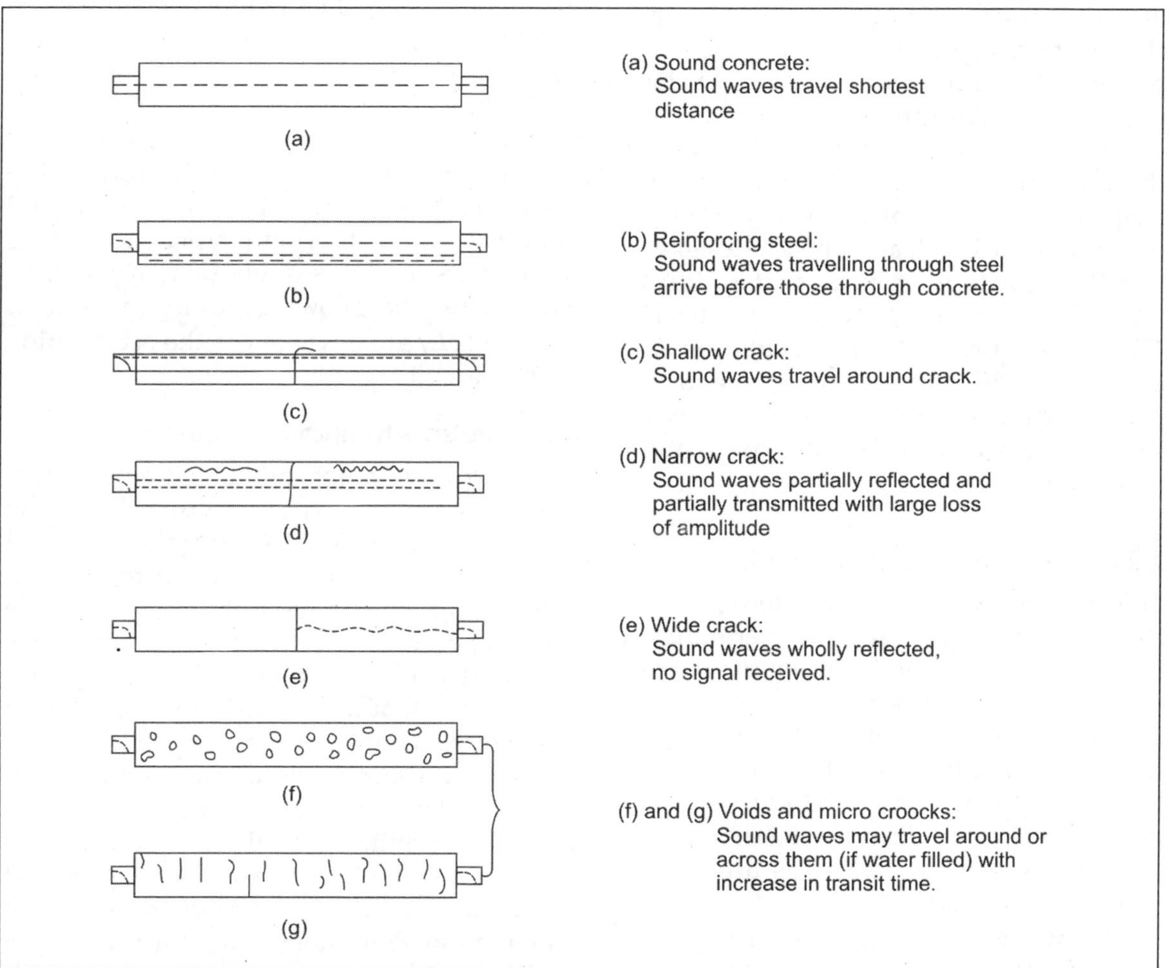

Fig. 7.6: Behavior of ultrasonic pulses in different concrete medium

Advanced Ultrasonic Testing Equipment

This category comprises the range of instruments that use sound or stress waves in order to determine the properties of concrete and other materials nondestructively. The first and most widely used system is V-meter, which utilizes the ultrasonic pulse velocity method for evaluating construction materials in the field. Transducers are available for a variety of frequencies from 24 kHz to 500 kHz. This unit has also been modified to suit the special needs of ceramic users and can be found as the ultra-pulse.

The **portable impact echo system (PIES)** is an advanced instrument for nondestructive detection of flaws and defects in a variety of civil infrastructure ranging from bridges, parking structures and buildings to dams, piles, tunnels, tanks and marine structures, etc. The **E-meter system** represents the state-of-the-art in bench top laboratory resonant frequency analysis of materials. This unit comes standard with a test bench designed to handle a variety of modes of vibration, including longitudinal, torsional and flexural. This line of products represents the most effective and efficient way to evaluate concrete and other materials in the field and utilizes the latest technology in order to guarantee accurate results.

7.2.3 Resonant Frequency Method

This method is based on the determination of the fundamental resonant frequency of vibration of a specimen. The equipment used for this is usually known as sonometer. Resonant frequency method is mostly used in the laboratory. The size of specimen in this test is usually limited to 150 mm × 300 mm cylinder or 75 mm × 75 mm × 300 mm prisms. The resonant frequency test is used for the following purposes:

- To study the deterioration effects on concrete subjected to repeated cycles of freezing and thawing

- To study the effects due to acidic and alkali reactions
- To determine the damage due to fire
- To calculate the dynamic Young's modulus of elasticity of concrete.

7.2.4 Dynamic or Vibration Method

This is an important nondestructive method used for testing concrete strength and other properties. The fundamental principle of this method is based on the velocity of propagation of sound through a solid material. A mathematical relationship, could be established between the velocity of sound through the specimen and its resonant frequency. From this relationship, the modulus of elasticity of the material is determined. For deriving this relationship the solid mediums are considered to be homogenous, isotopic and perfectly elastic. However, these relations are also applied to heterogeneous material like concrete. The velocity of sound, v in a solid material is a function of the square root of the ratio of its modulus of elasticity e and density ρ. Thus, the following formula can be used

$v = f\,[gE/\rho]^{1/2}$, where g is the acceleration due to gravity.

7.2.5 Pulse Attenuation Method

This is a wave propagation method in which electromagnetic waves, typically in the frequency range 500–1000 MHz are allowed to propagate through solids and the methodology is found to be useful for subsurface investigation in civil engineering structures and in particular concrete structures. The basic principle is that the attenuation properties of the electromagnetic waves are influenced by the electrical properties of the solid material tested. The dominant properties are the electrical permittivity, and thereby determines the signal velocity, and the electrical conductivity, which determines the signal attenuation. Reflections and refractions of the radar wave will occur at interfaces between different materials and the signal returning to the surface antenna can be interpreted to provide an evaluation of the properties and geometry of subsurface features. Limited use of radar has been seen in the investigation of tunnel linings in India. It is increasingly being recognized that there are many structural applications of the radar system as listed below:

- To determine major construction features
- To assess element thickness
- To locate moisture
- To locate reinforcing bars
- To locate voids, honeycombing/cracking
- To locate chlorides
- To determine size of voids
- To estimate chloride concentration
- To locate rebar corrosion.

7.2.6 Pulse Echo Method

Considerable developments have taken place in marking commercially available instrument on pulse echo technique and this has become popular through extensive research and practical applications. Originally, pulse echo technique was developed for pile integrity testing, and now it is popular for concrete structures and structural elements. An instrument system commercially known as DOCTER marked in Denmark is available with a field computer, related software and transducers. The pulse echo and oscilloscope signal techniques are shown in Fig. 7.7.

The system is found to be useful for:
- Testing the thickness and flaws in concrete
- Testing the wave speed on the surface
- Testing the depth of surface opening cracks.

7.2.7 Radioactive Method

The use of X-rays and gamma rays as non-destructive method for testing properties of concrete is relatively new. X-rays and gamma rays, both components of the high-energy

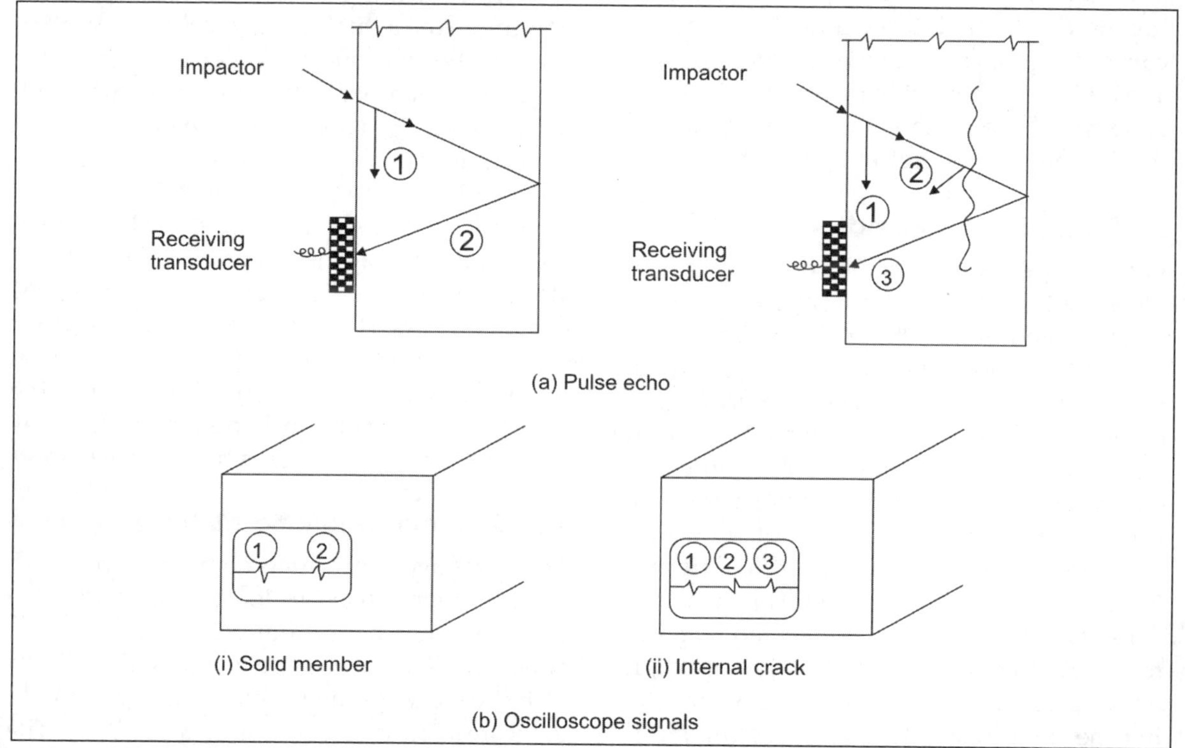

Fig. 7.7: Pulse echo method for crack location

region on the electromagnetic spectrum, penetrate concrete but undergo attenuation in the process. The degree of attenuation depends on the kind of matter traversed, its thickness, and the wavelength of the radiation. The intensity of the incident and the emerging gamma rays after passing through the specimens are measured. These two values are used for calculating the density of structural concrete members. Gamma rays transmission method has been used to measure the thickness of concrete slabs of known density. Gamma radiation source of known intensity is made to pass and penetrate through the concrete. The intensity at the other face is measured. From this, thickness of the concrete is calculated.

7.2.8 Nuclear Method

Use of nuclear method for nondestructive measurements of some properties of concrete is of recent origin. Two principle techniques have been reported, namely **neutron scattering method** for determining the **moisture content** of concrete and **neutron activation analysis** for the determination of **cement content.** These methods are not suitable for finding out the strength of concrete.

7.2.9 Magnetic Method

Battery operated magnetic devices that can measure the depth of reinforcement cover in concrete and detect the position of reinforcement bars is now available. The apparatus is known as **cover meter/pachometer**. This can be used for measuring the cover given in the lightly reinforced sections. Pachometer or cover meter is an electromagnetic device used for determining the locations of reinforcing steel embedded in concrete, and to determine the amount of cover over steel. The instrument consists of a search head connected by a cable to a metering unit, which may have a digital or analogue read-out. The purpose of locating the reinforcing steel, prestressing strand, cladding ties, etc. in concrete, is to avoid them during core sampling, direct pulse velocity

surveys, etc. and to make connections for electrochemical measurements. The knowledge of cover depth is essential if it is desired to obtain samples of the concrete at the level of the rebar for chloride ion analysis, pH measurements and the presence of carbonation.

It is also useful in determining the potential for corrosion and subsequent concrete deterioration, since it has been well established that structures in corrosive environments with inadequate cover get subjected to early deterioration. By this test, bar diameter, cover to reinforcement, spacing of the reinforcement, number of bars and any discontinuity in the bars can be detected. This test is performed using cover meter, which is based on electromagnetic theory.

Some constituents of concrete may affect cover meter readings and so it is always advisable to carry out a calibration check for the actual situation. This can be done by taking readings on bars at a range of different depths and either drilling down or breaking out to determine the actual cover. A calibration curve can then be constructed. When measuring the cover on a member with reinforcement in two directions at right angles to each other for such a slab, it is probably best to quickly determine the locations of the bars by moving the head across the surface and marking the location of the minimum reading. Some cover meters can also be used for assessing the diameter of the reinforcing bars. These instruments have different scales for different scale diameters. The bar diameter is assessed by locating a bar using any arbitrary scale and then recording the indicated reading using each bar diameter scale.

7.2.10 Electromagnetic Method

This category of products comprises the range of instruments used to locate and determine size and depth of steel reinforcing bar in concrete. The first type most widely used utilizes a low frequency electromagnetic field to locate ferrous objects within a structure. The second and latest technology utilizes ground penetrating radar to locate steel and other objects within a structure. The **HR rebar locator** shown in Fig. 7.8 and the data scan both utilizes a low frequency electromagnetic field to locate ferrous objects in a structure. By closely monitoring changes in the electromagnetic field the proximity of steel reinforcing bar can be determined. The HR is most economical model. It uses a simple analog meter to determine the strength of the electromagnetic field and correlate this to either the size of the rebar or the distance to the rebar. The **data scan** shown in Fig. 7.9 utilizes a more sensitive digital system to monitor the strength of the electromagnetic field. This allows the user more sensitivity in locating steel at greater depth and more closely spaced.

The **data scan MK-II** as shown in Fig. 7.10 utilizes the state-of-the-art in ground penetrating radar technology. By sending out

Fig. 7.10: Data scan MK-II

a high frequency pulse and observing the reflected electromagnetic waves features inside a concrete structure can be determined. The unit implements a number of different ways to visualize the data received, including a three dimensional image of the objects within the structure. This unit represents the current cutting edge technology in object location. The line of products represents the most effective and efficient way to evaluate steel reinforcing bar in the field and utilizes the latest in technology in order to guarantee accurate results.

Rebar Locator

It is rugged field instrument for finding the location, depth and size of reinforcement bars in place. The main features are as follows:

- Single unit construction; no physical strain for the operator and no probe cord to snag or break
- Post-mounted sensor with meter mounted separately for convenient viewing by operator in stand-up position
- Pinpoint accuracy for precise rebar location from direct reading, stable analog meter
- Detection up to 250 mm of cover with easy to read calibrated meter
- Light operating, weight less than 3.5 pounds (1.6 kg)
- Up to 8 hours continuous use between battery charges, improved electronic circuitry

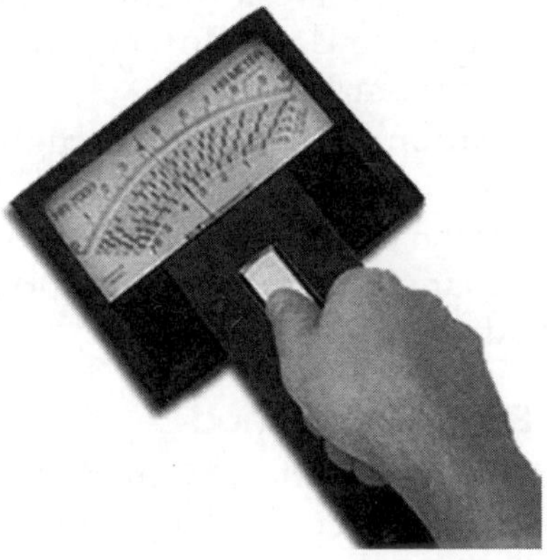

Fig. 7.8: HR rebar locator

Fig. 7.9: Data scan

Technical Specifications

Detection and orientation of rebars: The exact position and orientation of rebars can be measured quickly and accurately. Rebar-free areas can be identified for coring, grinding, resurfacing or insertion of new machinery mountings and under inspection as shown in Fig. 7.11.

HR rebar locator can be used to inspect new structures for compliance with specifications as well as old structures under modification.

Measurement of concrete cover: The amount of concrete cover over reinforcement bars is an important parameter on bridge decks, highways, columns and slabs. Concrete cover can be determined with an accuracy of +/–3 mm. Concrete thickness can be measured in tanks, pipes and other structures where the inner lining is steel and within 25 cm of the surface. Reinforcement bar size can also be estimated.

Location of ferrous metals: Locates any ferrous base material such as pipe, flues, wire and sheets embedded in concrete, masonry or wood. Identification is possible to a depth of 25 cm.

Location of prestressed cables: Locates the position of cables and lost tendon splices in pre or post tension concrete products.

Fig. 7.11: Rebar locator

Special mounting: The HR rebar locator can be mounted on an adjustable telescopic tube with handle. The sensor is mounted at the end of the tube and the meter placed within convenient operator viewing range. This enables areas to be scanned with the operator in a standing position.

Rebar Data Scan

Rebar data scan is a sophisticated reinforcement bar locator with an easy to read liquid crystal display. The features are as follows:

- Reinforcement bar cover can be detected to a maximum depth of 30 cm by reading directly from the high legibility liquid crystal display.
- Analog meter on probe facilitates rapid and easy location of rebar.
- Fast scanning of large areas in enhanced by an audible output via headphones.
- Built in memory stores over 1000 cover readings for late downloading to a personal computer.
- Calibration for rebars and high tensile steel tendons is available.
- The probe meter can be preset to react to any rebar within the value, ignoring those outside the range.

7.2.11 Electrical Method

Recently some electrical methods have been employed for determining the moisture content of hardened concrete, tracing of moisture permeation through concrete and determining the thickness of concrete payments. The accurate determination of the moisture content of hardened concrete is required in connection with creep, shrinkage and thermal conductivity studies. The fact that a dielectric property of hardened concrete change with changes in moisture content is made use of in this method. Electrical resistivity methods have been used to find out the thickness of concrete pavements. The method is based on the principle that the material offers resistance to the passage of an electric current. A concrete pavement has a resistivity characteristic that is different from that of the underlying subgrade layers. A change in the slope of the resistivity versus depth curve is used to estimate thickness of concrete pavements.

7.2.12 Acoustic Emission Technique

The acoustic emission technique has been used to study the initiation and growth of cracks in the concrete (locating delaminated concrete). Sounding of concrete with a hammer provides a low-cost, accurate method for identifying delaminated areas. When striking areas of delaminated concrete, the sound changes from "ping" to a hollow sounding "puck". Boundaries of delaminations can easily be determined by sounding areas surrounding the first puck until pings are heard. Hammer-sounding of large areas generally proves to be extremely time consuming. More productive sounding methods are available when working with horizontal surfaces. Chain dragging accomplishes the same result as hammer sounding. As the chain is dragged across a concrete surface, a distinctly different sound is heard when it crosses a delaminated area.

Automated electronic system is also available, eliminating the need for an operator to listen for differences in acoustic emissions. However, these methods have proved to give only a general picture of the areas of delamination. Consequently, they should be used only for general assessment, not for the detailed layout required for reconstruction.

Infrared **thermography** is a useful method of detecting delaminations in bridge decks. This method is also used for other concrete components exposed to direct sunlight. The method works on the principle that, as concrete heats and cools, there is substantial thermal gradient with in the concrete because concrete is a poor conductor of heat. Delamination and other discontinuities interrupt the heat transfer through the concrete. These defects cause a higher surface temperature than surrounding concrete during periods of cooling. The equipment can record and identify areas of delamination and indicate the depth of delaminations below the surface.

7.2.13 Radar Technique

Ground-probing or subsurface interface profiling radar can be employed in a number of different ways in the investigation of concrete structures. The technique has been used for measuring the thickness of members. For determining the spacing and cover to reinforcement, and for detecting the position and extent of voids. The technique is similar to the methods used in seismology except that radio frequency pulses are used rather than sound. The equipment generates electromagnetic pulses, which are transmitted to the member under investigation by an antenna close to its surface. The pulses travel through the member but are partially reflected at any interfaces where there are distinct changes in dielectric characteristics. Reflected pulses are received by a second antenna and processed by the equipment.

The surface of the member is marked out with transit lines with regular distance markers. The antenna is moved along the transit lines and a signal is inserted into the system each time a distance marked is passed. The radar signal is attenuated, or diminished in strength, as it passes through the member under investigation. The amount of attenuation is dependent on the conductivity of the material and the frequency of the signal used. Attenuation is increased in conductive materials and therefore it may be difficult to penetrate saturated or salt contaminated concrete. The velocity of propagation of the radar pulse is dependent on the dielectric constant of the concrete. This can vary quite widely depending on aggregate type, moisture content and other factors. There is a need, therefore, for depths determined by radar to be checked and calibrated by direct physical measurements at a few locations.

7.2.14 Radiography Method

Radiography employing X-rays or gamma rays is sometimes used to investigate defects such as honeycombing or poor compaction inside concrete members. The technique may also be used to determine the location of the reinforcement in congested sections or other sections, where the use of cover meter is not possible or is inappropriate. British standards 1881: Part 205 gives recommendations for the use of technique. Gamma rays sources are usually employed for concrete sections up to 500 mm thick. High energy X-rays are appropriate for greater thickness. The sources consisting of cobalt 60 or iridium 192 for gamma rays or a linear accelerator for X-rays is placed on one side of the member under investigation. A film, sensitive to X-rays, is placed in light-tight cassette in contact with the other side of the member. Intensifying screens of lead foil mounted on card may be placed on either side of the films. The foil emits electrons when it is subjected to irradiation and this causes an intensification of the image, thus reducing the explosive time. The British standards suggest thickness of both front and back screens for X-ray and gamma rays sources. Sources to film distances are also given for various source diameters to minimize lack of sharpness due to the configuration.

7.3 RECENT DEVELOPMENTS ON NDT INSTRUMENTS

For rehabilitation and repair of concrete structures, various test techniques are used for evaluating the input parameters required for the remedial measures for undertaking repairs. One of the major problems associated with concrete structures is the evaluation of the strength and serviceability of a structure during and after construction. As of now the acceptance of concrete quality is essentially based on the compressive strength of specimens (cylinders or cubes) obtained from standard tests. Situations, however, arise to test the concrete in structure. Such situations most frequently arise as a result of following:

- Doubt about the validity of standard test results

- Failure of these test results to meet the specifications
- Damage and deterioration of concrete due to various causes such as corrosion, fire, and improper construction and
- Expansion of an existing structure

Further, it is often realized that the standard test results are not easily traced after a period of completion of a structure. Under these circumstances, it is necessary to test the concrete in a structure and evaluate its condition for taking further action. Assessment of quality of *in situ* concrete during and after construction and periodical monitoring of structures call for several field testing techniques. Various testing methods have been developed in the recent past; a few like rebound hammer and core test, have been in use for a number of years and other like ultrasonic pulse velocity test and penetration and internal fracture test are more recent developments. These tests have been developed with a primary objective of quickly evaluating the condition of *in situ* concrete in structural members. While some of them are nondestructive, others are partially destructive. Each method has its own merits and demerits.

Various test instruments deployed all over the world are as follows:

i. **Schmidt concrete test hammer**: Nondestructive quality testing of concrete and other building materials in completed structures and prefabricated structural sections.

ii. **Digi-Schmidt tester:** Testing the strength of concrete, assessment of uniformity of concrete and structural components and detecting damaged zones.

iii. **Tico-ultrasonic concrete tester:** Portable instrument for accessing the quality of concrete either *in situ* or precast, provides information on wide variety of parameters, such as uniformity of concrete, cavities, cracks, honeycombing, defect due to fire and frost, modulus of elasticity and concrete strength.

iv. Profometer for rebars locater: Digital metal detector for reinforcement bar spacing, concrete cover depth measurement and reinforcement bar size determination. High accuracy depth measurement with spot probe up to 130 mm and depth probe up to 300 mm. Measure bar diameter to an accuracy of ±1 mm. Path measuring device with probe makes rebar visible through integrated cyber scan software.

v. Canin corrosion analyzing instrument: Nondestructive detections of corrosion in the reinforcement bars of concrete building elements. Large area measurements can be made using road and wheel electrodes and printouts can be taken directly to scale without **PC** via **RS232**.

vi. Resistivity meter: Measurement of the electrical resistance of reinforced concrete components for assessing the risk of corrosion.

vii. Dyna pulls off tester: Measures the surface strength of concrete. It also measures the adhesive strength of all kinds of applied coatings (mortars, plasters and plastic coatings) flexible and themoplastic coatings, paint finishes and coatings of metals and concrete, etc.

viii. Dyna extraction tester: It measures pull out force/displacement on anchored bolts and plugs.

ix. Torrent permeability tester: Fast reliable and nondestructive measurements of the permeability of concrete structures.

x. New concrete workability meter: Rapid, accurate measurement of the workability (water–cement ratio and slump) and temperature of fresh concrete for improved onsite quality control.

xi. Multicell surveyor: Monitoring steel corrosion in concrete using eight probes to record eight individual readings simultaneously.

xii. Half-cell surveyor: A single probe instruments that measures and records electrical activity in reinforced concrete.

xiii. Resistivity logger: A simple push and press instrument that measures resistivity of concrete, masonry and other materials and also stores the gathered data.

xiv. Logging cover meter: Effectively locates steel reinforcement bar and measures the depth of concrete cover over them, it also stores the data it has gathered.

xv. Bremor test: For finding out cement content of fresh mortar.

xvi. New rebar checker: Precession instrument that makes the detection of concrete cover over steel reinforcement bars rapid, simple and extremely accurate.

xvii. Port probe: Most accurate rugged and portable instrument for density and moisture measurement of soil, pavement, concrete and other construction materials by nuclear method.

xviii. AC-2R asphalt content gauge: Rapid verification of asphalt content of paving mixtures sampled at asphalt plant, lab or paving site prior to placement and compaction.

xix. Foundation pile diagnostic system: Used for integrity testing, pile driving monitoring, dynamic load testing, satanic load testing, structural vibration monitoring, hammer monitoring, sway monitoring, inclination monitoring, sheet pile monitoring and pile driving prediction.

xx. Doctor-impact echo system: For flaw detection or measurement of thickness of construction material such as concrete.

xxi. **Electrical time domain reflectometry (ETDR):** ETDR stress/strain sensing technique has been successfully used in geotechnical applications to detection rock deformation and long wall movement. Since durable sensor media that can survive harsh construction environment and has long service life can be used, the ETDR sensing method appears to be a practical technique for health monitoring applications of civil concrete structures. The applicability of using an embedded coaxial ETDR cable to sense structural deformation and detect load-induced crack damages of a concrete structure is investigated. Bending test was performed on small-scale concrete specimens with embedded coaxial ETDR sensors. The test results show that the deformation patterns of the concrete specimens are clearly represented and the location and magnitude of load-induced crack damages that pass the sensing line are evidently shown in the ETDR signal response of the sensor. The results of the current study demonstrate that the great potentials of the ETDR technique for the health monitoring application of large civil concrete structures.

Semidestructive Testing System

8.1 INTRODUCTION

Partially destructive test are used for assessing the *in situ* concrete strength. These tests cause localized damage, which do not cause any overall loss in the performance of the structure. All tests in this group are surface zone test which require assess to only one exposed concrete face. Compared to other tests the accuracy may not be as good but estimation of strength is immediately available and the testing is less disruptive and damaging. Some of the testing methods are available to assess the quality of concrete and damage of concrete under partial disturbance of the surrounding concrete.

The common methods in construction industry are given below:

- Penetration techniques
- Pull out and pull off tests
- Core sampling and testing
- Break off test
- Permeability test
- Half-cell potential survey
- Resistivity survey
- Carbonation and pH value test
- Chloride content test
- Abrasion resistance test.

In addition, some of the tests are used to find the quantity and quality of concrete ingredients and are used to help in the damage assessment of existing concrete.

These types of test are given below under chemical testing of concrete:

- Determination of cement content
- Determination of water content
- Determination of water–cement ratio
- Determination of free chloride content
- Determination of pH value of concrete

In addition, some of the special tests are available to assess the quality of concrete and rebar in corrosion damaged structures and fire damaged structures. These are also discussed in this chapter.

8.1.1 Penetration Techniques

The measurement of hardness by probing techniques was first reported during 1954, two techniques were used. In one case, a hammer known as **simbi** was used to perforate concrete and depth of bore hole was correlated to compressive strength of concrete cubes. In the other technique, the probing of concrete was achieved by blasting with **spit pins** and the depth of penetration of the pins was correlated with compressive strength of concrete. The accuracy of the test was found to be ±25%. Simbi and spit pins were more affected by the arrangement of coarse aggregate, than the test using rebound hammers.

During 1964 and 1966, a technique known as the **windsor probes** was advanced for testing concrete in the laboratory and *in situ*. The windsor probe is a hardness tester of the surface of the concrete. It is an equipment consisting of a powder activated gun, hardened alloy probes, loaded cartridges, and depth gauge for measuring penetration of probes. The probe is driven in to the concrete by firing of a precision powder charge cartridges. The exposed length is measured by calibrated depth gauge and this is correlated to the strength of concrete cylinder.

The **penetration resistance method** of assessing strength uses a hardened steel probe, which is fired into a concrete surface using a special gun and a standardized explosive charge. The surface is cleared of loose material and a steel plate is positioned adjacent to the probe. A steel cap is screwed onto the probe and the length of probe protruding from the surface is measured using a spring-loaded depth gauge shown in Fig. 8.1 and the depth of penetration can be determined. It is clear that the forces acting on the probe and processes occurring within the concrete are extremely complex. As the tip of the probe enters the concrete, it causes local crushing of the surface and creates a shock wave, which results in local spalling. As penetration continues, the forward motion of the probe is resisted by fraction along its sides, concrete is crushed at the tip and the concrete ahead of

the probe is compressed in a zone of indeterminate shape and size. The initial kinetic energy of the probe is dissipated by all of these processes and it is asserted that the depth of penetration can be calibrated against compressive strength. Cartridges are available at a choice of two power levels according to the strength of the concrete to be tested.

The use of the penetration resistance method is described in ASTM C803. The standard notes that penetration resistance testing can be used to obtain an indication of the development of strength in concrete, to investigate areas of deterioration or poor quality, and to assess the uniformity of concrete within a structure, but that the method is not intended to take the place of strength determination by conventional methods. The standard also notes that a relationship may be established between penetration resistance and strength for a particular concrete mix and particular test equipment and used to assess strength in structures. However, the relationship may be affected by curing conditions and the type and size of aggregate. The estimated strength values are unlikely to have an accuracy of less than ±20% at confidence levels of 95%.

Windsor HP probe system: For in-place testing of normal and high-performance concrete, windsor HP probe measures the compressive strength of concrete accurately and effectively, on site in the field. The windsor HP probe system rapidly and accurately determines the concrete compressive strength of a structure by driving a probe into the concrete with a known amount of force. Improved and enhanced over thirty years, this modern system is capable of measuring concrete with a maximum compressive strength of 110 MPa. It has been followed for use in the construction environment, yet refined to provide the user with a simpler system to operate. An electronic measuring unit has been added to help ensure proper test results,

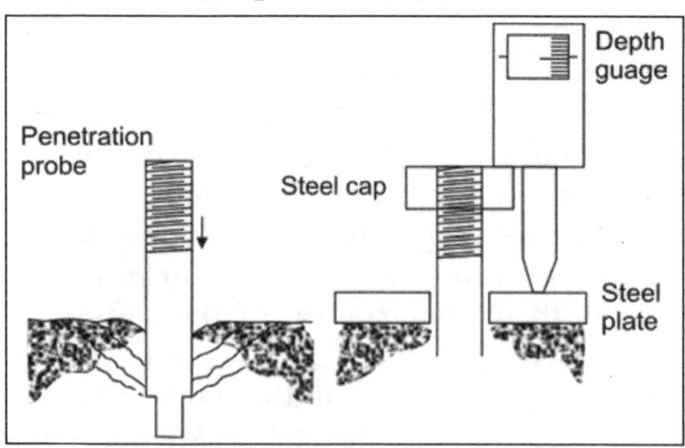

Fig. 8.1: Penetration resistance test

which can be recorded for later review or uploading to a personal computer.

Two probe styles are available one for lightweight, low density concrete with air filled aggregate and the other probe for more standard mix designs. Also, two standard power settings facilitate testing fresh concrete as well as mature mixes. The following are the features and benefits:

- New electronic measuring system enhances accuracy and efficiency
- Measurement to 17000 psi (110 MPa)
- Memory for data storage and uploading to PC
- Safe, since no accidental discharge and no recoil
- Fast and economical use
- Determines the developing strength of concrete; improves safety, ensures quality and reduces cost.
- Monitors the strength for rehabilitation as concrete ages
- Conforms to ASTM C803 and other official tests.

Further, it is safe and user-friendly. The windsor HP probe system does not require great skill to use and construction site supervisory staff or field technicians can obtain consistent results. In fact, among its users are contractors, engineers, and architects, testing laboratories, ready mix concrete producers, owners, managers and government authorities. This system has widespread use in testing concrete *in situ*, on conventionally placed, sprayed or precast concrete. The system is safe to use. The driver cannot be discharged unless it is fully depressed with some force against the actuating template, which rests against the surface being examined.

Probes: There is two-power settings available, low and standard power. The low power is used where concrete strength is less than 19.4 MPa. The newly designed silver probes can be used for high performance concrete with strength up to 110 MPa.

Procedure: A systematic test procedure is shown in Fig. 8.2.

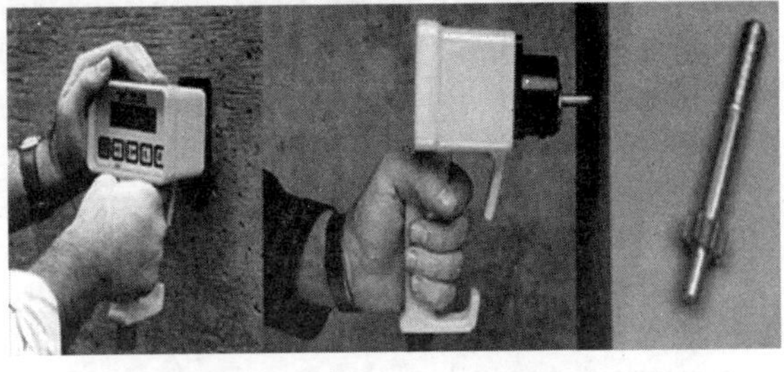

Fig. 8.2: Windsor HP probe test procedure

The actuating and measuring system is described below:

a. **Actuating:** Load the driver with a power load and probe suited for the type of concrete being examined. Place the driver firmly on the actuating template and fire. The locating template is then used to locate the probes at the corners of a fixed triangle. Normally, three measurements are required for consistent and statistically reliable results.

b. **Measuring:** The electronic measuring device is menu driven and programmed for selection according to the following parameters:
- Aggregate hardness
- Lightweight, normal, or HP concrete
- American or metric units

The three individual tests are automatically averaged and displayed on the LCD in accordance with ASTM procedure. This data together with time and date of the test are stored in the memory for subsequent uploading to a PC. An extractor is supplied to facilitate probe removal after the test. The standard windsor probe system includes a depth gauge and "strength chart" for determining the concrete strength. This is an economical alternative for many concrete test environments.

Technical: The windsor HP probe system is designed to evaluate the compressive strength of concrete in place. It is non-destructive and can be used with equal effectiveness on fresh and mature concrete. Equally accurate results are obtained on horizontal or vertical surfaces provided that the probe is perpendicular or at right angles to the test surface.

A hardened steel alloy probe is propelled at high speed by an exactly measured explosive charge into the concrete and its penetration measured. Each power load is guaranteed to have an energy level to give an exit muzzle velocity tolerance within $\pm 3\%$. The compressive strength of the concrete is directly related to the resistance to penetration of the crushed aggregate and cement matrix. The distance required to absorb the specific amount of kinetic energy of the probe determines this. The compressive strength of the concrete is empirically related to the penetration that varies with the hardness of the aggregate. This relationship is recognized by determining the Moth's scale of hardness of the aggregate and applying a correction factor to the penetration.

The combined contributions of both the aggregate and the cement paste to concrete strength are examined by the test. The accuracy of the inferred strengths has been examined in many independent tests and trials. The windsor HP results correlate well with the concrete strength determinations obtained by conventional means.

For most accurate test results ASTM recommends that a correlation be developed for the particular mix design being tested. Exact duplication of cylinder test results should not be expected. The probes measure the strength of the actual concrete in a structure rather than that of a sample compacted and cured under strict and somewhat artificial conditions, which do not necessarily represent those of the structure itself.

8.1.2 Pull out and Pull off Tests

Several tests are available which give estimates of compressive strength by measuring the force required to pull out embedded anchors or to pull off discs glued to the surface. Many of the pull out tests use anchors, which are placed into the concrete at the time of construction. In pull out test, either an insert is cast in the concrete or fixed into a hole, which is drilled into the concrete. Force required to pull out the insert is measured which is correlated with the compressive strength. Although the results relates to the surface zone only, the advantage is that a more direct measure of strength and a greater depth, compared to the surface, hardness test, and is available. Pull off test is based on the measurement of the *in situ* tensile strength of concrete by applying a direct tensile force. The method is especially useful in measuring the bond between the overlays. In this test, a metallic disk is glued either to the concrete surface to the surface of partial core. The force required to pull of the disc, causing tensile failure of concrete, is measured and correlated to the strength of concrete. The test requires the care in preparing surface and cause difficulty with damp surfaces.

A pull out test measures the force required to pull out from the concrete a specially shaped rod whose enlarged end has been cast into that concrete. The stronger the concrete, the more is the force required to pull out. The ideal way to use pull out test in the field would be to incorporate assemblies in the structure. These standard specimens could then be pulled out at any point of time. The force required denotes the strength of concrete. Another way to use pull out test in the field would be to cast one or two large blocks of concrete incorporating pull out assemblies. Pull out test could be performed to assess the strength of concrete.

Internal Fracture Test

It is the pull out test, which has been most widely used in the field. It was first developed for testing high alumina cement concrete components but further work has been carried out to produce correlation curves for Portland cement concretes. To carry out the test, 6 mm hole is drilled to a depth of 30–35 mm and the hole is blown out. Holes for individual tests should be at least 100 mm apart, should not be close to an edge and should be drilled away from reinforcement. It is usual to carry out six tests on a member. The average of the six maximum readings is used to read off the mean compressive strength from a standard calibration curve. The method allows concrete strength to be estimated within ±30% at the 95% confidence level. The test causes internal fracture of the concrete and there should be little damage to the surface. After the test the bolt can be sawn off and the surface made good.

Capo Test

The capo test is a modification of the **lok test,** which was developed in Denmark. The lok test uses cast-in inserts and confirms to ASTM C900. The capo test names based on the initial letters of the phrase "cut and pull out". To carry out the test, 18 mm hole is drilled to a depth of 45 mm. The hole is reamed out to form a 25 mm diameter groove at a depth of 25 mm. A ring insert is placed in the hole and expanded in the groove. The insert is connected to the hand-operated hydraulic jack and load-indicating gauge. Connecting to the test equipment is made by a threaded coupling rod 7.2 mm in diameter. The hydraulic jack produces a tensile force in the rod and bears on a reaction ring of internal diameter 55 mm on the concrete surface. The failure surface generated by the test is in the form of the frustum of core from the outer edge of the expanding insert to the inner edge of the reaction edge. The test can be stopped and the load released as soon as the peak is reached. This results in only a fine crack on the surface. Alternatively, the test can be continued until a plug of concrete is removed so that the disc can be recovered and reused.

Limpet Test

The limpet test is a pull off test. An aluminum disc, 50 mm in diameter is glued to the concrete surface at the test position using an epoxy resin or other suitable adhesive. The disc is connected to a mechanical loading device by a threaded rod. Load is applied by turning a handle on the side of the machine. Applied load is measured by integral strain gauge equipment with digital read-out. The applied force pulls the disc away from the surface with an attached thin layer of concrete. The concrete fails principally in direct tension. It is normal to carry out three tests on a member and calculate a mean pull off force. The compressive strength can be assessed using suitable calibration curves. Since the layer of concrete detached with the disc is usually only a few mm thick, the results can be strongly influenced by surface effects such as laitance, poor curing or carbonation. This difficulty can be overcome by core drilling using a 50 mm diameter barrel to a depth of 20 mm or greater, to penetrate concrete below the surface layer. The central section is not broken off and the aluminum disc is attached to it before pulling in the normal way. This technique is also useful for testing the bond of repairs or overlays. In this case, the core is drilled through the repair or overlay into the parent concrete.

8.1.3 Core Sampling and Testing

The core tests are performed under the following situations:

- When the standard 28 days cube strength test gives lower strength than acceptable and the primary aim of the core test is to ascertain whether the structural element is of adequate strength.
- When it is required to estimate the load carrying capacity of the structure for its safety under change of loading contemplated for the structure.
- To detect the segregation, honeycombing and to check the bond at construction joint or to verify the thickness of the pavement.

Core test serve as a good support test for any field investigation. Special electric or gasoline driven core drilling machines have been developed to cut cores from the structural members. Cores of 25 mm to 300 mm diameter can be cut including reinforcements. The cores can be tested for compressive strength, chemical analysis, petrographic examination and evaluating physical parameters. The strength testing of cores provide almost a direct evidence on the quality of concrete as it exists on the structure. The final results are influenced by the factors such as diameter of the core, slenderness ratio, core location, presence of reinforcement and curing condition.

The general procedure includes the following:

Core location and size: Cores for test should be normally taken at points where minimum strength and maximum stress coincide, for example, from the top surface near mid span for simple beams and slabs or any face near the top of lifts for columns or walls. In general, the accuracy decreases as the ratio of size of aggregate to core diameter increases, and 150 mm is regarded as the preferred size for aggregate of up to 40 mm, 100 mm cores must not be used if the maximum aggregate size exceeds 25 mm, and this should preferably be less than 20 mm for 75 diameter cores.

Testing: Each core must be trimmed and capped before visual examination, assessment of voidage, and density determinations.

Trimming: Trimming with a masonry or diamond saw, should give a core of a suitable length with parallel ends, which are normal to the axis of the core. If possible, reinforcement and unrepresentative concrete should be removed.

Capping: Cores should be capped with high alumina cement mortar or sulphur–sand mixture to provide parallel end surfaces normal to the axis of the core. Capping should be kept as thin as possible.

The honeycombing, cracks, aggregate distribution, drilling damages and other defects are easily seen on a dry surface. However, for the assessment of type of aggregate, size and characteristics a wet surface is preferable. Voids between 0.5 to 3 mm are classified as small, 3 to 6 mm sized voids as medium and greater than 6 mm as large voids. When the voids are interconnected, they are called honeycombing. The precise locations of the reinforcement present in the core must also be recorded.

Density determination: The method is as follows:

- Measure volume of trimmed core (V_u) by water displacement.
- Establish density of capping material (D_c).
- Before compression testing, weigh soaked/surface dry capped core in air and water to determine gross weight W_1, and gross volume V_1.
- If reinforcement is present, this should be removed from the concrete after compression testing and the weight W_s and V_s are determined.
- Calculate saturated density of concrete in the uncapped core from
$$D_2 = W_1 - D_c (V_1 - V_u) - W_s / V_u - V_s$$
- If no steel is present, W_s and V_s are both zero.
- Estimated excess voidage
$$= [(D_p - D_a)/(D_p - 500)] \times 100$$

where, D_p = the potential density based on available values for 28-day old cubes of the same mix

D_a = the actual density from core samples.

Factors influencing the core compressive strength are as follows:

- Moisture and voids
- Length/diameter ratio of core
- Diameter of core
- Position of cutout concrete on structure
- Direction of drilling
- Effect of age
- Concrete characteristics
- Reinforcement
- Method of capping.

Compression Testing

Compression testing will be carried out at a rate of 15 N/mm^2/min in a suitable testing machine and the mode of failure noted. If there is cracking of the caps, or separation of cape and core, the result should be considered as being of doubtful accuracy. Ideally, cracking should be similar all round the circumference of the core, but a diagonal shear crack is considered satisfactory except in short cores or where reinforcement or honeycombing is present. Estimation of an equivalent cube strength corresponding to a particular core result must initially account for two main factors. These are:

- The effect of the length/diameter ratio (correction factor)
- Conversion to an equivalent cube strength (conversion factor based on general relationship);
  Cylinder strength = 0.8 × cube strength
  Horizontal drilled cores;
  Estimated cube strength = $2.5f_\lambda/(1.5 + 1/\lambda)$
  Vertical drilled cores;
  Estimated cube strength = $2.3f_\lambda/(1.5 + 1/\lambda)$
  where f is the measured strength of a core with $1/d = \lambda$.

Due to the presence of reinforcement, the measured strength of concrete is found reduced up to 10%. Thus, as far as possible reinforcement from the core must be avoided. In case the reinforcement is present in the core, the measured core strength should be corrected as follows:

$$\text{Corrected strength} = \text{measured strength} \times [1 + 1.5\,(\phi_r\, h/\phi_c\, I)]$$

where,

ϕ_r = diameter of reinforcement bar

ϕ_c = diameter of the core

h = distance of bar axis from the nearest end of core

I = uncapped length core

In case core has multiple bars, then

$$\text{corrected strength} = \text{measured strength} \times [1 + 1.5\,(\phi_r\, h/\phi_c I)]$$

IS:516-1959 describes the method of preparation of core after drilling and the procedure of test. According to IS:456-1978 the concrete in the member represented by the core test shall be considered acceptable, if the average equivalent cube strength of the core is equal to at least 85% of the cube strength of the grade of concrete specified for the corresponding age, and no individual core has the strengthen less than 75%.

Advantages of core testing are as follows:

- The structural integrity of the concrete across the full cross-section may be affected.
- The ratio of length/diameter other than standard ratio 2 will have its effect on strength.
- Capping of both ends of the core is also liable to affect the strength of concrete.
- Existence of reinforcement will also have its effects on the strength of concrete.

Petrographic Examination

Petrography is a section of geology dealing with description and classification of rocks, but petrographic techniques can also be used to inspect samples of concrete in the laboratory. A standard practice is described in ASTM C856 and a description of microscopic method is given in Concrete Society Technical Report No. 32. Areas in which microscopic methods have been found to be useful include the following:

- Identification of mix ingredients including coarse and fine aggregate type, cements type, pulverized fuel ash and ground granulated blast furnace slag
- Assessment of mix proportions and air void content
- Investigation of various durability problems including carbonation, alkali-aggregate reaction, fire damage and chemical attack.

Petrographic techniques can be used in a quantitative way to assess the original mix proportions. Two methods using either point counting or linear traversing of polished sections are commonly employed. The basic principles of both methods are described in ASTM C457 in relation to the determination of air void content and bubble size and spacing. In each of the methods the polished specimen is placed on a hand or motor-driven stage which can be moved in two directions at right angles. Using either of these methods, the volumetric percentages of each of the components can be determined. With the measurement of some additional parameters the cement content, water/cement ratio and other details of the original mix can be assessed.

8.1.4 Break-off Test

A test, which measures the force required to break off a core, has been developed in Norway. The test was originally developed to monitor the early strength of concrete, and utilized a plastic cylindrical insert at the time of casting to from the cores. However, it can also be used to drilled cores. A core of 55 mm diameter is drilled to a depth of 70 mm, the annular hole on the surface is enlarged to form a circular group 10 mm wide and 10 mm deep to accept the loading device, as shown in Fig. 8.3. The load is applied transversely at a depth of 5 mm from the surface. The flexural, tensile stress developed at the base of the core causes it to rupture and the core breaks off. Figure 8.3 shows the break-off test.

8.1.5 Permeability Test

Several types of apparatus have been developed for site use in measuring properties related to permeability. When the test is undertaken on a vertical surface, a means of keeping the cap firmly in contact with the surface has to be provided. Water is introduced into the cap to give a pressure head of 200 mm using a filter funnel. A second port in the cap leads to a horizontal capillary tube. The rate at which water is absorbed into the concrete surface is determined by closing the connecting to the reservoir and measuring the movement of water surface in the capillary tube during a fixed time period. On site, the test is usually carried out 10 minutes after filling the cap with water. A diagram of the apparatus is given in Fig. 8.4.

The second test was developed at the building research establishment, UK in the early 1970 to enable measurements of either air permeation index or water absorption of concrete to be made. This is commonly known as the modified Figg's method. In this test, a hole is drilled into the concrete surface and after thorough cleaning, it is plugged from outside surface by polyether foam and then sealed with a catalyzed silicon rubber. After hardening of rubber, a hypodermic needle is pushed through the silicon plug as shown in Fig. 8.5. For air permeation, a mercury filled manometer and a hand vacuum pump is attached to the needle. The pressure within the system is firstly reduced to a standard value and then after isolating the pump, the time for the pressure to raise a standard value is recorded. This time is taken as a measure of the air permeation index of concrete. For water absorption, a water head of 100 mm is used. Water is forced into the assembly using a syringe and one minute after contact between water and concrete, the syringe is shut off. The time for the meniscus in a capillary tube to

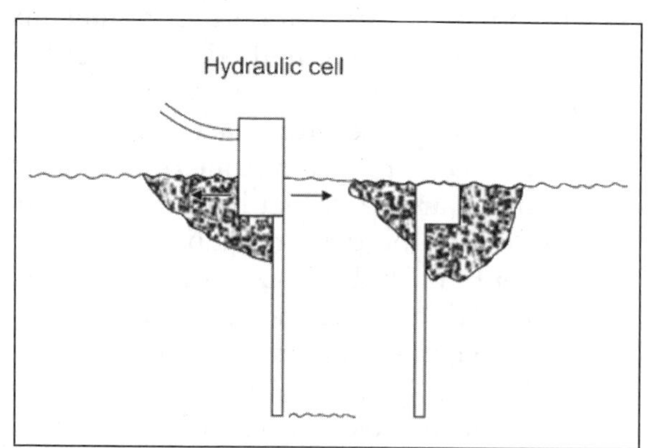

Fig. 8.3: Break-off test

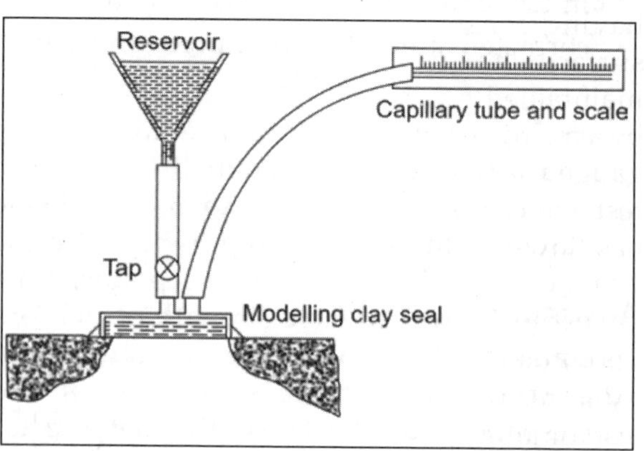

Fig. 8.4: Initial surface absorption test

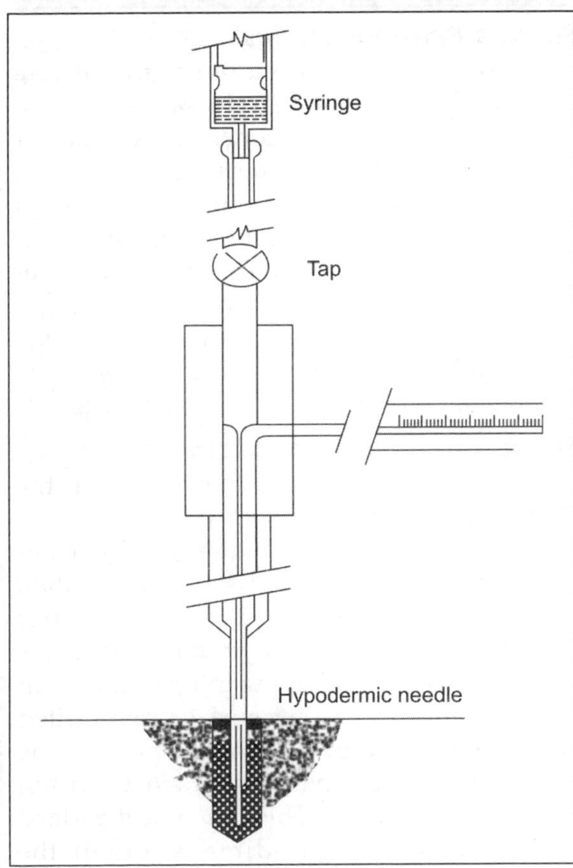

Fig. 8.5: Figg's apparatus

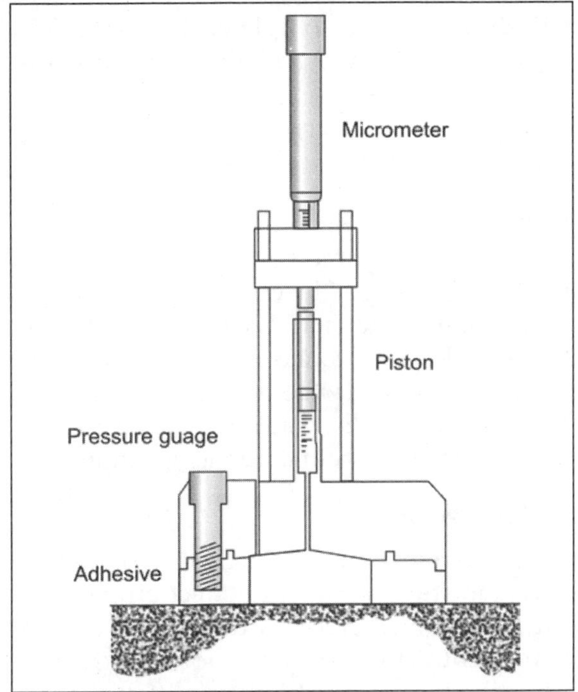

Fig. 8.6: Clam apparatus

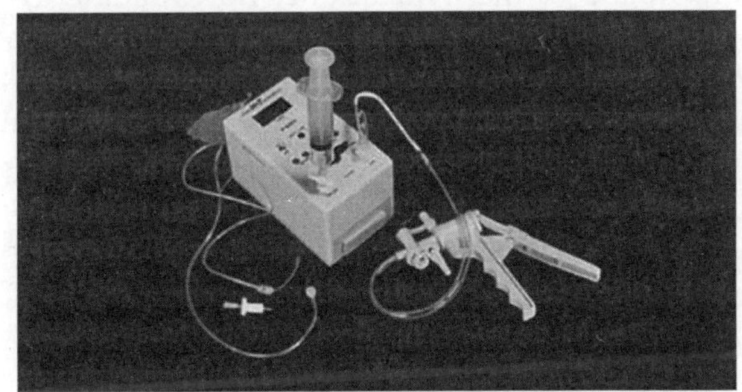

Fig. 8.7: Poroscope plus concrete air/water permeability tester

travel 150 mm is recorded and is taken as a measure of the water absorption index of concrete.

A third portable instrument called the clam uses a circular steel ring bonded to the concrete surface. The main body of the instrument having constant pressure is applied to the surface shown in Fig. 8.6. The pressure is maintained by means of a piston controlled by a micrometer screw gauge. Readings on the gauge are noted at the beginning and end of a test, which allows the volume of water, which has flowed into the concrete to be calculated.

Advanced Permeability Tester

Poroscope plus concrete air/water permeability tester is shown in Fig. 8.7. This is a field test for air and water permeability of concrete using the Figg technique.

The main features are as follows:

- Both air and the same instrument measures water permeability of concrete.
- Permeability both at the concrete surface as well as within the concrete mass can be determined.
- Porosity in sealants and surface mortars can be checked.
- The test is nondestructive (only a small plugged hole required) and can be completely carried out on site.

- Each test can be completed in only a few minutes and gives reliable reproducible results.

The test enables meaningful concrete durability prediction to be made.

Internal Test

A hole 10 mm diameter and 40 mm deep is drilled and plugged leaving a cylindrical test void 10 mm diameter and 20 mm high situated 20 mm below the concrete surface. The time required for air and water to permeate through the test material to the void is used as an index to determine the quality of the concrete under test.

Air Permeability

The air permeability test is always done first since, moisture has a large effect on permeability. Connect the air outlet tube on the instrument to the luer connector on the top of the hypodermic needle. Connect the hand operated vacuum pump to the air connection on the top of the instrument and evacuate to greater than 55 kPa. The instrument times in seconds for the vacuum to fall from 55 kPa to 50 kPa. This time is the Figg number and is a measure of the air permeability of the concrete.

Water Permeability

Connect the water outlet tube to the luer socket on the top of the hypodermic and ensure that the fine plastic inner tube is of sufficient length to reach the bottom of the test cavity. After filling the syringe with distilled water connect it to the water inlet on top of the instrument. The water is then forced into the cavity and the air displaced out through the outer tube through the overflow tube which is 10 cm above the surface of the concrete. The cavity is filled when water starts to flow out the overflow tube. The instrument flow sensor and timer then automatically measured the time taken for the water meniscus to travel a distance of 50 mm and this time in seconds is displayed on the LCD display of the instrument. The time in seconds is Figg number for water permeability.

Surface Permeability Test

Measurements are carried out at the surface by clamping a stained steel chamber on the smooth surface of the concrete. An exactly dimensioned cup grinding wheel is used to smooth the sealing surface of the concrete if necessary. A measurement of the time required for related amounts of air and water to permeate through the concrete is used as an index of the surface conditions. This time can then be used to determine the condition of any concrete sealant or surface mortar.

Surface Technical Specification

A stainless steel surface chamber with the same surface area and exactly twice the volume of the hole used in the internal test is now used as the void for this test. The method of sealing the surface chamber to the concrete eliminates the possibility of variation in the test due to sealants seeping into the chamber, or voids along the sealing surface. The surface chamber is sealed to the concrete by grinding a smooth donut in the surface with the cup wheel provided. This cup wheel is sized to exactly match the dimensions of the surface chamber. A pair of o-rings mounted concentrically in the surface testers 'flange' is then used to seal the chamber to surface. The two o-rings eliminate the possibility of a surface void in the material being tested defeating the test. After clamping the surface chamber to the surface a strong seal is now provided with no variation in volume.

This surface chamber is now used as the void for testing porosity of the surface. Rather than the walls of the hole being the tested surface the surface that the chamber is sealed against is now the surface tested. This provided a check for water and air penetration through concrete sealants, surface mortars and any other methods used to seal construction material surfaces.

The surface chamber has been designed to easily accommodate attachment to the instrumentation. By first performing the air test as outlined in the internal test and then the water test the instrumentation will provide the time required for the chamber to lose 5 kPa vacuum or once filled with water, 0.01 ml water. With the surface tester attached to the instrument both a Figg number and direct indices for air and water surface permeability can be established.

Technical Information

The ingress of air and moisture into the concrete can cause corrosion of the steel reinforcement and lead to a deterioration in concrete strength. Therefore, measure of the ease of movement of the liquids and gases through the surface layer of the concrete is a better method of assessing the soundness and expected life of concrete than strength alone. Permeability is recognized as being the most important parameter in assessing concrete durability. The air permeability test involves measuring the time then for air to flow into a known volume of a sealed, evacuated chamber in the concrete reducing the vacuum from –55 kPa to –5 kPa. This time is a measure of the air permeability of the concrete.

The water permeability test utilizes the same sealed chamber in the concrete, which is completely filled with water, and the total time in seconds for a volume of 0.01 ml of water to escape is taken as a measure of the water permeability of the concrete. The moisture content of the concrete has a major effect on permeability. For example, fully saturated concrete is almost impermeable to air and results in extremely long times in the water permeability test. For effective testing the concrete should be dry and near surface moisture content measured. Permeability test results have shown that there is a good correlation with both water/cement ratio and compressive strength of the concrete.

8.1.6 Half-cell Potential Survey

Corrosion being an electrochemical phenomenon, the electrode potential of steel rebars with reference to a standard electrode undergoes changes depending on corrosion activity. A systematic survey on well-defined grid points gives useful information on the presence of probability of corrosion activity. The grid points used for other measurements, namely, rebound hammer and UPV can be used for making the data more meaningful. The common standard electrodes used are:

- Copper–copper sulfate electrode
- Silver–silver chloride electrode
- Standard calomel electrode

To perform the half-cell potential measurements, electrical connections have to be made with reinforcement and with concrete surrounding it. Electrical connection with the reinforcement is made simply by making a hole in the concrete to expose part of the reinforcement and connecting wire to it. Electrical connection with concrete is made by electrolyte in the $Cu/CuSO_4$ reference electrode which wets both the concrete surface and the conductor to which another wire can be attached. Thus, for taking half-cell potential readings, the half-cell, i.e. $Cu/CuSO_4$ reference electrode is contacted to the surface of the concrete via a sponge soaked in contact fluid and the wires attached to the reinforcement and reference electrode are connected to a high impedance millivoltmeter which gives half-cell potential readings. The measurement consists of giving an electrical connection to the rebar and observing the potential difference between the rebar and a reference electrode in contact with concrete surface. A simple field arrangement is shown in Fig. 8.8.

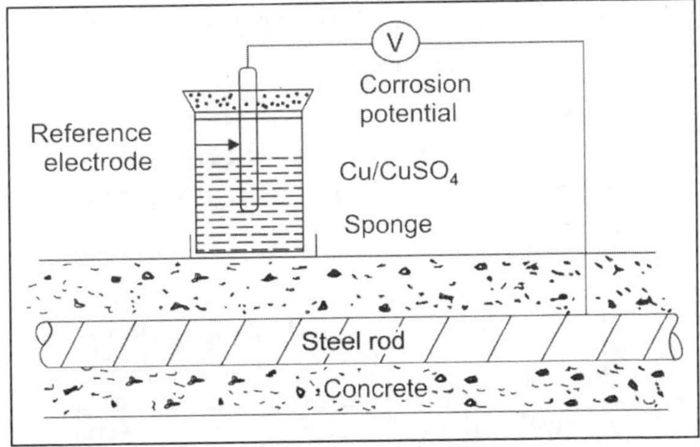

Fig. 8.8: Typical setup of potential measurement in rebars

Generally, the potential value become more and more negative as the corrosion becomes more and more active. However, less negative potential values may also indicate the presence of corrosion activity, if the pH values of concrete are less. Table 8.1 shows the general guidelines for identifying the probability of corrosion based on half-cell potential values covered by ASTM C876-1991.

However, these interpretation criteria should not be normally utilized under the following conditions:

- To evaluate reinforced steel in concrete that has been carbonated/and have highly variable moisture or oxygen content at the level of steel
- To evaluate galvanized reinforcements which have a range of active/passive potentials completely different of that to bare steel.

Advanced Testing Systems

The **ohmcor** and **cormap** systems represent more economical methods of evaluating steel reinforcement corrosion. The cormap systems use half-cell potential mapping in order to identify areas of probable corrosion. The ohmcor system is typically used in conjunction with the cormap system in order to verify the presence of steel reinforcement corrosion. This older technology is still widely used in the field.

Cormap system: A simple economical method for identifying areas of probable rebars corrosion. The major advantages are:

- Easy to use
- Detachable electrode extension pieces facilities measurements in hard to reach locations

- High impedance digital meter is designed for tough field conditions
- Economical.

Technical Specifications

Corrosion, which is an electrochemical process, occurs in concrete when oxygen and moisture are present. The actual corrosion is an exchange of energy within different sections of the uncoated reinforcing steel. The relative energy levels can be determined in relation to a reference electrode with a stable electrochemical potential. By connecting a high impedance voltmeter between the reinforcing steel and a reference electrode placed on the concrete surface, a measurement can be made for the half-cell potential at the location of the reference cell. This then is a measurement of the vicinity of the reference cell. The reference cell is copper in copper sulfate solution. By taking half-cell potential measurements a fixed distance apart a grid of half-cell potentials can be quickly made and thus areas delineated with a high probability of corrosion of the reinforcing steel. To analyze the results, the measurements made with cormap can be plotted on a grid and lines of possible corrosion activity. For example, the following guide is listed in ASTM C876 using a copper/copper sulfate half-cell:

- For readings of –350 mV and greater, there is a 95% chance of active steel corrosion.
- For readings –200 to –350 mV, there is a 50% chance of active steel corrosion.
- For reading less than –200 mV, there is only 5% chance of active steel corrosion.

The method is particularly useful for following:

- Bridge decks
- Parking garages
- Concrete piers and docks
- Substructure
- Tunnel lining
- Foundations

Cormap II system: Advanced system for corrosion potential data acquisition and analysis, allowing the user to quickly identify areas of probable corrosion in the field is shown in Fig. 8.9.

Table 8.1: Corrosion risk based on potential readings

Potential readings	Corrosion
More negative than –350 mV	>90%
–200 mV to –350 mV	50%
More positive than –200 mV	<10%

Fig. 8.9: Corrosion analyzer equipment, cormap II

Features and Benefits

Ruggedized electronics allow rapid analysis of data in the field or office and conforms to ASTM C876 standard method of half-cell potential of uncoated reinforcing steel in concrete. Electrode is designed for use on horizontal, vertical and inverted positions. Temperature and humidity sensors facilities inclusion of environmental condition in data analysis.

For steel reinforcing bars in concrete, corrosion is an exchange of ions from the steel to the concrete. This chemical exchange of ions produces rust (FeO_2). It also produces areas of concrete where there is a larger concentration of negative ions due to the corrosion process of the steel reinforcing bar than areas where there is no corrosion. This larger concentration of ions creates a small electric voltage potential. By measuring and mapping the voltage potential found in the concrete we are able to determine rapidly the presence of corroded steel reinforcement without costly and time consuming demolition of concrete.

This is done by recording the voltage between the rebar and half-cell, which is mapped across the surface of the concrete. Areas of rust with high corrosion will exhibited significantly lower voltage than areas without corrosion, thus areas of corroding steel reinforcing bar in concrete can be rapidly found. There is no need to know the exact position of the steel reinforcing bar or the amount of cover, the presence of the steel is that is required. However, the voltmeter has to be connected to an exposed piece of the rebar network, and because the concrete is being tested, any material on the surface should be removed.

Half-cell Reference Electrode

The cormap II system comes complete with rugged half-cells designed for the tough construction environments. Porous ceramic tips are used in order to provide long life and eliminate problems from clogs in Cu/CuSO4 half-cell. The specially shaped tip has also been designed to allow the half-cell to take readings in the vertical, horizontal or inverted position. They also have a semi transparent full view window that allows liquid level observation without removal of sealed ends and still protects the half-cell from damage by sunlight. The Ag/AgCl electrode uses an easily replaceable cell. This also allows reading to be taken in the vertical, horizontal or inverted position. It also makes maintenance a simple matter.

Instrumentation Unit

They fully integrated data acquisition and analysis unit has been designed for the rapid analysis of data in the field or office. A large amount of data are normally generated, interpretation of this information can be very difficult. Employing the simple to use menu driven cormap main unit, data can not only be collected quickly and easily, but it can also be analyzed directly in the field on the graphic display. The unit produces a symbolic map of the structure, where symbols represent various half-cell potential voltage levels previously acquired. This symbolic map can then be interrupted like a contour map where an area of high potential represents areas most likely to be corroding. Not only can this information be acquired and analyzed but the unit will also read the general environmental parameters of temperature and relative humidity. All this data can also be stored and uploaded to a PC. This allows the user to include the data in subsequent reports and spreadsheets for further analysis.

Specifications:

Instrument weight:	6 lbs (2.75 kg)
Ship weight:	15 lbs (6.8 kg)
Instrument dimension:	4.5″ × 8.5″ × 10.5″ (115 mm × 225 mm × 267mm)
Battery:	12 volt 4–10 hours continuous operation
Display:	320 × 240 pixels backlit for daylight use
Storage:	over 5000 readings
Operating temperature:	0–5°C
Temperature reading range:	–273 to +130°C
Temperature accuracy:	+/–0.5%
Humidity reading range:	0–100%
Humidity accuracy:	± 5%

8.1.7 Resistivity Mapping

The corrosion of a specific length of reinforcement is dependent on the algebraic summation of the electrical currents originating from the corroding sites on the steel and flowing through the moist surrounding concrete to noncorroding sites. Hence, the electrical resistance of concrete plays an important role in determining the magnitude of corrosion at any specific location. This parameter is expressed in terms of resistivity in ohm centimeter or kilo ohm centimeter by four probe resistivity meter shown in Fig. 8.10.

The method essentially consists of using a 4 probe technique and the probes are inserted into the concrete to a depth of 6 mm with a spacing of 50 mm between the probes in which a known current is applied between two outer probe and the voltage drop between the inner two elements is read off allowing for a direct evaluation of resistance R. Using a mathematical conversion factor, resistivity is calculated as $p = 2Ra$. The factors, which govern the resistivity values are:

- Continuants of concrete
- Chemical contents of concrete
- Type of pore structure of concrete

Table 8.2 indicates the general guidelines of resistivity values based on which areas having probable corrosion risk can be identified in concrete structures.

Advanced resistivity meter

RM-8000 ohmcorr resistivity meter to assess corrosion currents in concrete is shown in Fig. 8.11.

Table 8.2: Corrosion risk based on resistivity values

Resistivity ohm cm	Corrosion probability
Greater than 20,000	Negligible
10,000–20,000	Low
5,000–10,000	High
Less than 5,000	Very high

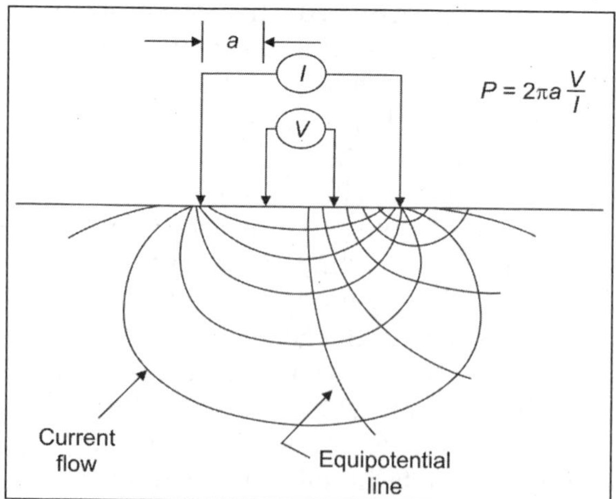

Fig. 8.10: Resistivity testing for concrete

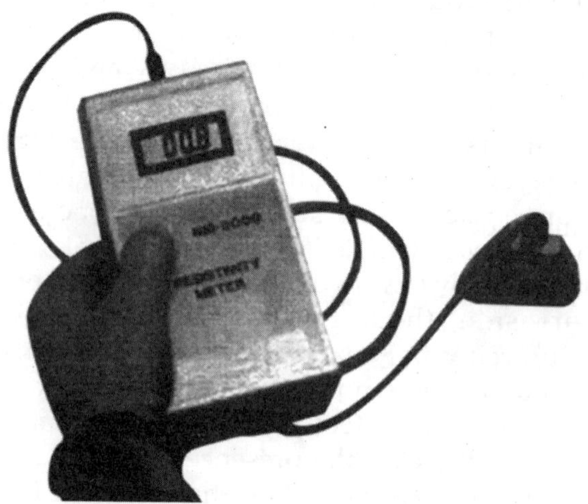

Fig. 8.11: Ohmcorr resistivity meter

Features

- Assesses damaging corrosion currents in concrete
- Economic and easy to use
- Direct digital readout of resistivity. Measuring from two small holes avoids the problems and errors of surface measurements.
- Used in conjunction with cormap system to produce resistively plots.

The ohmcorr, when used in conjunction with the cormap system provides an economic and sound means of diagnosis of corrosion in reinforced concrete.

Technical Information

The electrical conductivity of concrete is an electrolytic process that takes place by the movements of ions in the cement matrix. This ionic movement will take place when contaminants such as chloride ions or carbon dioxide are introduced into the cement mortar matrix. A highly permeable concrete will have a high conductivity and low electrical resistance. Because resistively is proportional to current flow, the measurement of the electric resistance of concrete surface should be avoided. The resistivity meter, ohmcorr, has two probes spaced 5 cm apart, which are placed in two holes drilled to a depth of 8 mm and filled with conductive gel. The concrete resistively is displayed on an LCD, when the control switch is activated. Table 8.3 correlates arrange of values versus the possible rate of corrosion of the reinforcement bars.

8.1.8 Carbonation Test and pH Value

Carbonation test can be carried out using simple phenolphthalein test. Phenolphthalein spray is an indicator of pH value and as carbonation has its effect of pH of concrete, the change in colour indicates the extent of carbonation. The test is conducted by drilling a hole on the concrete surface to different depths up to cover concrete, spraying with phenolphthalein, and observing the colour change. Uncarbonated surface exhibits pink colour while carbonated concrete exhibits no change in colour. The depth of carbonation is estimated based on the change in colour profile.

The pH value can be determined by collecting broken samples or from core samples by analyzing in the laboratory from water extracts and pH meter.

8.1.9 Chloride Content Test

Chloride content can be determined by collecting broken samples or from core samples. The test consists of powdering the sample, obtaining the water extracts and conducting standard titration experiment for determining the water-soluble chloride content, which is expressed by weight of concrete or by weight of cement if the mix ratio is known. One recent development for testing for chloride content includes the use of chloride ion sensitive electrodes. This is commercially known as *rapid chloride test kit* and the test consists of obtaining powdered sample by drilling, collecting the sample from different depths (every 5 mm), mixing the sample (of about 1.5 gm weight). With a special chloride extraction liquid, and measuring the electrical potential of the liquid by chlorideion sensitive electrode, with the help of a calibration graph relating electrical potential and chloride content, the chloride content of the sample can be directly determined.

Table 8.4 gives qualitative guidelines for identification of corrosion prone locations based on pH values and chloride content.

Advanced Equipment

The poroscope and chloride field systems allow the user to rapidly evaluate mechanisms

Table 8.3: Risk of corrosion due to resistivity level

Resistivity level ($k\Omega$ cm)	Possible corrosion rate of reinforcement rebars
<5	Very high
5–10	High
10–20	Moderate to low
>20	Insignificant

Table 8.4: Corrosion prone based on pH chloride content

Test results	Interpretations
1. High pH value greater than 11.5 and very low chloride content	No corrosion
2. High pH value and high chloride content (0.4 to 0.6% by wt. of cement)	Corrosion prone
3. High pH value and high chloride content greater than 0.6% by wt. of cement	Corrosion prone

that cause corrosion. The poroscope quickly and simply evaluates concrete permeability. The chloride field system will rapidly evaluate field chloride levels without expensive and time consuming laboratory tests.

Features

- Fast: Results within minutes at the site.
- Economical: Low coat per sample compare to laboratory testing.
- Accurate: Results are comparable to laboratory testing.
- Covers wide range from 0.002% to 2% chloride by weight.
- Automatic compensation for changes in ambient temperature.
- Digital display for direct reading of lbs/cu.yd. and percentage of chloride by weight.
- Correlates to ASTM C114 and meter conform to AASHTO T260.

Technical Specifications

The determination of the chloride ion concentration in concrete is essential in assessing the need for maintenance on, for example, bridge decks and parking structures. The test can also be used to ensure that materials used in new construction are free from potentially harmful chloride ion levels. With this method, the concentration of acid soluble chlorides is measured. In most cases, this is equivalent to total chloride concentration. A sample of powder is obtained by drilling and careful quartering. Then an accurately weighed 3 gm sample is dissolved in 20 ml of extraction liquid which consists of a precise, measured concentration of acid. For sampling wet concrete a 3 gm sample of mortar (i.e. without coarse aggregate) is used. The chloride ion reacts with the acid of the extraction liquid in an electrochemical reaction. An electrode, with integral temperature sensor, is insured into the liquid and the electrochemical reaction measured. A uniquely designed instrument converts the voltage generated by the chloride concentration. The instrument automatically applies the temperature correction and correlates to ASTM C114. Once the sample is obtained, test results can be determined and read in less than five minutes to avoid contamination, the electrode should be thoroughly washed with deionizer water after each test. Replacement packs, containing twelve bottles of extraction liquid each for one time use, are available. Five calibration liquids each with known concentrations are supplied with each pack. These liquids are used to establish the calibration curve, and to check that the system is functioning correctly. Calibration is not required for each use and the instrument will indicate when calibration is required. The calibration liquids are colored to avoid confusion between them and with the extraction liquid. All equipment necessary to complete the chloride test is supplied in standard size carrying case.

8.1.10 Abrasion Resistance Test

ASTM standard C779-89a prescribes three test procedures for abrasion resistance of horizontal concrete surfaces. In procedure A, the revolving disk test, the abrasive action is provided by the flat faces of three (60.3 mm diameter) cold-rolled steel revolving disks, each attached to motor driven vertical shafts which also revolve about a vertical axis. The abrasive grit used in this procedure is silicon carbide. The comparison of measurements of average depth of wears of representative surfaces (specimen size 305 mm × 305 mm × 95 mm) at 30 min and 60 min, exposure to abrasion will indicate the relative abrasion resistance of these surfaces. The other two procedures are *dressing wheels abrasion test and ball bearing abrasion test.*

8.1.11 Alkali–Silica Reaction Test

Alkali–silica reaction damage is assessed by *ASR detect equipment.* Simply apply each of the two regents to the broken surface of a concrete core drilled in a suspect structure and rinse off the excess. On ASR contaminated concrete, the resultant stain reveals the presence of ASR. The stain distribution shows the extent of ASR in the concrete, and their proximity to different components of the aggregate gives clues to the source of trouble. The two gels that are identified, one staining yellow, the other pink indicate the stage of ASR progression. Yellow signals that degradation has begun. Pink warns that degradation is advancing. typically, ASR occurs in cracks and these cracks often cut through the aggregate and usually do not follow the aggregate-paste boundaries. ASR tends to fill air voids. The ASR detect is shown in Fig. 8.12, where field test to detect alkali–silica reaction is shown.

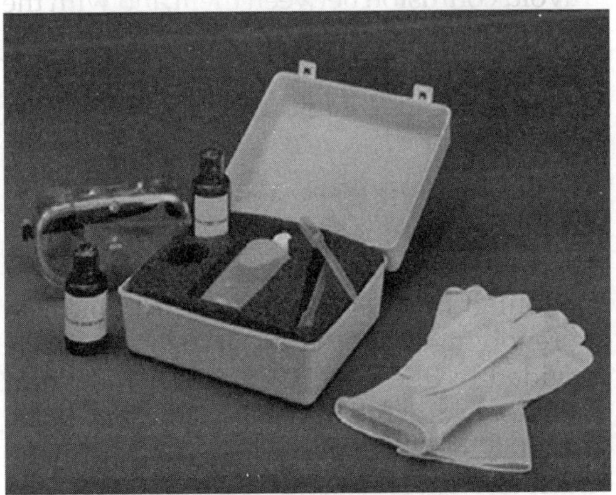

Fig. 8.12: Simple coloured dye field test

Features

- Test can be carried out completely on site.
- Minimal operator training and no special equipment required.
- Utilizes only two environmentally safe days.
- Identifies ASR in concrete and differentiates ASR from other causes of degradation.
- Results obtained in less than five minutes are easy to interpret. Economic, fast and easy to use.

Advantages

In contrast to the two established methods of ASR detection—petrography analysis and uranyl acetate analysis, ASR detect has numerous benefits. Because the reagent stains are clearly visible even before the treated sample dries, a complete diagnosis is possible in less than 5 minutes. ASR detect systems are relatively inexpensive. Petrographic analysis requires shipment to a laboratory, adding time and raising the costs substantially per sample, also the uranyl acetate reagent is almost prohibitively expensive. ASR detect is simple enough to use in the field. Its reagent stains are visible to the naked eye and are distinctive enough to be recognized and interpreted by anyone with minimal training. Petrographic analysis requires specially trained technicians working in a well-equipped laboratory. Therefore, the use of ASR detect can considerably reduce the cost of diagnosis by reducing drastically the need for petrographic analysis.

The ASR detect reagents present minimal danger to either human health or the environment. Uranyl acetate is radioactive and contains a heavy metal and therefore, has the potential to cause health and disposal problems. ASR detect provides information not only about the presence but also the severity of ASR. Its affordability allows an engineer to analyze enough samples to obtain an accurate diagnostic picture of an entire structure. The high cost of a petrographic analysis permits only a limited number of samples to be examined. Concrete tested with pink gel, yellow gel and pink yellow gel are shown in Fig. 8.13 and showing beginning and advanced stages of ASR degradation.

Applications

ASR detect is both a practical and a scientific tool. Its principal application is analyzing existing concrete structures. By identifying ASR deterioration in its earliest stage, ASR detect facilitates the problem being identified when remediation techniques can be applied; for example, treating the concrete with a lithium bearing solution to inhibit further deterioration is advanced, ASR detect provides a clear picture of the extent and depth of the damage. As a scientific tool, ASR detect can be applied for improving the understanding of where, how and why ASR occurs. That understanding is basic to developing ASR preventatives that allow high-alkali cements or poor-quality aggregates to be used in concrete mixes without risking the development of ASR.

Technical Specification

One of the primary causes of premature concrete deterioration is alkali-silica reaction (ASR). ASR causes concrete to deteriorate when sodium and/or potassium from the cement attacks silica rich components in the aggregate, producing gels that expand and eventually crack the structure. Most gels contain cations (positively charged atoms or, molecules) that readily exchange with other cations in solution. ASR detects two reagents react with cations found in the two gels associated with ASR. The first reagent exchanges sodium with the potassium found in some ASR gels and then reacts to a form a bright yellow precipitate. The second reagent reacts with calcium-rich ASR gel to form a bright pink satin. In concrete containing ASR,

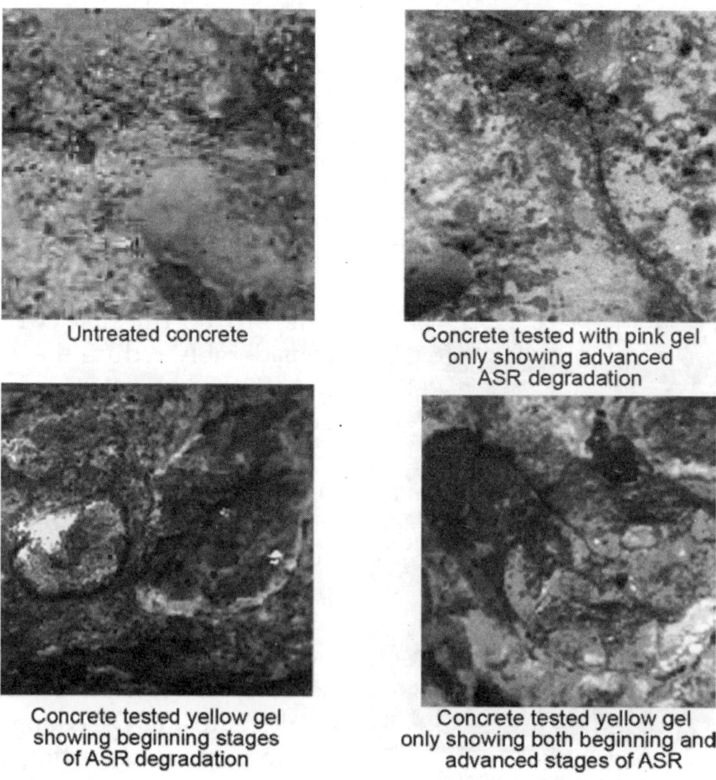

Fig. 8.13: Beginning and advanced stages of ASR degradation

the result is brightly colored surface showing the presence of the targeted gels; concrete with no ASR is unaffected.

8.2 CHEMICAL TESTING OF CONCRETE

Chemical testing of hardened concrete is mainly carried out to identify the causes of deterioration, and in the context of rebar corrosion, mainly consist of the determination of original water and cement contents, chloride content, carbonation depth and pH of concrete. Chemical testing of *in situ* concrete can be carried out in the laboratory through proper sampling of concrete from the structure under consideration.

Determination of Cement Content

The cement content of the powdered sample of concrete can be determined following the standard procedure recommended in ASTM C1084-87. In this procedure percentage silica in the concrete or percentage calcium oxide in the concrete is determined by soluble silica subprocedure or calcium oxide subprocedure. The cement percentage by weight is expressed as

$$C = Cs \text{ or } Ce$$

Where

Cs = (% silica in the concrete/% silica in the cement) × 100 (assume 21% if not known) and

Ce = (% calcium oxide in the concrete/ % calcium oxide in the cement) × 100 (assume 63.5% if not known)

Knowing the cement % (C) by weight, the cement content of concrete can be obtained as:

Cement content = $C \times D/100$

where D is density of concrete.

The cement of concrete has important bearing on its corrosion resistance properties and it affects the rebar corrosion because high alkalinity of the concrete, which resist the depassivation of rebar, is mainly due to the cement. It has been demonstrated by fried land that concrete with cement content less than 235 kg/m^3 can have a highly enhanced rate of rusting even when the w/c ratio is proper, and increasing the cement content to 300 kg/m^3 improves the product nature of concrete; however beyond this cement content the improvement in productive nature of concrete is not proportional. Various codes of practices have recognized the above fact and IS:456-2000 restricts the minimum cement content for reinforced concrete as 250–360 kg/m^3 for mild, moderate and severe exposure respectively.

Determination of Original Water Content and w/c Ratio

Determination of original water content and w/c ratio of the hardened concrete is covered in BS 1881 Part-124; 1988. The original water content is the amount of water present in the concrete mix at the time of setting. It is the sum of the capillary porosity of the concrete (Q) originally filled with water at the time of setting; and the combined water of hydration present in the prepared concrete sample. The capillary porosity of concrete can be determined using unfractured core sample. The sample is oven dried at 105°C for at least 16 hours cooled to room temperature in the desiccators, weighed, and immersed in a liquid of known density in vacuum desiccators. The pressure is reduced, causing the air from the capillaries to be evolved. The vacuum is then released and the sample kept immersed in a liquid for further 5 min, the weight of liquid observed can then be calculated and the capillary porosity can be determined as

Q = Wt. of observed liquid density (1.33) × Wt. of dry sample × 100

If aggregate control samples are not available as in case of diagnosis of old RC structures, it may be assumed that the combined water of hydration of concrete is typically 0.23 × C, where C is the percentage cement content. Further, if the aggregate absorption value is not known, then only the original total w/c, which includes the water absorbed by the aggregates at the time of setting unlike the free w/c, which excludes the water absorbed by the aggregates at the time of setting, can be quoted and is given as following:

The original total $w/c = (Q/C) + 0.23$

The w/c of concrete plays different roles in the process of rebar corrosion in different exposure of concrete. When the RC structures are immersed in some aggressive solutions, it is the permeability of concrete, which is a function of a water–cement ratio, affects the corrosion of rebar.

Determination of Free Chloride Content

The researches have furnished recently a rapid, accurate method for determining the water soluble chloride (free Cl ions) in hardened concrete samples obtained from the site. This rapid method involves the extraction of water soluble free chloride ions, present in concrete samples, by boiling. After boiling the powdered sample for ten minutes in water, undissolved matter is filtered out, and the filtrate acidified to a pH value of 3 before titration against standard mercury nitrate solution using diphenyl carbasone as an indicator. This titration gives an accurate estimate of the free chloride, percent by weight of concrete. IS:456-2000 limits the total amount of chloride in the concrete at the time of placing to 0.15% by mass of cement.

8.3 DIAGNOSTIC METHODS OF CORROSION DAMAGE

Diagnosis of the distressed structures is the first step in the process of their repairs and rehabilitation to identify the real causes of distress. Its importance lies in the fact that rehabilitation carried out without comprehensive diagnosis, will not be able to cure the real causes of distress; and thus rendering the repaired structure shows signs of cracking, spalling or rust staining, or any other sign that might indicate corrosion distress; the first task should be to find out how serious it is, and what has caused it, and how it can be set right.

The corrosion of reinforcing steel in RC structures is one of the most commonly occurring causes of distress in many structures even in places situated far away from marine environment or sea shore. It is a well-known fact that the corrosion of reinforcing bar in RC structures occurs mainly due to two reasons:

- Reductions of alkalinity of concrete at rebar level due to carbonation and/or penetration of other acidic gases such as sulphur dioxide.
- Presences of aggressive anions such as free chloride ions which can lead to the de-passivation of rebar.

In an urban environment such as one encountered in industrial zones of Delhi, the reduction of alkalinity of concrete may be either due to carbonation owing to atmospheric carbon dioxide or due to penetration of polluting industrial acidic gases. The penetration of external chloride into the concrete is more prevalent near the sea shore; however at places situated far away from marine environment, presence of free chloride ion in concrete is likely to be through addition of $CaCl_2$ during the time of construction as an accelerating admixture or through impurities in aggregates and/or mixing water.

A large number of corrosion distress cases have been reported in recent years. Some examples are corrosion distress of RCC bridge across little Rann of Kutch, corrosion distress of prestressed concrete Thana Creek bridge, and corrosion distress in buildings located in various parts of Northern India, most of which were constructed in the sixties and seventies, whereas a few cases refer to relatively recent construction. The cause of corrosion in RC structures under marine or chloride bearing industrial environment has been well understood which is due to presence of large amount of chloride in concrete made available through penetration from outside; but the cases of corrosion distress in the RC structures away from marine or chloride bearing industrial environment, such as those encountered in Northern India, draw attention towards the need of pinpointing the real cause of rebar corrosion, which may possibly be more than one.

8.4 INVESTIGATION STRATEGIES

A visual survey of the affected structures provides valuable information to ensure that whether the corrosion of rebar is really a cause of distress or there is some other cause of distress. This survey consist of a careful investigation of the structures for any sign of distress, such as cracking, cracking and spalling, cracking and rust staining, and simply rust staining. If visual inspection of structure suggests that the cause of distress is the corrosion of rebar only, the next step is to make a careful examination of the structure and carryout detailed test which will positively identify the cause and extends of distress, and allow some predictions to be made about the future life of the structure. The prediction of the future life of the structure needs measurement of the rebar corrosion rate and assessment of the loss of sections from the reinforcement or prestressing steel due to rusting. Measurement of the corrosion rate is useful for life prediction in terms of the residual time during which the cover concrete would crack or spall due to rusting; whereas the assessment of the loss of section from the rebar or prestressing steel is useful to predict the life in terms of the residual time during which the strength of structure is impaired.

After the initial sign of corrosion has been confirmed from visual observation, a thorough preliminary investigation may be carried out to locate the rebar and to estimate the cover depth. This may be followed by half-cell potential mapping and concrete resistivity mapping to isolate the corroding and noncorroding regions. The causes of rusting can then are ascertained by chemical testing of concrete both *in situ* and in the laboratory using cored samples obtained from corroding regions. The rate and extent of damage may be estimated through electrochemical testing techniques and through chemical cleaning of randomly collected samples of rusted rebar.

8.4.1 Detailed Test and Inspection Techniques

A considerable range of test and inspection techniques is available for use *in situ* during inspection, or on samples removed from the structures for laboratory investigations. Test and inspection techniques can be subdivided as follows:

- Those dealing with structural integrity and location of reinforcement, e.g. pachometer or cover meter service.
- Those dealing with prestressing steel or reinforcing steel serviceability and condition, e.g. electrochemical and non-electrochemical measurements.
- Those dealing with concrete quality and composition namely chemical testing of samples obtained by crushing of cores collected from site, for determination of chloride content, pH cement content and original w/c ratio, etc.; additionally the core samples obtained from the site can be tested for estimation of potential cube compressive strength and absorption test, etc., in order to ascertain the durability quality of concrete.

8.4.2 Determination of Structural Integrity and Location of Reinforcement Visual Surveys

The first visibility of reinforcing steel distress in a conventional RC structure is frequently the appearance of delamination of the outer concrete layer and rust staining on the concrete surface. Obviously, this only occurs where the surface can be visually examined. A later visual stage occurs where cracks which might have developed at the subsurface, reach the surface and spalling of the concrete ultimately occurs. Hairline cracks in concrete running parallel to the rebar may be due to shrinkage or they may be the first sign of corrosion distress. That cracks developed, whether due to rusting or due to shrinkage, can be differentiated on the fact that the delamination with rust staining on concrete surface appears due to rusting, rather than shrinkage, but this evidence is not always apparent. There are many tools which may be used to detect delamination or surface fracture planes parallel to the concrete surface. These devices range from simple chain drags or lightweight striking hammers, involved in surrounding technique, to more sophisticated devices such as delamtect. Almost any surrounding device can be used to locate hollow areas are delamination caused by corrosion of rebar. Ultrasonic reflective method to determine the existence and extent of sub-surface delamination and the depth of visible surface cracks can be alternatively used. The suitability of various delamination detection techniques/devices in different situations is presented in Table 8.5.

Visual survey should be carried out by experienced personnel capable of making careful observations and record findings in a systematic manner, paying attention both to areas exhibiting distress and areas which are visually sound for comparison. The visual survey is normally used as the basis of confirmation of the reinforcement corrosion

Table 8.5: Suitability of delamination detection techniques/devices

Sr. No.	Delamination detection techniques/devices	Suitability
1	Simple chain drag	For locating delaminated areas during repair operations
2	Automated delamtect	For surveying large numbers of bridge decks or other horizontal surfaces such as parking garage floors, if record of the area of delamination is required
3	Surrounding technique using lightweight striking hammer	For upper surfaces of bridge decks and also on vertical surfaces
4	Ultrasonic reflective technique using soniscope	May be modified so as to be of use even on submersed surfaces

distress and for any on site testing non-destructive testing and sampling.

Pachometer or Cover Meter Surveys

Pachometer or cover meter is used to locate reinforcing steel embedded in concrete, and to determine the amount of cover over steel. The purpose of locating the reinforcing steel, pre-stressing strand, cladding ties, etc. in concrete, is to avoid them during core sampling, direct pulse velocity surveys, etc. and to make connections for electrochemical measurements. The knowledge of cover depth is essential, if it is desired to obtain samples of the concrete at the level of the rebar for chloride ion analysis, pH measurement and the presence of carbonation. It is also useful in determining the potential for corrosion and subsequent concrete deterioration, since it has been well established that structures in corrosive environments with inadequate cover get subjected to early deterioration.

8.4.3 Determination of Steel Serviceability and Condition

Measurement of reinforcement corrosion in concrete structure needs a method which can determine simply, accurately and non-destructive way not only whether corrosion of reinforcement is taking place but also its intensity and the rate of damage. The most widely used *in situ* methods are half-cell potential mapping, and the resistivity testing of concrete. Among the other methods the linear polarization or polarization resistance method is considered to be the most suitable for determining the rate of corrosion. Furthermore, the gravimetric test and visual observation can also be performed on rebar samples, obtained from site, to assess the percentage weight loss of rebar and its geometrical features after rusting.

Half-cell Potential Mapping

The half-cell potential is also called as open circuit potential or rust potential or corrosion potential. The test procedure is discussed under the previous headings. Half-cell potential data is normally presented as iso-potential contour plots, known as half-cell mapping which can be used to delineate anodic and cathodic areas of reinforcement. It is important to note that the potential mapping can provide information on the extent of corrosion and its probability but rates of corrosion cannot be inferred from absolute values of potential. In general, where corrosion is slow, potential difference rarely exceed 100 mV, where as in reinforced concrete undergoing significant corrosion potential differences over 200 mV are common. This method is widely used when assessing maintenance and repair requirements, with the advantages of being nondestructive, inexpensive, simplest to perform, and able to survey the whole structure quickly. It is particularly valuable in isolating the regions in which corrosion may cause future difficulties, from those in which it has already occurred but with no visible evidence at the surface of concrete. This helps in restricting the corroded regions of structure for detailed investigation. However, this method is difficult to perform when contaminants are present in concrete.

Resistivity Testing of Concrete

The procedure of the test is explained in the previous headings. Like half-cell potential mapping, resistivity measurements, particularly using surface contact probes, can be a rapid scan technique that provides useful information of rebars corrosion. However, resistivity testing is normally used only in conjunction with detailed half-cell potential mapping. For *in situ* usage, variations in moisture conditions are likely to have the major influence upon results. Furthermore, it should be prone in mind that resistivity measurements generally relate to the concrete cover to the steel and not to the resistance of the concrete immediately next to the steel where corrosion is occurring. It is generally accepted however that for practical purposes the depth of zone of concrete affecting the

measurement will be equal to the electrode spacing and a value of 50 mm is commonly adopted. The method is therefore only indicative and other factors, such as oxygen availability at the cathode or anode–cathode ratios, may also influence the corrosion rate of the reinforcement.

Corrosion Rate Measurement

In reinforced concrete structures, determination of actual rate at which the reinforcement is corroding assumes larger importance. One method is known as linear polarization resistance (LPR) method for the on site study of corrosion rates of steel in concrete. The fundamental principle of linear polarization is based on the experimentally observed assumption that for a simple model corroding system, the polarization curve for a few millivolts around the corrosion potential obeys a quasi-linear relationship. The slope of this curve is the so called polarization resistance (R_p):

$$R_p = \left(\frac{\Delta E}{\Delta I}\right)_{\Delta E-0}$$

From this slope (Fig. 8.15), the corrosion rate can be determined

$$I_{\text{corr}} = \frac{B}{R_p}$$

where, B is a constant which is a function of the Tafel slopes and determined from the formula given below:

$$B = (B_a \cdot B_c)/2.3\,(B_a + B_c)$$

The value of B usually lies between 13 and 52 mV depending on the passive and active corroding system. For on site measurements, the testing system consists of a potentiostat, counter electrode, reference electrode, and the reinforcement as working electrode. This system is schematically illustrated in Fig. 8.14. A typical plot of linear polarization curve is shown in Fig. 8.15. It is necessary that, for measurements in concrete, the potentiostat should have electronic ohmic compensation or otherwise, the value is to be obtained by calculation or by separate experiments.

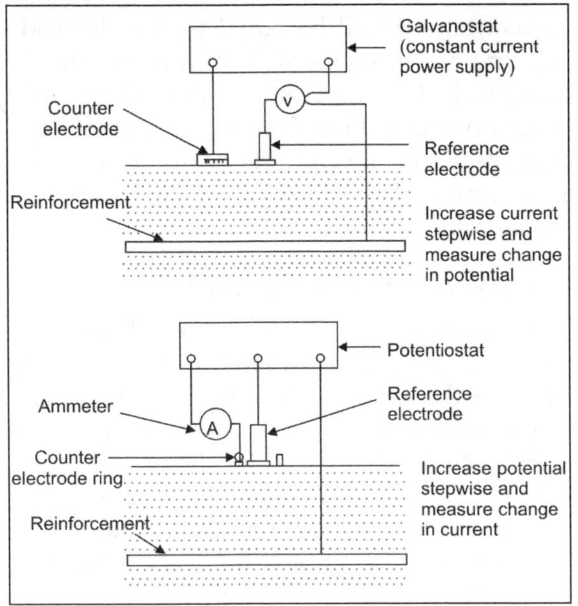

Fig. 8.14: Typical setup of linear polarization resistance method

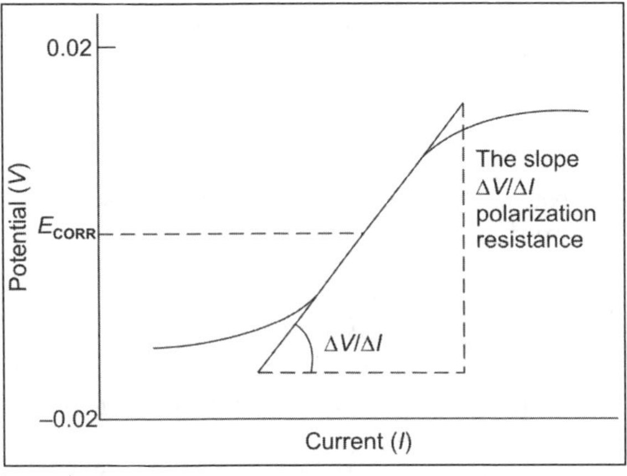

Fig. 8.15: Typical polarization curve

The application of commercial instrument known as Gecor-6, for measuring the corrosion current in reinforcement in an actual structure. This works on the principle of LPR technique.

This category comprises the range of instruments that evaluate parameters related to the corrosion of the concrete or the steel reinforcing bar within the concrete. The leading instruments in this field are Gecor systems. These represent the latest research and technology in determining the corrosion rate of steel reinforcing bar in concrete. The Gecor-8 is most sophisticated system using a closed loop modular confinement system in order to confine the linear polarization test from outside parameters. The Gecor-6 is a more economical method of determining corrosion rate measurement. These will allow the end user to accurately locate areas of high corrosion without expensive demolition work and to estimate the service life of the structure.

Gecor systems for corrosion analysis of concrete steel reinforcement bars is the world's most advanced system for analyzing corrosion of concrete steel reinforcement bars. Gecor-8 represents the latest technology in steel reinforcing bar corrosion rate determination. It combines state-of-the-art embedded microprocessor systems and computerized flash technology with the world's leading research in reinforcing bar corrosion rate analysis.

The Gecor-8 features are brought out below:
- A method for rapidly mapping a structure's corrosion rate.
- An advanced modulation confinement technique for precise corrosion rate analysis.
- New corrosion rate analysis techniques for use in very wet or submerged structures.
- New methodology for measurement of cathodic protection efficiency and more.

The rapid mapping technique of the Gecor-8 allows the engineer to quickly classify areas of a structure, while the unit's built-in programming can graphically analyze this data and produce contour maps of suspected areas. Further, our advanced modulation confinement technique accurately measures the true polarization resistance of the steel reinforcing bar. Combined with the latest advances in interface systems and database storage technology, the Gecor-8 is an easy-to-use, reliable, automatic and intelligent corrosion rate analysis system. A system that can save an engineer's time, money and effort in evaluating structures for subsequent rehabilitation.

Gravimetric Test and Determination of Geometrical Features of Rebar Sample

Gravimetric test and measurement of the reduced diameter of rusted rebar can be carried out by cleaning the rebar samples using Clark's solution as per ASTM G1-81. Composition of the Clark's solution is as follows: One litre HCl of sp. gr. 1.19, antimony trioxide 20 gms and stannous chloride 50 gms. After proper cleaning, the difference in weight and reduced diameter of bar can be measured. The difference in weight can be used to determine the average corrosion rate of rebar. No differential or instantaneous corrosion rates can be measured, but only a mean can be measured during the period of test. When pitting is present, it must be accompanied by visual observation. The measurement of the reduced diameter of the rebar can help in determining the residual strength of a structure.

8.5 DETERMINATION OF CONCRETE QUALITY AND COMPOSITION

Testing of Drilled Cores

The test procedure of core drilling and sampling is already explained in the previous headings. The only way to get detailed information about the *in situ* concrete is by obtaining and testing drilled cores from the distressed structures. However, obtaining the cores is relatively expensive and time consuming. The cores drilled from concrete structures can be tested in compression in addition to chemical testing to measure the compressive strength of concrete; they can be tested in splitting to measure the tensile strength of concrete; they can be weighed to measure its density; sliced to measure its permeability; and examined petrographically. After these physical tests on cores, they can be crushed to analyze the concrete chemically to determine cement content and type,

chloride content, water–cement ratio, pH, aggregate type and grading. Further, if a core is cut through the reinforcement, its condition can be examined too by electrochemical and nonelectrochemical methods. However, normally coring through reinforcement should be avoided, because it can seriously weaken the structure besides rendering it unsuitable for compression testing. Corrected absorption value of 3% for 75 mm dia. core sample indicates good concrete, whereas the same between 3–5% indicates moderate concrete and absorption higher than 5% indicates a poor concrete.

Chemical Testing of Concrete

These tests include:

- Determination of cement content
- Determination of original water content and w/c ratio
- Determination of free chloride content
- *In situ* determination of depth of carbonation
- Determination of pH value of concrete

The above test procedures are already explained in the previous headings.

8.6 SYSTEMATIC ASSESSMENT OF FIRE AFFECTED STRUCTURES

The assessment of damage caused by fire in a building is of utmost importance, as it is a key to any subsequent repair work and its economic and practical viability. Damage caused by fire can be of three categories named damage beyond repair, which is more difficult and requires close examination of each structural member. The strategy generally adopted in this situation is to determine the extent of damage and residual material strength of the structural member either directly or through indirect methods. Direct estimation of the material strength of the member is usually by means of nondestructive method which involve estimation of peak temperature and the peak temperature profile attained in the material during the time of fire, from such known temperature history of the material, a reasonable estimate of their residual strength can then be made from available chart etc. Yet another way of estimating the temperature history of material is through theoretical calculation. By estimating the probable value of the air temperature attained during fire in the room, from heat transfer calculation and using charts, one can estimate the temperature profile in structural sections. The steps in the procedure generally adopted, is the estimation of residual strength and degree of damage, and lastly stabilize check of the building.

8.6.1 Preliminary Investigation

A prompt and careful preliminary investigation of fire damaged structure is necessary to discover any urgent requirements for temporary propping before cleaning up and safe detail investigation. The evidence of the preliminary survey should also enable the engineer to identify the structural element to be examined first. The investigation helps in selecting investigational techniques and test methods which would be best suited to the damaged structure in question. On the basis of preliminary investigations, a damage classification of structure can be done as given in Table 8.6.

Table 8.6: Damage classification of fire damaged concrete

Sr.No.	Damage level	Investigation requirement
1	Repairable damage	Requires close examination of affected structural member by means of the best suitable investigational technique and test
2	No damage	Decision regarding this part of structural member can be made during preliminary investigation just by visual observation
3	Damage beyond repair	Decision regarding this part of structural member can be made just by observation such as deflection, etc. of the member concerned

8.6.2 Detailed Investigation

Detailed inspection and investigation may be taken up once the structure has been propped up and found reasonably safe. This process involves a range of technique starting from simple thorough visual inspection to highly sophisticated methods. Each individual method has its own advantages and disadvantages and a number of methods are used together in conjunction to arrive at worthwhile conclusion.

Detail Visual Inspection

This includes visual survey of colour, texture and damages of surface concrete and collection of evidential debris, resulting in identification of zones based on extent of damage, approximate assessment of gas temperature in the room and temperature attained by the surface concrete. Concrete is known to undergo a set of irreversible chemical and physical changes when subjected to high temperature regime. Concrete made from non-igneous aggregates usually undergoes a permanent color change through oxidation on heating. It has been stated that below 300°C of temperature, colour of concrete is normal, progressively between 425°C–600°C it may be pink hue depends to red. Between 600°C–1000°C colour changes back through grey buff. However, this colour change depends upon type of aggregate, for example, calcareous and igneous crushed aggregate are less susceptible to this effect.

When concrete is subjected to high temperature, the evaporable water present in concrete is lost first, followed by the commencement of dehydration and breakdown of cement gel at about 180°C. At about 300°C inter layer water at C-S-H and some chemically combined water is lost. Dehydration of $Ca(OH)_2$ begins at about 500°C and complete decomposition of C-S-H is attained at 900°C. The binding material of concrete thus loses its binding property with the imposition of high temperature. Aggregate in the concrete however is relatively inert but some aggregate themselves may disturb due to explosive expansion of free and combined water or gases, for example decarbonation of limestone, etc. Expansive changes in aggregate on heating may even result in spalling. At about 300°C surface crazing in concrete may be observed. Cracking due to expansion, due to excessive tensile stresses may become apparent at 600°C, but if the temperature attained during the fire is 800°C and so, saucer like pieces will break away from concrete resulting in concrete spalling. The change in the colour and appearance of concrete subject to different temperature is given in Table 8.7.

Direct Strength Assessment

Essentially this type of test includes non-destructive tests, core test and direct load test. Nondestructive testing can be performed on selected zones identified from damage classification. Commonly used NDTs are Schmidt hammer test and ultrasonic pulse velocity test, etc., however their uses may be limited by the available surface condition, especially in case of UPV which requires polished surfaces. Besides UPV cannot be used for slabs and very thick structures. Rebound hammer test however is relatively more useful. Other methods such as windsor probe, lok test, etc. are yet to attain sufficient popularity in this context. Correlating strength with rebound number and UPV essentially requires a calibration that can be obtained through limited number of core test. Correction for commonly affecting factors such as moisture content, carbonation, etc. must be duly applied.

Table 8.7: Changes in the colour and appearance of concrete		
Temperature of specimen	*Colour*	*Surface appearance*
Room temperature	Cream colour	Sound in texture
200°C	Slight pinkish	No cracks visible
400°C	Whitish grey	Very fine few cracks
600°C	Dove grey	Wide cracks with spalling

Core Test

One positive way to confirm suspicion about the quality of *cast in situ* concrete is to take core from it. The relevant properties are determined by the standard test procedures, where detailed results of compressive strength, modulus of elasticity, hardness, etc. are required. Cores which intersect reinforcement steel provide an opportunity of examining the steel and its condition of bond with cement paste. In order to see the comparison between the undamaged and damaged concrete, cores are taken from undamaged concrete also. Core test can also be helpful in obtaining the calibration for other NDTs. This method provides an in-depth investigation of the damage occurred to concrete during fire. However, taking out more and more cores will make the already weaken structure more weak.

Direct Load Test

Method is very expensive and time consuming. For *in situ* test whether the member under test is actually subjected to the assumed test load during fire is difficult to predict due to load sharing effect. All the direct methods of assessing residual strength of concrete provide information regarding average potential cube compressive strength and are not indicative of strength profile across the section. Direct load testing again may not be possible where the loading may damage the structure permanently or where the structural system is too stiff to show any deflection.

Indirect Test

For indirect assessment of residual strength, its relationship with the temperature attained must be known well ahead of the test. Thermogravimetric test (TG), dilatometric test, differential thermal analysis (DTA), thermoluminescence test (TL), etc. belong to this category.

Thermogravimetric Test (TG)

This test together with dilatometric test was used to determine the temperature attained during fire. As the concrete undergoes irreversible chemical changes during fire there is weight loss due to the process of dehydration of C-S-H and dihydroxylation of calcium hydroxide and is indicated by thermogravimetric test. This process can be considered as a process of chemical stabilization. Thus, the concrete which has not been subjected to fire at all is likely to show weight loss at a slow rate initially below 100°C may become rapid beyond 300°C up to a temperature of 800°C when the dehydration process is complete. A concrete which on the other hand has been subjected to a temperature of say 500°C, if tested within a reasonable period from the date of fire is unlikely to show any weightloss till 500°C, beyond which weightloss pattern will be identical as that of unaffected concrete. The unaffected concrete can be treated as a reference specimen with which thermogravimetry curves of fire affected concrete may be compared to ascertain the temperature. The quantity of test samples required is of order of 500 mg of ground material, therefore if concrete sample is collected from different depths, approximate temperature profile can be established through this test.

Dilatometric Test (DT)

The principle underlying this test is similar to that mentioned above for TG test. During the process of dehydration irreversible shrinkage of concrete takes place and which may be detected through DT test. In this test again unaffected concrete is used as reference where the shrinkage takes place rapidly beyond a temperature of about 150°C after initial expansion, till about 850°C and line expansion versus temperature represents dilatometric curves. The fire damaged concrete on the other hands demonstrates thermal expansion monotonically till the temperature to which it has been subjected during fire, beyond which the shrinkage follow reference curve for unaffected concrete. Thus, by comparison of the dilatometric curve of fire affected concrete with that of unaffected concrete, one can establish the probable temperature to which concrete was subjected to. The sample material originally used was 7.62 cm × 1.3 cm × 1.3 cm square representing the test specimen. The rate of temperature rise at which the test was performed as 5°C/min. Details of this test together with TG test and their use for assessing the temperature of fire damaged concrete. It must be recognized that this method provides only indicative value of temperature and error can be anything of the order of ±15 to 25%.

Differential Thermal Analysis (DTA)

This technique is based up on measurement of temperature change of small sample of powdered concrete accompany the irreversible physiochemical transformations at a temperature when subjected to a heating region in a furnace. Usually the rate of change is compared with an inert sample, whose rate of temperature rises is controlled to be as nearly uniform as possible. The sign of the measured temperature change increase or decrease in relation to the reference determines whether the change is exothermic or endothermic and indicated by peaks in the measurement. The exothermic/endothermic peak temperatures are characteristics of particular physiochemical changes such as loss of water of crystallization, dehydro oxidation and decarbonation, etc. The DTA can be performed for an unaffected sample and expected peaks can be identified. The absence of peak at a desired temperatures for the concrete subjected to fire excludes the possibility of the expected physiochemical changes there by giving an indication that this temperature was already attained by the sample during the fire. DTA curve therefore can throw light on the temperate history of the sample and an approximate assessment of the temperature profile of the concrete section can be made when drilled sample from different depths are analysed. Sample weight may be of the order of 16–20 mg depending up on the instrument, and rate of temperature increases in the furnace may be of the order of 20°C/min.

However, literature on application of this method to post fire assessment of temperature of concrete is not many.

Thermoluminescence Test (TL)

The test is useful in finding out the temperature history of concrete exposed to temperature of 300°C–500°C. Like the other indirect tests, this test also requires very small amount of sample and is sensitive to the thermal exposure experienced by concrete rather than just maximum temperature attained. TL is the emission of visible light which occurs when certain minerals are heated. The intensity of emission of light versus temperature curve for a particular material depends upon its thermal and radiation history. Essentially TL is a result of presence of deflects in crystal in materials like quartz, feldspars, etc. When energy is supplied, the vibration of lattice causes escape of electron or ion trapped as defect in the crystal which become free to defuse and may be recombined with another ion resulting in radioactive light emission called thermo-luminescene. If the sample has already been subjected to heating up to temperature under consideration then this emission will be absent at that particular temperature during the test. This forms the basis of use of the test. Temperature versus TL curve of the fire affected samples may be compared with that of unaffected samples, and from the comparison, the duration of exposure to the given temperature and the temperature itself for the sample can be estimated. Using cored sample from the cross-section of the structural element, the profile of temperature thus can be obtained. The equipment required for this method is somewhat complex and drilled sample used may be washed in concentrated acid solution to remove mineral which can give spurious results. Usually it is TL of quartz up to 500°C. However, this test is sensitive to thermal exposure rather than maximum temperature and is sensitive to exposure of concrete to ultraviolet radiation.

9

Destructive Testing System

Damage in structure takes place due to natural or man-made causes. Even structures built in recent past have shown extensive damage and distress due to various causes, giving rise to problems both investigation and repair. Studies on durability of various construction materials and methods of repairing and renovating the damaged structures are increasingly attracting the attention of engineers and researchers. Research efforts have resulted in the development of investigation procedures and instruments and repair materials and techniques. Concrete is the ace card for the construction industry. Civil and structural engineers design concrete structures for an anticipated life of about 100 years. But due to various factors durability of concrete became a matter of serious concern and today repairs to concrete structures are given importance and different methods are being developed for repairs.

Before going into details of this method, it is necessary to discuss the probable causes of decay of concrete and its testing to evaluate the degree of deterioration before applying this method. Diagnosis of the damage is of greatest importance. Before remedies can be correctly prescribed the illness must be diagnosed and before an accurate diagnosis is possible the doctor has to have a thorough knowledge of the disease, its various symptoms and treatments. Similarly concrete repair expert needs similar knowledge in this field, so that the repairs prescribed for the distressed/ troubled structure are long lasting and successful. One must know the cause for the deterioration and distress of the existing structure rather than just the symptoms of each. A very methodical approach is required for this purpose. Data collection and analysis at each of the following stages will lead to methods and stages of repairs.

9.1.1 Purpose of Assessment

During a disaster an engineer may be tasked with a variety of functions, some of which may not be within the purview of his/her expertise. One of these could be the assessment and monitoring of the structural stability of a reinforced concrete structure. It is the promise of this instruction that this can be performed in a cursory fashion by a nonstructural engineer (in the absence of a competent structural engineer), within a relatively static environment. Therefore, the purpose of this instruction is to provide a fundamental understanding that will allow a nonstructural engineer also to perform a cursory safety assessment/monitoring of a reinforced concrete structure.

Reinforced Concrete Structures

Reinforced concrete comprises two basic materials, steel and concrete. The two materials work in a synergistic fashion when constructed properly to provide composite components which have very strong structural characteristics. Reinforced concrete is commonly used in structures designed for heavy use and long life, such as governmental and institutional buildings and public works structures. Concrete has a great capacity to support compressive loads. However, concrete has only a limited capacity to support tensile loads. Steel has a great capacity to carry both compressive and tensile loads. However, it is expensive to use as a single structural element and is prone to degradation (e.g. rusting) in certain environments. Therefore, in order to construct structural elements which are long lasting, economical, and have both compressive and tensile capacity, steel is combined with concrete to harness the compressive strength of concrete and the tensile strength of steel. Beams are horizontal structural components that support floors, ceilings, roofs, or decks (i.e. bridge and parking decks). The loads carried by a beam are primarily perpendicular to the longitudinal axis of the beam. As load is applied to a beam, it generally tends to cause bending of the beam, with the center of the unsupported section being forced away from the applied load.

This bending creates compressive forces (compression) in the upper depth of the beam and tensile forces (tension) in the lower depth of the beam. A typical reinforced concrete beam is designed to allow the compressive forces to be carried by the concrete material and the tensile forces to be carried by the steel. Columns are vertical structural components that support beams and other structural elements. The compressive loads carried by columns are primarily parallel to the vertical or longitudinal axis of column. As a uniform concentric load is applied to a column, compressive loads occur within the column's cross-section, and are distributed between the concrete and longitudinal steel reinforcement. When an eccentric load is applied to a column bending will occur in the column, with outer sections of the column being in tension and inner sections being in compression. The tension and compression loads within the column are carried by vertical reinforcement and concrete. Columns are usually designed with a ring of structural steel around the outer perimeter to provide confinement of the concrete in case of column failure. Concrete structures are built using cast-in-place, precast, or a combination of either procedure. Structures built prior to 1975 were built using nonductile concrete methods and hence have more structural damage due to unforeseen events. Cast-in-place buildings are erected using temporary forms that concrete is placed into and then forms removed. Precast buildings elements are generally formed off-site and then shipped to be assembled on site using some type of anchoring device. Cast-in-place structural failures will occur throughout the concrete member and /or its connections, while failures in a precast building member will typically occur at the joints and/or connections.

9.1.2 Rapid Structural Safety Assessment

The objective for the rapid structural safety assessment is to quickly inspect and evaluate the concrete structure and determine if the damaged structure is unsafe for personnel within the building and rescue personnel accessing the building. Two primary concerns need to be considered when performing this assessment of the structure that has sustained structural damage. This includes a quick evaluation of the building structural components (e.g. beams, columns, decking, etc.) and of the building nonstructural components (e.g. structural debris, partitions, ceilings, glass, pipe anchoring, electrical/mechanical equipment anchoring, etc.). If there are any visual signs of structural and/or nonstructural damage, then the specific buildings need to be isolated, secured, and marked as unsafe. The on-scene commander should be informed and the area remains in this unsafe condition, until a structural engineer proves otherwise. The rapid structural damage assessment would note the major failures within the structure including the major structural elements of beams, columns, and roof and floor decks. Typical failures would be found at the connections of the major structural elements, or at elements that no longer have adequate vertical support (e.g. unsupported roof and floor decks that are now cantilever elements). Indications would include cracking, spalling (i.e. loss of concrete from an exterior surface), and/or complete loss of all or part of a structural element. The on-scene commander should be notified immediately of the risk, and the area secured and marked unsafe.

The rapid nonstructural damage assessment would note the major failures within the building structure envelope including such items as structural debris, partitions, ceilings, glass, piping, and electrical/mechanical equipment. Concrete is a brittle material and, therefore, has a tendency to fragment into small, dense, hard pieces with rough edges. Many of these fragments may be precariously lying near or hanging from the exposed building edges and some fragments may be barely attached to exposed reinforcement steel. Settlement or shifting of the damaged structure may cause fragments to fall resulting in serious harm to personnel and/or additional damage to the remaining structure. The issue of potential harm to personnel from concrete fragments and other building materials (such as glass, and other items that may have been in and around the structure) is exacerbated by activities that may be occurring in and around the remaining structure by rescue personnel operating throughout the building, along with

rescue and/or press rotary-wing aircraft operating around the building. These activities may tend to cause accidental (e.g. physical interaction with something causing debris to fall) or inadvertent (e.g. vibration or rotor wash from rotary-wing aircraft causing debris to fall) mishaps. Typical failures in other nonstructural elements would be found where they originally were attached and/or secured. Failures would include anchoring tensile or shear failure, nonstructural element damage, and potential of future damage due to gravity loads and/or inadequate support bracing. The on site commander should be notified immediately of the risk, and the area secured and marked unsafe.

Depending on the structural and non-structural elements within the building matrix and their relation to rescue/recovery activities, there may be varying importance on the ability of these elements to provide adequate support. For example, failure of a column at ground level may cause failure of the remaining structure above those columns, or failure of a beam above an area where rescue/recovery workers are operating may at least cause collapse of a floor (and its contents) into the rescue/recovery area. Therefore, the rapid structural safety assessment should focus on those remaining structural elements of greatest importance to the remaining structure. In addition, a notebook, pen or pencil, spray paint (for making annotations on or around the compromised structure) and a camera (if available) will be needed for the assessment. The locations of any structural and non-structural elements should be documented via a written description and an associated sketch or photograph (digital) for use in more detailed assessment/analysis following the rapid structural safety assessment. Personnel performing the rapid structural safety assessment in and around the remaining structure should wear appropriate personal protective equipment (PPE). At a minimum PPE should include a hard hat, steel-toed boots (with steel shanks), gloves, and respiratory protection from dust. The specific site conditions may dictate further PPE. The rapid structural assessment should be performed in the following order:

- Review the entire outside of the structure.
- Enter the building only if necessary to determine extent of damage.
- Determine the degree of damage found in the structural and nonstructural elements.
- Secure all areas that need to be isolated and post UNSAFE signage.

9.1.3 Monitoring

A monitoring program can be established during the rapid assessment effort. Monitoring of potentially dangerous debris, unsecured equipment, structural cracks and failures until any mitigation/removal effort is complete may be simple periodic surveillance of the affected areas. The primary mode of failure for nonstructural elements within a compromised structure is due to the potential future movement of these elements. All unsecured elements need to be documented and recorded so that any future movement of these elements can be clearly established. Two primary modes of failure for structural elements within a compromised structure are shear and buckling. Shear failure implies failure generally along a plane (e.g. the failure of a deck slab that is no longer properly supported, could fail in a clean shear break). Buckling failure is a compressive failure that results in the element collapsing (e.g. the lateral displacement of a column).

Failure events may occur suddenly (e.g. the clean break of a cantilevered floor or roof section with no structural steel in the upper depth) or could be subtle as remaining structural elements adjust to the loading from the remaining structure. This adjustment may be radically different from the original design loads of the structural elements, as well as their potentially different structural composition. For example, they may have cracking and/or spalling affecting the carrying capacity of the element and/or the characteristics of the materials comprising the remaining structural elements may have changed. Therefore, monitoring is critical to expose subtle changes that may have the potential to result in structural failure. Monitoring individual structural elements for possible shear failure can be performed simply by marking a straight line perpendicularly across cracks of concern using a pencil or fine-tip permanent marker. Any slippage along the crack (possible shear failure) would then be indicated by a disjunction of the straight line. The length of cracks also should be monitored by drawing an arrow on the affected structural element with the tip of the arrowhead placed at the clearest discernible termination point of the crack. The date and time that the arrow was placed also should be clearly noted on the structural element such that its association with the arrow is unambiguous. This will document any extensions of cracks which may indicate possible weakening of the affected structural element with the potential end result being buckling failure. This data could then be utilized to make decisions on necessary shoring or bracing efforts by engineers with structural experience.

Monitoring of the overall structure can be accomplished by establishing a matrix of witness marks or survey marks on the exterior of the remaining structure. A line of markings can be made vertically through a line of columns and/or horizontally through a line of beams. This can be accomplished with the use of a transit, and bright-colored spray paint. It is desirable to have clear visibility of all markings from a common transit station that can be permanently established, without having to be periodically broken down during the monitoring effort. The intent of the placement of the witness marks is to allow monitoring of the various sections of the remaining structure relative to the others. This allows a relatively easy, fast, and safe means of determining if a given section is moving relative to the rest of the remaining structure and may illuminate significant and possibly imminent structural issues. The monitoring effort should be constant with data logging at regular time intervals. The interval will be dependent upon available resources, the stability of the remaining structure, and the dynamics of the environment and rescue/recovery efforts within and around the remaining structure.

9.2 INVESTIGATION OF DAMAGE

The investigation of reasons for damage to structures is largely a matter of gathering information by observation, studying records and asking questions, supplemented if necessary by a certain amount of testing, and then interpreting the information thus obtained. A systematic investigation of concrete structures is highly important and essential in order to:

- Find out the cause of damage
- Assess the condition of the structure in its damaged state
- Formulate recommendations for repair and restoration.

The general procedure for an investigation can be briefly summarized as given below:

- Documentation of damage
- Visual observation
- Measurements on geometrical parameters such as deflections and deformations in vertically and horizontally
- Experiments for evaluating material properties and behavior, these includes type of testing
- Interpretation and analysis of test results
- Analysis of the building in damaged state taking into account the test results obtained during investigation
- Formulation of repair measures
- Post-repair evaluation.

9.2.1 Observation

The texture of a concrete surface may suggest the possibility of chemical attack by a general softening, leaching of the matrix or in the case of sulfate attack, and whitening of the concrete. Rust stains often indicate corrosion of reinforcement but they may be caused by contamination of the aggregates with iron pyrites. If cracked concrete is broken out, the state of the crack surface gives useful information. General flaking of an exposed concrete surface suggests frost damage. In fire damaged structures, the colour of the concrete gives an indication of the maximum temperature reached. When concrete is cracked the crack pattern can be informative. A mesh pattern suggests drying shrinkage, surface crazing, and frost attack or in rare cases, alkali-aggregate reaction. Exudations from cracks may be the result of water passing through the concrete and washing out calcium salts, or from alkali-aggregate reaction. Relatively straight cracks usually indicate excessive but fairly uniform tensile strain, and they should be studied in relation to the probable stress pattern in the member concerned.

Cracks caused by unidirectional bending will be widest in the zone of maximum tensile stress and will taper along their length, while cracks caused by direct tension will be of roughly uniform width. Continuing movement at a crack often produces crumbling at the edges. Pop outs in concrete are usually associated with particles of coarse aggregate just below the surface. Highly absorptive particles may expand if serve frost occurs while they are saturated. Occasionally, coarse aggregates are contaminated with particles of lime or clay that expand with weathering.

9.2.2 Questioning

Records of mix proportions, sources of materials, cube test results weather conditions, etc. may be available, particularly for recent built structures but, even with these, reliable information may be difficult to obtain. It is always useful to ask questions of as many as possible of the people who were concerned with the design or construction. Their recollections may not be completely accurate and their accounts of events may conflict but, by questioning, one can often find out what actually happened as distinct from what should have happened. This applies not only to work on site during construction but also to subsequent use or misuse of the structure.

9.2.3 Assessment Procedure

The first key step of investigation is to make a diagnosis of the defect or the failure. It is basically an inductive procedure and it requires abundant caution. Figure 9.1 shows

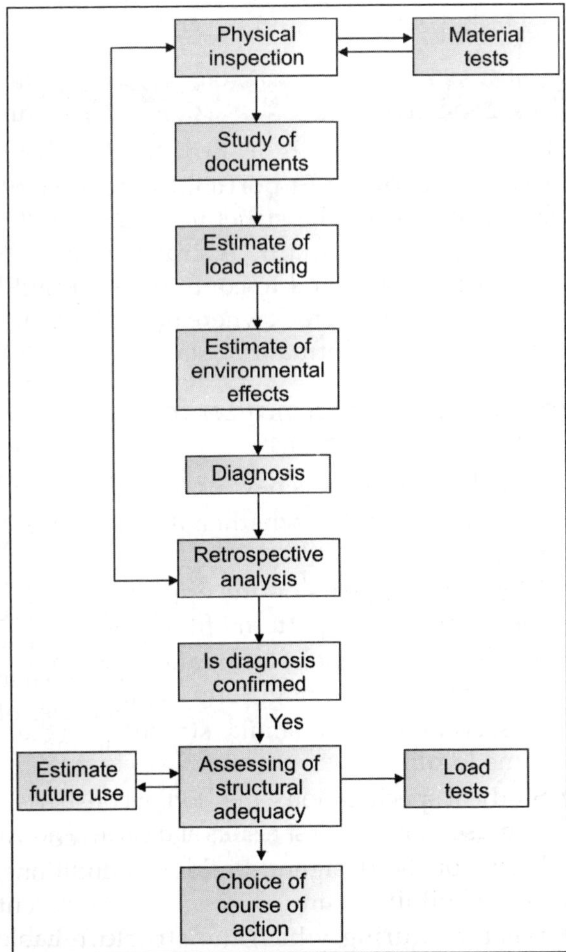

Fig. 9.1: Assessment procedures for damage

the flow chart of damage assessment procedure. Some important steps are:

- Visual inspection
- Study of available documentation
- Estimation of actual loads and environmental effects
- Diagnosis.

Visual inspection: Any damaged structure, as a first step, requires an extensive visual inspection followed by documentation of the details. A perusal of the documentation, supported by photographs will reveal all possible evidence of structural defects and damage.

Study of available documentation: Study of available documentation will give some idea on the history of construction, original quality, analysis and design methods with assumption made, and the type of materials used. It is unfortunate that the relevant details are not readily available in many cases. A comparison of adjacent buildings also helps in a proper diagnosis.

Estimation of actual loads and environmental effects: It is generally found that there is, in majority of the cases, the loads acting on a structure will be much different from the loads assumed in design calculations. Therefore, cracking or any other damage may, sometimes, be attributed to the fact that these loads or a certain load combinations were not considered in the analysis and design. Environment effects are likely to be different from those assumed or not considered at the design stage. Effects of temperature changes, or a hostile atmosphere would impose serviceability or durability problems. Environmental changes will result in undesirable effects in foundations.

Diagnosis: In any investigation, diagnosis of the cause or causes of damage is of prime importance and is difficult too. A proper and reliable diagnosis can be made only by the conducting a systematic investigation using proven test methods and experienced personnel.

9.3 TESTING SYSTEM OF HARDENED CONCRETE

Concrete as material has high adaptability to satisfy many aspects in structure, such as functional, economy, maintenance, aesthetic acceptability, protection against corrosive environments, protection against fire, resistance to cyclic loading, explosion resistance and control over deflection, etc. The concrete structure that are to be constructed as per specifications, during its execution need testing of constituent materials of concrete, tests on mixed concrete when it is fresh, and finally when it has hardened. These tests on concrete are very important as their properties give durability. The structures which have already been built do exhibit distress after lapse of some years during the service life. There are number of testing system available and they could be used on structures to assess the quality of concrete and steel. Testing system of hardened concrete can be divided into three categories:

- Nondestructive testing system (NDTS)
- Partially destructive testing system (PDTS)
- Destructive testing system (DTS)

Nondestructive testing systems: To assess the quality of concrete in its damaged state without any disturbance of surrounding concrete.

Semi/partially destructive testing systems: To assess the quality of concrete in its damaged state with partial disturbance of surrounding concrete. These two categories are already covered in earlier chapters.

Destructive testing systems: To assess the quality of concrete with complete disturbance of concrete (loaded up to failure). The range of instruments is typically considered to be two parts. The first are nondestructive field tests of compressive strength. The second are tensile field tester systems to either determine the tensile strength of an overlay or bond material, or tensile strength of anchors embedded in the concrete. The first group is pure nondestructive testing where the strength of the material is determined by correlation to another parameter more easily available and readily apparent. This is typically the hardness of the concrete or the resistance to penetration by either a pin or probe. The windsor probe, windsor pin and rebound hammers all fall within this category. These are widely used standard tests and as such have seen use throughout the world.

The second set of instruments is our concrete tensile testers. These have been optimized to both test the strength of the anchors and repair overlay material. They can be used to test until failure or to simply verify that the material will not be affected by a specific amount of force. A number of considerations were taken into account when designing this line of products; include viscous damping of the resultant failure backlash, portability, and ruggedness.

9.4 EVALUATION OF CRACKS

The cracks in a structure are broadly classified in two categories: Superficial cracks and structural cracks. The structural cracks may be active and dormant. A crack where a movement is observed to continue is termed active, whereas the crack where no movement occurs is termed dormant or static.

The following information may help in diagnosing the cracks:

- Whether the crack is new or old
- Type of crack, i.e. whether it is active or dormant
- Whether it appears on the opposite face of the member also, pattern of the cracks
- Soil condition, type of foundation used, and sign of movement of ground, if any
- Observations on the similar structures in the same locality
- Study of specifications, method of construction used and the test results at the site, if any
- Views of the designer, builder, occupants of the building if any
- Weather during which the structure has been constructed.

From the above discussion, it is evident that cracking is a complex phenomenon. As in the case of a medical practitioner prescribing medicine without thoroughly examining the patient, it is difficult for a repair engineer to advocate any repair technology without making a thorough investigation. Before proceeding with repair, the investigations should be made to determine the location and extent of cracking, the causes of damage, and the objectives of repair. Calculation can be made to determine stresses due to applied loads. For detailed information, the history of the structure, structural drawings and specifications, and construction and maintenance records should be reviewed. The objectives of repair include restoration and enhancement of durability, structural strength, functional requirements and aesthetics.

The evaluation of cracks is necessary for the following purposes:

- To identify the cause of cracking
- To assess the structure for its safety and serviceability
- To establish the extent of the cracking
- To establish the likely extent of further deterioration.
- To study the suitability of various remedial measures
- To make a final assessment for serviceability after repairs.

Apart from visual inspection, tapping the surface and listening to the sound for hollow areas may be one of the simplest methods of identifying the weak spots. The suspected areas are then opened up by chipping the weak concrete for further assessment. The comparative strength of concrete in the structure may be assessed to a reasonable accuracy by nondestructive testing and by the tests on the cores extracted from the concrete. The commonly used nondestructive tests are the rebound hammer test and ultrasonic pulse velocity test.

9.5 DESTRUCTIVE TESTING SYSTEMS

The most common destructive test is load test and is used to assess the strength of concrete structural elements. Load testing is sometimes used as an alternative method of assessing structural capacity.

Load tests are usually carried out for one of the following reasons:

- There are still doubts about the satisfactory performance of the structure under load after a survey and local testing.
- It is difficult or impossible to determine adequate information about the structure and its materials.
- Verification of structural analysis in cases where the complexity of the structural form does not lend itself to rigorous analysis.

Deficiencies in detail, material or construction are suspected and such deficiencies would mean that the normal procedures or assumptions on which structural analysis is based were not appropriate. After confirming that the reinforcement did not go into plastic, it is decided to conduct full scale load tests on the beams, so as to determine the load carrying capacity of the cracked beams and compare it with original design strength before any remedial measures are suggested. Accordingly load tests are conducted on most distressed beams.

The following aspects are set for the testing programme:

- To conduct the load test on beams up to 1.25 times the designed live load
- To monitor the deflections and recovery of the beams during incremental loading and unloading
- To compare the actual deflections with that of theoretical deflections.

9.5.1 Assessment of Existing Concrete Structures

It has been learnt that concrete structures require a closer inspection, not only immediately after construction but also periodically at a regular interval. The quality control measures during construction, generally, consist of workability tests on fresh concrete and cube compressive strength of concrete samples, after some specified days of curing. It is a well-known fact that the results of the above tests do not reflect the true quality of the concrete, existing in the concrete structure because the quality of a concrete structure depends on many factors such as method of mixing, transporting, placing, compacting and curing of concrete. While concrete members with certain amount of imperfections can satisfy the requirements relating to strength and serviceability, such concrete may not satisfy durability requirements. Assessment of quality of concrete is necessary to ensure that the quality of execution is satisfactory and to identify any deficiencies so that they can be rectified. This can be achieved only by conducting some *in situ* tests on the structures besides visual inspection. The *in situ* tests are nondestructive tests and partially destructive tests. These tests measure indirectly the strength of concrete except in the case of core test, where direct evidence on the condition of concrete and a measurement on compressive strength are possible. These testing methods are in use for evaluating existing concrete structures with regard to their strength apart from assessment and quality control of hardened concrete.

9.5.2 Direct Load Test

This is a method of assessing the strength of the *in situ* concrete member. In most cases this test is performed for the proof of structure capacity, not for suspect or critical location. The principal aim in this testing is to demonstrate satisfactory performance under an overload, above the design working value. This is usually judged by measurement of deflections under this load which may be sustained for a specific period. The selection of specific members or portion of a structure to be tested will depend upon the general features of convenience as well as the relative importance of strength and expected load effect at various locations. Selection of member may often be assisted by nondestructive methods coupled with visual inspection to locate the weakest zones or elements.

There are certain inherent problems in this test. This method is very expensive and time consuming. In *in situ* test whether the member under test is actually subjected to the assumed test load during fire is difficult to predict due to load sharing effect. All the direct methods of assessing residual strength of concrete provide regarding average potential cube compressive strength and are not indicative of strength profile across the section. Direct load testing again may not be possible where the loading may damage the structure permanently or where the structural system is too stiff to show any deflection.

9.5.3 Actual Load Test on Structural Element

A live load of 1 ton/sq m was taken for imposed load calculation. Floor finishing load and partition wall load was also considered. Thus, imposed load was taken as 1.25 times live load as per IS: 456-2000 code recommendation. The load was imposed on the beam as udl spreads over an area of 55.5 sq m. Thus, a total load of 84 tons was arrived at and imposed in 12 stages of 7 tons each at an interval of 15 minutes. The load of 7 tons was imposed with the help of 200 sand bags weighing 35 kg each as first step and increased in steps up to 84 tons. Dial gauges and electrical strain gauges were fixed at different locations. Separate platform was provided to read and record the deflections from the dial gauges, strain from strain indicator during loading and unloading.

Deflections for the above mentioned loading steps were recorded. Total time for reaching maximum load was 26 hr and the maximum load was kept for more than 43 hr during which creep effect was monitored and the time taken for unloading was 32 hr during which deflection recovery was monitored. The recorded deflections and theoretical deflections were compared.

SHORT ANSWER QUESTIONS

1. Define a systematic investigation.
2. Name the general procedure for investigation.
3. What are the satisfactory aspects of concrete structures?
4. What are the testing systems of hardened concrete?
5. Name some nondestructive testing methods.
6. Name some partially destructive testing methods.
7. Differentiate nondestructive, semi-destructive and destructive testing methods.
8. What is the principle of rebound hammer test?
9. What are the factors that affect the readings of rebound numbers?
10. What are the applications of rebound hammer test?
11. How will you assess the quality of concrete from rebound numbers?
12. What is the object of ultrasonic pulse velocity test?
13. What are the types of waves generated in UPV instrument?
14. Relate compression wave velocity and dynamic modulus of elasticity in elastic medium.
15. How do you evaluate the strength of concrete by UPV test?
16. What are the factors that influence the calibration charts?
17. What is the use of pulse velocity measurement?
18. How do you measure the pulse velocity?
19. What are the parts of pulse velocity measuring equipment?
20. What are the factors that influence the pulse velocity measurements?
21. Write down the criteria adopted for concrete certification by UPV.
22. What is the principle of pulse attenuation method?
23. Which is dominant property in the pulse attenuation method?
24. Write down the structural application of the radar system by pulse attenuation method.
25. What is the need for pulse echo technique?
26. What is the use of echo system?
27. What is the use of pachometer?
28. What are the determinations by electrical methods?
29. How do you identify the delaminated areas under acoustic emission technique?
30. What is the use of thermography test?
31. What is the principle of thermography test?
32. How will you measure the hardness by probing technique?
33. Define windsor probe technique.
34. What is meant by pull out test?
35. What are the situations when core test is performed?
36. What is the location of core taken?
37. What is the specified size of core?
38. What is the requirement for number of cores?
39. Write about the core drilling equipment.
40. How is the core prepared for testing?
41. How will you obtain trimming and capping?
42. What are the identifications by visual examination of core test?
43. Write about the density determination from core sample?
44. How will you assess the voidage of concrete from core sample?
45. How do you conduct compression testing on core samples?
46. What are the factors that influence the core compressive strength?
47. How does the reinforcement modify the strength of concrete in core test?

48. How will you estimate the cube strength from core strength?
49. What is the codal representation of core test?
50. What are the disadvantages of core method?
51. What is meant by acceleration curing test?
52. What are the methods available for accelerated using test?
53. What is the determination of bond strength?
54. What are the uses of resonant frequency test?
55. Write about dynamic or vibration methods.
56. Derive the mathematical relation between velocity and modulus of elasticity.
57. What are the properties evaluated from cores?
58. How do you assess the quality of concrete from the core test?
59. What are the tests available to find concrete corrosion?
60. Explain the phenolphthalein test.
61. How can we determine pH value in concrete?
62. How will you conduct the chloride test?
63. Give the qualitative guidelines for identification of corrosion prone based on chloride content and pH value.
64. What are the standard electrodes used in half-cell potential survey?
65. What is measurement by half-cell potential survey?
66. Give the guidelines for identifying corrosion based on half-cell survey.
67. Define resistivity.
68. What are the factors that govern the resistivity value?
69. Give the guidelines of corrosion risk based on resistivity values.
70. Define load tests.
71. Define TNS tester.
72. Explain in short the internal fracture test.
73. How will you conduct epoxy grouted bolt test?
74. Explain in short the determination of cement content in concrete.
75. Discuss in short the determination of water content in concrete.
76. How will you find out the sulfate content in the concrete?
77. Define abrasion resistance test in concrete.
78. Explain the rapid chloride test.

THEORY QUESTIONS

1. Explain assessment procedure for evaluating damages in structure.
2. Discuss about the evaluation of different types of crack.
3. Name the various testing system used in the assessment of distressed concrete structures.
4. Differentiate nondestructive testing, partially destructive testing and destructive testing methods.
5. Explain clearly the rebound hammer test and its limitations.
6. Explain clearly the testing procedure of ultrasonic pulse velocity testing of concrete.
7. How are the electrical, magnetic and nuclear menaces used to assess the property of concrete?
8. Explain the radar system used to assess the damages in concrete structures.
9. Explain the pulse echo system used to assess the quality of concrete.
10. How will you assess the fire damaged concrete structures by NDT methods?
11. Discuss about the penetration techniques in concrete structures.
12. How do you conduct pull out and pull off test?
13. How do you conduct core sampling and testing?

14. Explain *in situ* permeability test on concrete.

15. Explain the testing methods to assess the corrosion damages in concrete structures.

16. How do you conduct chloride penetration test and pH value test?

17. Explain the determination of cement content, water content and water–cement ratio in concrete structures.

18. Explain the use of half-cell potential survey in the damage assessment of concrete structures.

19. Explain about the destructive testing methods.

20. Discuss briefly the load test on existing concrete member.

Part 3
Repair Materials

10. Materials for Repair
11. Function of Repair Materials
12. Special Types of Repair Materials
13. Selection and Evaluation of Materials

10

Materials for Repair

Essential parameters of repair materials for concrete are:

- Low shrinkage properties
- Requisite setting/hardening properties
- Workability
- Good bond strength with existing substrate
- Compatible coefficient of thermal expansion
- Compatible mechanical properties and strength that of the substrate
- Should allow relative movement
- Minimal or no curing requirement
- Alkaline character
- Low air and water permeability
- Aesthetics to match with surrounding
- Cost effective; durability; and non-hazardous/nonpolluting.

Wide range of materials for repair of concrete is available differing in cost and their performance. The application range covers the following:

- Materials for surface preparation
- Chemical rust removers for corroded reinforcement
- Passivators for reinforcement protection
- Bonding agents
- Structural repair materials
- Nonstructural repair materials
- Injection grouts
- Joint sealants
- Surface coatings for protection of RCC.

Though the above materials are being marketed under their brand names, yet these could be classified in the following categories:

- Premixed cement concrete/mortars
- Polymers/latex modified cement additives for mortars/concrete/cement slurry
- Epoxy resins
- Chemical for corrosion inhibitor, removal of rust.

- Retarding admixtures
- Water reducing admixtures
- Air entraining admixtures
- Superplasticizing admixtures

Polymer modified mortars and concrete (PMM/PMC): Materials used in polymer modified systems are the same as those employed in normal mortar and concreting operations but for latex/polymers, which is used as modifier.

The physical and mechanical properties of polymer modified mortars/concretes are:

- Better workability
- Water retention property is more
- Better resistance to bleeding and segregation
- Increased resistance to crack propagation
- Noticeable increase in tensile and flexural strength
- Chemical resistance more
- Temperature effect: Strength depends on temperature
- Dry shrinkage is less
- Water proofing quality or permeability increases
- Improved adhesion or bond strength
- Better abrasion resistance
- Durability and nondegradability: Generally these materials are nondegradable after total polymerization takes place.

Fields of Application

- Structural repairs to RCC
- Ultrarapid hardening polymer modified shotcrete
- Polymer ferrocements
- Antiwashout under water concrete
- Protective anticorrosive and water proofing coatings
- Bond coats.

Precautions for Use of Epoxy

Epoxies are generally toxic in nature and these require a lot of care in their handling. The special care to be taken are:

- Should not come in contact with the skin, worker to be provided with rubber gloves.
- Utensils/equipment used for mixing resin and hardener should be cleaned immediately after their use.
- Pot life of mixed epoxy is very limited, half to two hours. Should be applied within pot life period.

- The epoxies generally used as bond coat between old concrete and repaired concrete, should not be used as exposed environment.
- Epoxies have much higher bond strength than other polymers, but they are much costlier than others.

10.3 ESSENTIAL PARAMETERS OF COATING

- Should possess excellent bond to substrate
- Be durable with a long useful life
- Little or no colour change with time
- Should have maximum permeability to allow water vapour escape from concrete substrate
- Should have sufficient impermeability against passage of oxygen and carbon dioxide from air to concrete
- Should be available in a reasonable range of attractive colours.

10.4 REPAIR MATERIALS (SPECIAL)

Shotcrete: This is a method of applying a combination of sand and cement, which mixed pneumatically and conveyed in dry state to the nozzle of a pressure gun, where water is mixed and hydration takes place just prior to expulsion. The material bonds perfectly to properly prepared surface of masonry and steel. In versatility of application to curved or irregular surfaces, its high strength after application and good physical characteristics, make for an ideal means to achieve added structural capability in walls and other elements.

Epoxy resins: These are excellent binding agents with high tensile strength. There are chemical preparations, the compositions of which can be changed as per requirements. The epoxy components are mixed just prior to application. The product is of low viscosity and can be injected in small cracks too. The higher viscosity epoxy resin can be used for surface coating or filling larger cracks or holes. The epoxy mixture strength is dependent upon the temperature of curing (lower strength for higher temperature) and method of application.

Epoxy mortar: For larger void spaces, it is possible to combine epoxy resins of either low viscosity or higher viscosity, with sand aggregate to form epoxy mortar. Epoxy mortar mixture has higher compressive strength, higher tensile strength and lower modulus of elasticity than cement concrete. Thus, the mortar is not a stiff material for replacing reinforced concrete. Epoxy is a combustible material. Therefore, it is not used alone. The sand aggregate mixed to form epoxy mortar provides a heat sink for heat generated and it provides increased modulus of elasticity too.

11

Function of Repair Materials

Patching materials currently available appear to have been formulated to meet the demands of a repair schedule and conditions. Minimum downtime has dictated the development of fast setting, high strength-developing materials; poor accessibility has led to the development of high flow, self-leveling materials. While some materials have unique characteristics which offer significant advantages in particular situations, others, when considered against more conventional materials, appear to offer borderline advantages for their higher costs. The success of any patch will depend largely on overcoming the tendency of the concrete/mortar or polymer patching material to shrink after placement, and on securing a bond to the substrate concrete. A variety of methods have been used to overcome shrinkage and to promote a better bond. The list of requirements for a patching material is given as follows:

- It should be as durable as the surrounding material.
- It should require a minimum of site preparation.
- It must be tolerant of a wide range of temperature and moisture conditions.
- It must be chemically compatible with the substrate.
- It should have a similar colour and surface texture to the surrounding material.

11.1.1 Cementitious Materials

Portland cement concrete and mortar offer a number of advantages as patching materials, like thermal movement similar to the existing concrete, similarity in an appearance, cost, ready availability, and familiarity. Concrete is most often used for complete replacement of sections and deep cavities extending beyond the reinforcing bars, while mortar can be used for small cavities. Special proprietary mixtures are available that are based on Portland cement and contain hard mineral metallic aggregate to increase the resistance of floors to heavy traffic.

Most of these products use high early strength cement and contain an expansive agent that causes the concrete and mortar to expand either in the plastic stage or after it has hardened. The expansion produced in the plastic state establishes intimate contact with the substrate before it hardens, thus completely filling the space and promoting a good bond. The expansion produced in the mortar or concrete is usually obtained by the use of an expansive agent, such as aluminium powder, coke powder, anhydrous calcium sulfoaluminate, or calcium oxide.

Some of the disadvantages of these materials are that high shrinkage and a tendency to crack may result, if great care is not taken to keep the water content low and provide proper curing. Also, in cold weather, the rate of strength is too slow. High early strengthen cement with a mixture of super plasticizing and nonchloride accelerating admixtures can have a more rapid setting time and rate of strength gain. Commonly used admixtures for this purpose include naphthalene or melamine formaldehyde sulfonates and nitrates, calcium formate, sodium sulfate or carbonate.

11.1.2 Polymer Concrete and Mortar

There are two broad classifications of polymer concrete and mortar. Polymer concrete are materials containing latex emulsions used with hydrating Portland cement, and the second are systems without Portland cement binder. The first class of materials includes latex modified mortars and epoxy emulsions. The second group includes epoxy, polyester, mortars, and concrete. Polymer concrete comprises a blend of coarse and fine aggregates held together in a polymerized matrix. Some of the materials are discussed below.

Latex Modified Mortars

Latex modified mortars include mixtures of Portland cement, sand and latex admixtures, such as styrene, butadiene, polyvinyl acetate, acrylics, and epoxy emulsions. The latex is usually used at a level less than 20% by weight of cement in the mixture. Addition of the latex improves the permeability, bond, tensile, and flexural strengths. Such mortars can be placed in sections ranging from 12 to 50 mm thick in both horizontal and vertical applications, such as column and wall repair. The chief advantage of the system is its good workability and ease of application when compared to other similar systems. When the patched area is shallow and extends across a crack or joint that is subject

to movement, a flexibilized epoxy should be used to accommodate the anticipated movement. However, if a patch is more than 9 mm deep, the patching material should not bridge the crack.

Epoxy Patching Compounds

Epoxy patching materials are usually epoxy mortars. They are generally prepackaged systems containing resin binders consisting of two components that are added to a selected blend of aggregates, which may be contained in the prepackaged system or purchased separately. A typical mortar mix consists of 1 part of mixed epoxy resin and hardener to 3 parts concrete sand. Due to their strong adhesive character, these materials can be placed in very thin layers, which will resist heavy impact loading. Some of the commonly used epoxy compounds for this purpose include black coal tar epoxies, the cream-colored petroleum based epoxies and a variety of polysulfide epoxies which range in colour from clear amber to an opaque grey.

Epoxy patching materials present a major advantage over cementitious materials because of their rapid cure and faster rate of strength development, high resistance to aggressive chemicals, high bond strength, good resistance and long-term durability. Normally, traffic can be resumed over a patched area within 24 hours and with a more rapid curing system, the down time can be reduced to 4 to 5 hours. Due to the high cost of epoxy materials, it is too costly to place an epoxy mortar layer more than 9 mm thick. If extensive resurfacing and/or large patches are required, a thin layer of epoxy or latex cement slurry can be used to bond new regular Portland cement mortar to the existing concrete substrate.

Polyester Resins

Patching compounds based on unsaturated polyester resins have not been used as widely as epoxies. They generally harden by a reaction between the base resin and small amounts of catalyst. Accurate control of proportions and mixing are therefore more difficult than with the two component epoxies. Typical polyesters include unsaturated polyesters, such as methyl methacrylate. Polyester patches cure faster than epoxy materials and are less sensitive to lower temperatures. Vinyl ester patching compositions combine resiliency, impact resistance, and excellent chemical resistance. Typical application areas are floors, trenches and pickling and plating tanks.

Acrylic Concrete and Mortar

The most widely used patching materials of this type are acrylic monomers. Acrylic concrete and mortar contains neither water nor Portland cement. Instead, the aggregates are held together by an acrylic polymer. Two types of monomers are usually used: Methyl methacrylate (MMA), which has been used for the last more than 20 years, or high molecular weight methacrylate (HMWM), a relatively new material. All components of the product can be mixed and placed, or the monomer may be poured over a patch area filled with pre-placed aggregate. One commercial, acrylic concrete system consists of two components, viz. liquid acrylic and a package of premixed fine aggregate, promoters, initiators, and pigments.

Polymer concretes are used for patching sections that are generally larger, or more extensive than those suitable for repair with epoxy mortars. A range of properties, such as rapid strength gain, low permeability, and good abrasion resistance is achievable by custom design of the mix. The unique features of rapid strength gain and high ultimate strength in conjunction with its excellent durability characteristics make this material suitable for a wide array of applications. Bridge slabs, parking garage decks, industrial warehouse floors, and tanks are potential areas where this repair material could be used.

11.1.3 Quick Setting Compounds

These compounds are used as repair materials in areas which are heavily trafficked to be achieved with the least disruption to traffic, or, in the case of industrial floors, to production.

Also, faster setting and curing materials are required during cooler periods, or in areas which are subjected to continued cold temperatures, such as food freezer floors. A number of unique and specialized patching materials have been developed for such special repair applications.

High Alumina Cement Containing Compounds

High alumina cement patching compounds with accelerated set time and strength gain can be obtained by mixing this cement with the appropriate proportion of Portland cement. High alumina cements are three to four times as expensive as other cement, but it produces a faster rate of strength gain and is more resistant to sulfate attack.

Magnesium Phosphates

These are fast-setting, rapid strength developing materials. One of the most widely used commercial products is Set-45. In the two-component package, dry magnesia is mixed with the liquid phosphate in small quantities and work very rapidly. It usually produces a high strength, low permeability patch with a good bond to many surfaces. However, water will affect hardening; even small amounts produce severe strength reduction. Some evidences indicate that permeability may be high enough to jeopardize freeze-thaw durability. It is probable that poor mixing, resulting from an increase in the viscosity of the phosphate binder under low temperature conditions, produces a relatively more porous material than under normal temperature.

Magnesium phosphate products cost about the same as epoxy resin mortar. Due to its ability to cure and harden at temperatures at or below freezing, it appears to be best suited for repair of silver spall and pop outs or for emergency work. Typical applications include repairs to bridge decks, commercial freezer rooms, airport runways, industrial floors, highway payments, truck docks, ramps, and stairways. Other features which make it a suitable patching material are the similarly of its coefficient of thermal expansion and linear shrinkage value to that of Portland cement concrete. The use of these materials greatly reduces the duration of interruption of the use of the structure under repair.

Molten Sulfur

Molten sulfur has been used either on its own or in combination with tar-based additives as a quick hardening patch for concrete. It has also been used to impregnate deteriorated concrete and as a binder for concrete.

Calcium Sulfate Based Materials

Many of the proprietary, rapid-setting patching products are calcium sulfate based. Most contain Portland cement in varying amounts, some also contain chlorides. They gain strength very rapidly and can be used at any temperature above freezing. Many have not proved to be reliable when exposed to moisture and freezing weather. These materials can also promote sulfate attack on the surrounding concrete.

11.1.4 Bituminous Materials

Bituminous materials have the lowest cost of all patching materials, are readily available, and are easy to apply. However, when they are of poor quality, and carelessly applied, there is some evidence that they hold water. Poorly graded aggregate and unconsolidated patches are particularly prone to water pickup, which accelerate deterioration of surrounding concrete. Bituminous patches probably find their best use for repairing payment prior to the application of an overlay. Hot-mixed, densely-graded asphalt concrete is widely used for patching. Asphalt emulsions, both anionic and cationic, have also been used with some success for patching. Other bituminous materials used for patching include cold-mixed cutback asphalts, tars and rock asphalt.

11.1.5 Grouts

A grout is any flowable, plastic, or packable material that can be used to fill the space between the underside of a machine or column and the foundation on which the unit is to rest, and then harden there to support the unit. In addition to this principle role, they also find wide use in the repair of deteriorated structures. Typical applications include void filling, and as a mortar for filling large cracks. The most important requirement of a grout is that it completely fills the space in which it is placed. Therefore, it must have negligible shrinkage and remain stable in place, without cracking, delaminating or crumbling. There are a variety of grouts in use today which can be classified as follows:

- Nominal cement/sand mixtures
- Gas-forming grouts
- Metallic aggregate grouts
- Inorganic sulfoaluminate cement or expansive additive-based grouts
- Fiber-reinforced grouts
- Organic resin grouts.

Nominal Cement/Sand Mixtures

Ordinary cement/sand grouts are the lowest in material cost. However, they require high water and cement content to achieve peaceable consistencies that shrinkage and cracking promotes on hardening. Dry packing with a stiff grout mix overcomes some of the problems. The use of cement/sand grouts has declined sharply with the advent of specially formulated nonshrink grouts.

Gas-forming Grouts

These grouts contain expansion producing ingredients which react with the cement liquor to generate gas bubbles. The gas expands the grout to help compensate for any shrinkage that occurs in the plastic state before the grout has hardened. Vertical and lateral restraint is required to achieve specified strength and volume stability in the hardened stage. Aluminum powder and finely divided carbon are used to produce the gas producing reaction. One of the disadvantages of these products is that they are quite temperature sensitive.

Sulfoaluminate Grouts

Most of the expansion produced by sulfo-aluminate grouts occur after the grout has set. The expansion can be produced by shrinkage-compensating cement or an anhydrous sulfoaluminate expansive additive used with Portland cement. The expansion producing reaction occurs when the anhydrous calcium sulfoaluminate is converted to the hydrated ettringite. Sulfoaluminate cement-based materials are less prone to strength reduction due to increased w/c ratios than the Portland cement-based materials. The initially achieved expansion could well be lost, if there is no moist curing after the grout is placed.

Fiber-reinforced Grouts

These grouts may have polypropylene, steel, or glass fibers dispersed in either a Portland cement or a shrinkage compensating mortar. The fibers impart much improved impact resistance and flexural strengths. The stress required to produce cracking is significantly increased and the integrity of the grout after cracking is preserved, thus preventing a catastrophic type of failure.

Metallic Aggregate Grouts

These contain iron fillings to impart tough-ness, and an oxidation catalyst, such as ferric chloride, that produces expansion to compen-sate for settling and drying shrinkage. To develop their full strength and stability such grouts require rigid confinement. They are not recommended for areas that are exposed to wet and dry cycling and stray electrical cur-rents. Most of the cementitious grouts are available as one-component products that re-quire only the addition of water in the field. Storage life, under proper conditions of low humidity and temperature, is 12 to 18 months.

Polymer Grouts

Polymer grouts are generally used for equipment bases or concrete repair, especially in areas subjected to chemical exposure, freeze/thaw, impact stresses, or when a fast turn around time is important. Most grouts are supplied as three-component materials having:
- A liquid resin component
- A fluid-reactive curing agent component
- A dry filler component.

In some cases, two-component products are packed with the curing agent mixed with the aggregate. Some speciality system formulated for underwater applications are available. The composition of these products may consist of resin and hardener systems which are tolerant to moisture. They are usually pumped or poured. Since such materials are unique products, most products are sold as pre-packaged units which avoid the need for field proportioning and the consequent errors and problems that result from it. Although epoxy grouts generally have good chemical resistance to oils and chemicals, there are some applications where the chemical environment may be too severe for epoxy grout. Under such conditions, it is more advantageous to use a vinyl ester based grout. To prevent potential penetration of the corrosive agent to the concrete, a glass-reinforced monolithic surfacing is installed prior to the application of polyester, vinyl ester, and epoxy.

Because of the more costly installation procedure and the higher shrinkage, polyester, vinyl ester, and epoxy should only be used in situations where rigorous chemical exposure conditions dictate their use. Properties of these grouts include high compressive strength of 3 to 4 times that of standard concrete, and high tensile bond strengths. One of the most important performance factors which distinguish the service capabilities of organic polymer grouts from those of cementitious grouting materials is their excellent adhesion to variety of construction materials.

Acrylate and Urethane Grouts

These grouts are used in subgrade injection to repair sewer leaks, control groundwater movements, and stabilize soil. The chemical systems used for such purposes include acrylamides, polyacrylamides, urethanes, and acrylates. These grouts have very low viscosities and can be controlled to set almost instantly after predetermined elapsed times. Some, like urethanes, are even activated or catalyzed by water itself. The grouts are true solutions; their constituent materials dissolve completely or combine with water to provide a very thin mix. Since their particle size is less than one third the size of the void being filled, they penetrate completely and prevent fracture of the void being filled.

11.2 RESURFACING MATERIALS

Resurfacing materials are protected concrete structures under certain exposure conditions from spalls, cracks, and disintegrates. Resurfacing materials are referred to as coatings and toppings. Protective coatings can greatly reduce the effect of deteriorate conditions, and significantly improve the durability characteristics of the concrete. A variety of coatings are available, and some are tailored for greater chemical resistance, while others are formulated to resist wear and erosion. Unfilled and partly filled coatings less than 30 ml are described simply as coating, while filled system with thickness in excess of 30 ml, and ranging up to 10 mm are described as toppings.

11.2.1 Protective Coatings

When considering a coating for concrete, the specifier or buyer must take into account the properties of the concrete and location of the area to be coated in the structure. The properties of concrete which affect the successful application and performance of a coating are:

- Porosity
- Moisture content
- Presence of contaminants on the surface

Porosity will affect coverage rats, interface adhesion, and "pin holing" of the coating. A porous surface will absorb greater quantities of the resin from the surfacing material in comparison to a surface that is dense and of low absorptivity.

A coating of high viscosity may be absorbed to a smaller extent than those of low viscosity. It is a good practice to use priming or first coat which may be tailored to meet a dense or porous substrate. To ensure proper bonding of the coating to the concrete, it is important that the concrete have low moisture content and be surface dry. When concrete contains a significant amount of water, the moisture may subsequently move to the surface and blister or disrupt the coating film. It is possible, however, to apply latex coatings to slightly damp surfaces because these coatings are water-based emulsions.

The condition of the concrete surface to be coated is of prime importance. A concrete surface that contains residual form-release agent, curing agent, or oil, grime and laitance may not accept the surfacing or not coat the surface uniformly. The factors due to location of the concrete in the structure that need to be considered include the following:

- Continuous or intermittent exposure of the coating to moisture
- The degree of wear encountered in service
- The temperature fluctuations to which it will be subjected.

In some concrete, there is continued ingress of water from an exterior source. This occurs in tanks, tunnels, flatwork on grade, laundry rooms, and improperly constructed basement walls and other such locations. The concrete may contain water soluble salts which can be carried to the surface of the concrete to produce efflorescence. The crystal pressures of such salts, deposited just below a coated surface, can disrupt the film and pry it away for the wall. Therefore, the application of a coating to the surface on which efflorescence might appear is not an effective preventive measure. The degree of wear and traffic anticipated in a particular location may influence the choice of the coating, e.g. a coating for a hospital floor would be expected to be less wear resistant than one intended for a factory floor. The following factors are considered in the application of resurfacing material.

- **Climatic conditions:** Coating is best done under favorable climatic conditions because the viscosity of coatings increases as the temperature decreases causing difficulties in application. Coating should not be done below an ambient temperature of 10°C or above 30°C.

- **Concrete temperature:** Some two-component coatings will perform poorly if the concrete temperature is cold, even though the coating is warmed. Therefore, the air and concrete temperature must be above 10°C.

- **Moisture content of surface:** For most polymer based paints, the provision of a dry surface on which a coating is to be applied is imperative. Water based coatings may be safely applied to damp surfaces.

- **The thickness and number of coats required:** This will depend on the texture of the surface and the intended application. A purely decorative application may need a thickness and number of coats to obtain colour uniformity. For abrasion resistance the thickness and the number of coats are required.

- **Coverage rate:** Rough and porous surfaces often require two coatings. The first one is usually thinned to allow for penetration or a compatible primer coat may be used. In estimating coverage rate for a given film thickness, allowances should be made for losses caused by spillage and overspray. Resurfacing of a concrete surface requires thorough cleaning or removal of the existing coating. If the old coating is in good condition, then compatibility between the old and the new coating should be determined.

Types of Coatings

Many different wall and floor coatings are available which exhibit varying degrees of chemical resistance, physical durability, and ease of application. Some of the more widely used products are as follows:

- ***Bituminous coatings and mastics:*** Asphalt or coal tar coatings may be applied cold or hot. Two coats are usually applied to surface dry concrete. A thin priming solution is applied first and when the primer has dried to a tacky state, it is ready for a finishing coat. Multiple coats at right angles to each other are usually applied to avoid pinholes. Bituminous mastics may be applied cold or as heated fluids. Cold mastics are cutbacks or emulsions produced as thick, pasty materials which contain finely powdered, siliceous mineral fillers and asbestos fibers. Emulsions are slower drying, more permeable, and less protective.

- ***Polyesters and vinyl esters:*** These coatings are two- or three-part systems consisting of an unsaturated polyester resin, peroxide catalyst, and possibly a promoter. Catalyst and promoter are mixed separately into the resin, and the amount of catalyst must be carefully controlled because it affects the rate of hardening. Fillers, such as silica flour and glass fibers are often used to reduce shrinkage, and the coefficient of expansion, and to compensate for brittleness. Selective modifications to resins have produced new, less rigid types, which retain moderate to good chemical resistance. Vinyl esters are different from general polyesters, because the resins are characterized by an acrylic termination of epoxies. They have excellent resistance to oxidizing chemicals, improved handling characteristics, rapid cure, and low cost, and are widely used in the food, chemical process and paper industries for the repair of floors, trenches, storage tanks, scrubbers, and waste treatment facilities.

- ***Urethanes:*** These coatings may be one- or two-part systems. Most one-part systems are moisture cured and must be used on dry surfaces to prevent blistering during the curing period. The two-part system is available in two types—catalyzed and polyol cured. Polyol cured coatings are the most chemically resistant of the urethane coatings but require the greatest care in application.

- *Epoxies:* Epoxy coatings have excellent chemical resistance to most chemicals. They have a better alkali resistance than acid resistance and bond well to concrete. The coatings are resilient and abrasion resistant. They are generally two-package systems composed of epoxy resin and a curing agent. Common curing agents suitable for room temperature curing are amines, polyamines, amine adducts, polyamides, polysulfides and tertiary amines. The coating properties are dependent on the type and amount of curing agent used. Generally, three coats must be applied to eliminate pinholes; fillers like silica flour or glass flake may be used to bridge pinholes. Fiber-reinforced epoxy coatings may be used on concrete that may undergo thermal movement.

- *Neoprene:* These coatings may be one- or two-part systems. One-part systems are generally used in thinner films. The two-part system has lower chemical resistance, cures slowly at room temperature, and usually requires a holding period between mixing and application. Since the materials are sensitive to moisture in the concrete, coating should not be done until the concrete is at least 28 days old.

- *Coal tar epoxy:* Most proprietary products are two package systems. A combination of coal tar, filler, solvent, and epoxy resin may be in one package and one of the commonly used curing agents in the other. The coal tar, filler, solvent, and curing agent may also be blended together to make up one package and the epoxy in the other package. To ensure proper cure and chemical resistance, the correct mixing ratios must be used, and stipulated mixing times be adhered. Three main types of coal tar epoxy coatings are recognized, the classification being according to epoxy resin content, namely, high epoxy resin content coatings for dry film thickness, medium epoxy resin content coatings for concrete pipe linings, and low epoxy resin content materials for nonsag application. The first type requires a primer and its thickness is achieved in two coats, while the latter two materials require no primer and can be applied in a single coat. Spray application gives the best coverage.

Acrylics: These materials are available as solvent-based products and as the more recent 100% solid resins. Solvent-based acrylic coatings have excellent wetting characteristics, a fair degree of chemical resistance, and good colour retention characteristics on outdoor exposure. Because of their good intercoat adhesion, they are easily recoated. Because of their low viscosity, these materials are ideally suited for sealing and impregnation of cracks. They can be used at temperatures as low as 10°C, because they cure rather rapidly at this temperature. At normal temperatures, treated surfaces are reportedly ready for use after only 1 to 3 hours of curing. Therefore, they can be used for the repair of plant floors without causing prolonged work stoppages. Floors treated with acrylic monomer type materials are skid resistant, hygienic, seamless, easy to clean, and show good resistance to gasoline, oils, alkalis, dilute mineral acids, and organic acids, such as lactic acid. They are reported to develop excellent intercoat adhesion as well as bond to older, previously cured acrylic coatings even after years of service, thus achieving easy and effective repairs.

11.2.2 Intumescent Coatings

These coatings are specially formulated fire resistant paints applied on structural elements for the purpose of increasing their fire rating. These products are finding increased use in the repair of fire damaged buildings where they are used for fire protection of steel work. The coatings can be brush applied up to a thickness of 3 to 4 mm for thin coatings and up to 30 mm for thicker coatings. These coatings look like ordinary coatings, but are chemically formulated to expand 25 to 30 times their original volume when exposed to temperature of 200°C and above. Most system consist of three components, a primer, intumescent base coat, and an external seal coat usually based on chlorinated rubber to retain colour fastness and protect the base coat from wear and tear.

11.2.3 Resin Based Toppings

When an industrial floor is at an advanced stage of deterioration, heavy duty, monolithic, resinous toppings are probably the best all round materials for resurfacing eroded and spalled floors. In general, the polymer resins, curing agents, catalysts, and fillers used are the same for both the high-build coatings and seamless flooring compositions. Most of the products in use consist of thermosetting resins and special sand fillers that are generally applied as a single 3 mm to 6 mm thick layer. Some of the desired properties provided by these toppings include superior abrasion and chemical resistance in comparison to concrete, rapid cure and a tenacious bond to clean sound concrete.

Nearly all topping are three-component systems consisting of a resin, curing agent and aggregate fillers. The resins used for heavy duty, monolithic toppings are epoxies, polyurethane, polyesters, polyacrylate, and phenolic materials. In addition to binding the aggregate and providing adhesion to the concrete, the resins provide chemical and impact resistance. Epoxies, polyesters, and polyurethanes account for most resins used in heavy duty monolithic toppings. In addition to lowering the cost and providing an abrasion resistant, textured surface, a well-graded aggregate provides other advantages, such as increased compressive strength and reduced curing shrinkage. It also reduces the coefficient of thermal expansion to a value closer to that of concrete, increases toughness, and improves trowel ability. Some of the materials used as heavy monolithic toppings are brought out below.

Epoxy Floor Toppings and High-build Floor Coatings

These products have excellent resistance to wear and impact, are completely waterproof and impervious to grease, oils, detergents, and many chemicals. Because the epoxy binder used is usually 100% solid they show negligible shrinkage. Two types of topping are currently used: A sand-filled epoxy mortar and a glass fiber-reinforced, multicoated system.

Epoxy Mortar

Such systems contain pure or modified epoxy resins of low viscosity blended with silica sands and fillers with mesh sizes ranging from 40 to 200. The blend produce a trowel able mix which is applied at a thickness in the range of 3 to 6 mm. Epoxy/amines combinations have excellent solvent and alkali resistance and good resistance to acids except for strong oxidizing agents. Epoxy polyamides have less chemical resistance than epoxy polyamines, but their physical properties are better, they are tougher, more flexible, have better adhesion, and are also more tolerant of moisture. The polysulfide epoxies are less widely used because they are far less chemically resistant and are more expensive. Since they are more flexible than the other two, they find use in some floor joint sealant applications.

The aggregate fillers are generally silica, quartz and emery. The ratio of aggregate to epoxy resin binder will vary, based on mesh size of aggregate and desired consistency anywhere from 3:1 to as much as 8:1. The finer mesh sizes produce a smooth finished surface but the higher resin demand by the large, specific surface area of the filler makes the system quite expensive. A coarser mesh size of 30 produces a more skid resistant surface, loads the resin matrix more fully and proves better troweling properties. A mixture of these mesh sizes is more widely used where colour is required, chemically resistant pigments may be ground into the epoxy resin or curing agent, or they may be dry blended with the aggregate.

Epoxy mortar compounds and coatings are used in many areas other than on floors. They are used to line silos, build and repair airport runways, line cooling tower pans, and provide nonskid areas on roadways, bridges ramps, and docks. A typical floor is applied in following three steps:

- Priming of the concrete surface with a low viscosity epoxy resin to ensure proper wetting of the surface and sealing of the pores.
- Troweling of a 6 mm sand-filled mortar over the primer.
- Sealing of the above topping with a low viscosity epoxy resin to fill in any remaining pores and ensure that a complete barrier system is presented to splash chemicals.

Glass Fiber-reinforced Multicoated System

The glass fiber-reinforced multicoated system is a heavy duty chemical resistance system used in locations where the exposure conditions involve higher concentration of chemicals and longer periods of exposure. The thickness of the system can range from 0.5 mm to 6 mm depending on the severity of the in-service conditions. The thicker coats provide a surface where discontinuities or pinholes are minimized or eliminated so that an effective barrier system, which protects concrete against acids and other aggressive chemicals, is produced. The system is built up in four steps as brought out below:

- The surface is primed with a low viscosity epoxy
- A 50 ml thick coat of the filled epoxy is then placed.

- A woven glass fabric is then applied.
- After that the epoxy is partially cured.

Polyurethanes, Seamless Floor Toppings

Many of the synthetic monolithic flooring compounds, termed seamless flooring, are either epoxy- or polyurethane-based products. Polyurethane products result from the chemical reaction between isocyanate groups on the liquid resin, and hydroxyl groups of the curing agent. They are characterized by excellent resistance to chemical attack, impact, abrasion, good adhesion, and flexibility.

These flooring toppings can be one- or two-part systems. A one-part system may be moisture cured. They are usually supplied at a solid contain of 40 to 50% in a blend of solvents. In flooring applications, a base coat of the higher solid content material is first applied. While the base coat is still wet, silica or emery aggregate is sprinkled or blown into it. The aggregate may be applied in one or two applications and a sufficient amount of time must be allowed for the base coat to develop a gel so that the aggregate particles do not settle to the bottom of the coating. Two types of the two-part system are available; catalyzed and polyol cured. Catalyzed toppings have a limited pot life after mixing and cure rapidly. Solid content ranges from 55 to 60% and the solvent is xylol. These materials are the most chemically resistant of the urethane floor toppings, but they require the greatest care in application. Most of the seamless flooring compositions can be readily applied by brush, spray or roller.

Limitations of toppings made from thermosetting resins are:

- High coefficient of thermal expansion
- Ultrahigh cohesion and bond strength
- Measurable curing shrinkage
- Small, but measurable, liquid absorption.

11.2.4 Surface Hardeners

The surface hardeners and overlays are used for the repair and upgrading of industrial floors, where loaded vehicles, such as forklift truck traverse the floor surface. A number of materials available that, in specific instances, will improve wear resistance or chemical resistance, reduce dusting, or improve the appearance of concrete floors.

Surface hardeners: Generally refers to a material used to upgrade floor wear resistance, reduce dusting and increase chemical resistance. Two main categories in current use are sprinkled shake hardeners and liquid hardeners.

Shake hardeners: Shake hardeners produce a special finish that is obtained through the incorporation of selected natural or metallic aggregate into the surface of a newly poured floor. There are many commercial products sold for this purpose, the required characteristics being high aggregate to cement ratio, good finishing characteristics at low cement factor, and capability of being incorporated heavily enough to make a substantial difference in the composition of the floor surface. Two types of aggregate are used in shake hardeners, viz. mineral and metallic. Metallic floors are generally used in floors where there is continuous steel wheel traffic and point impact. Floors containing mineral aggregate tend to be harder, thicker, and will not rust.

Liquid floor hardeners: The liquid hardeners are usually floor sealing liquids, which by their reaction with certain cement hydrates, $Ca(OH)_2$, adhesive character and sealing ability, are able to bind the floor surface and reduce the rate of wear. More recently, dilute solutions of both solvent and water based emulsions of resins, such as epoxies, urethanes, and methyl methacrylate is finding wider use for this purpose. Solvent and water based epoxies and urethanes consisting of 6 to 20% solid solutions are generally used to improve both wear resistance and resistance to oil and grease penetration. Two coats of polyurethanes, epoxy, or methyl acrylates are often used to prevent dusting on concrete floors. Because of their adhesive character, these polymeric materials bind the surface well, providing better abrasion resistance than that obtained by the use of inorganic hardeners.

11.2.5 Overlays

Concrete overlays may be applied as a second stage of construction on a new floor or deck, as preventive maintenance on a deck that has been open to traffic for a short time but was built without a deck protective system, or in the rehabilitation of existing deteriorated floor slabs or decks. The terms overlay and topping are often used interchangeably in the above mentioned applications. Materials currently available for this method of repair can be categorized as the follows:

- Normal or high early strength Portland cement concretes and mortars
- Latex modified concrete
- High alumina cement concrete and mortar
- Fiber-reinforced concrete
- Silica fume concrete
- Resin based mortars.

The materials mentioned above are limited to the rehabilitation of industrial floors and bridge decks. The general requirements for the above systems include the following:

- Low permeability to water and deicing chemicals
- Adequate skid resistance
- High abrasion resistance
- Insensitivity to minor variations in ambient temperatures
- Sufficient working period to setting time to approximately 15 year service life
- Sufficient flexibility to avoid cracking.

Several potential advantages of concrete overlays are as follows:

- The overlay can be tailored to provide the required thickness of high strength concrete and maximum durability at the deck surface.
- Properly proportioned and consolidated concrete is effective in reducing the penetration of chloride ions.

- The overlay is an integral component in the load carrying capacity of the deck.
- Use of a concrete overlay permits vapour exchange between the concrete and environment, preventing the build-up of vapour pressure that occurs beneath an impermeable membrane.
- The overlay has a compatible thermal coefficient of expansion with the base concrete and absorbs less solar radiation than asphalt overlay.
- A smooth riding surface can be provided because minor irregularities in profile can be corrected and dead load deflections are minimal.
- High quality aggregate can be incorporated into the concrete mixture to provide good wear and skid resistance at a modest additional cost.
- In repair work, the overlays will fill in areas where concrete was removed without the need for a separate placing operation. Work can proceed while part of the deck remains open to traffic.

Normal or high early strength Portland cement, concrete, and mortar overlays: Typical overlay mixtures may consist of 1 part Portland cement, 1 part fine aggregate, 1.5 to 2 parts coarse aggregate by volume. The aggregates for such heavy duty overlays should be selected from clean, hard and irregular or rounded material, such as trap rock, granite, emery corundum, or synthetic products like silicon carbide or heat treated aluminium oxide. Separate bonded overlays, 50 to 76 mm thick are applied after the base structural slab has hardened. The overlays are placed as stiffly as possible. Their performance depends on their bond to the base slab. For this purpose, the base slab should be roughened with a coarse wire broom before it fully hardens to leave the coarse aggregate partially exposed. After the slab hardens, it should be blown clean of dust and debris, and dampened. Just prior to placing the topping, a grout consisting of a 1:1 mixture of Portland cement and sufficient sand mixed with

sufficient water/latex mixture to give a thick slurry dries out. Curing, of the freshly placed topping, is a very important step in securing dense surfaces which will not craze or dust. The success of the overlay is dependent upon good quality control procedures.

Latex-modified Concrete Overlays

Latex-modified concrete contains the same ingredients as conventional concrete with the addition of polymer latex. The water of suspension in the emulsion hydrates the cement, and the polymer enters the structure of the concrete, thereby providing supplementary binding due to the adhesive and cohesive properties of the polymer. This result in concrete having good durability and tensile and flexural strength properties that is well suited to use in overlays. The structural properties of this type of concrete can vary depending on the type and amount of latex, the type of aggregate, the cement content, and the w/c ratio. The choice between latex-modified mortar and latex-modified concrete is determined by the intend thickness of the overlay. Thickness of less than 25 mm is usually constructed from mortar and overlay greater than 25 mm from concrete.

High Alumina Cement Mortar and Concrete Overlays

High alumina cement mortars and concrete have the ability to provide high early strengths at normal temperature or cure and develop higher strengthens at lower temperatures faster than Portland cement concretes. In addition, such concretes are more resistant to heat, acidic conditions, and sulfate attack than Portland cement-based concrete. Consequently, their chief use is in locations which are subjected to chemical attack, such as pulp and paper factor floors, refractory conditions, such as steel rolling mills, and when repairs are done under cold weather conditions.

Fiber-reinforced Concrete Overlays

Several types of fiber have been investigated, but most of the work has focused on the use of steel and alkali resistant glass fibers. More recently, polypropylene fibers have been used as a crack-arresting measure in flat slab applications. The addition of steel fiber to concrete results in substantial increase in tensile and flexural strengths. Most of the high way applications of steel fiber-reinforced concrete have been in payment overlays where the materials high flexural strength and resistance to dynamic loads can be used to advantage for pavement rehabilitation projects, the high tensile and flexural strengths of the material overlay, and its resistance to reflective cracking, offer good potential. The following overlays are also in use nowadays:

- The silica fume concrete overlays
- Polymer concrete overlays
- Epoxy concrete overlays
- MMA (methyl methacrylates)

11.3 SEALING MATERIALS

There are various forces that can act alone or together to force water through concrete. These are hydrostatic pressure, capillary action, wind driven rain, a difference in a vapour pressure between the two sides of the concrete or some combination of these. Water finds its way most readily through porous built or sealed. Some of the materials used to seal concrete surfaces and joints to prevent the ingress of moisture into structure are described subsequently. Sealants are flexibilized polymeric materials which are used for mainly two purposes:

- To plug irregular gaps between two rigid surfaces
- To provide a dynamic bridge across the gap between two surfaces.

In this capacity, they perform three functions as brought out below:

- Accommodate joint movement
- Prevent the infiltration of the joint
- Prevent the ingress of water into the structure.

11.3.1 Sealant Types

Some of the common types of sealants used in industrial and domestic buildings and other engineering structures include acrylic, latexes, butyls, solvent release acrylics, polysulfide, urethanes and some flexibilized epoxies.

Oil-based sealants: Oil-based sealants are easily applied caulks that adhere well to most surfaces and require no special joint preparation. They are used in nonmoving joints and cracks. The selected oils and fillers that comprise the product slowly harden as the oils dry or oxidize. Some of the better oil based products have costs similar to acrylic latex based sealants.

Mastics and thermoplastics: Mastics are highly filled, inexpensive materials that are used in joints where the anticipated movement does not exceed 5%. The materials used as mastics are asphalt, rubber asphalts and pitch or coal tar which becomes plastic when heated. Thermoplastics are upgraded mastic type materials which show slight improvement in elasticity and chemical resistance when compared to mastics.

Latex-based sealants: Latex-based sealants are sealants that cure by water evaporation to a resilient material that can accommodate slight movement. The surface usually dries rapidly enough to be appointed 30 to 60 min after application. Due to their quick curing capability and good adhesion to a variety of surfaces without primer they find wide use for residential construction applications.

Butyl sealants: Butyl sealants are soft, low modulus, that cure by solvent release to a rubber, capable of sealing joints which experience small movement. They have excellent adhesion to wood, metal glass and concrete without primers. Although they possess good water resistance, they are not suitable for immersion service.

Solvent-based acrylic sealants: Solvent based acrylic sealants cure by solvent release. They are versatile construction sealants providing a high performance seal for many building construction joints without the need for primers. Acrylic have excellent adhesion.

Polysulfide sealants: The two-component systems are high performance products used to seal joints certain wall construction and for caulking expansion joints in masonry, concrete and glazing applications. They have a history of long-term performance having been in use for over 25 years. Polysulfide elastomers have good elastic qualities, good recovery, fast cure, very good adhesion, good resistance to weathering and ageing, and reasonably good chemical resistance. Two-component product cures by a chemical reaction between the liquid polymer base and activators component. The resulting synthetic rubber seals accommodate movement of 5.25% of joint width. They bond well to glass and metal but need a primer for full bond to porous surfaces, such as concrete, especially for immersion service conditions. The one-part system cures/by absorption of moisture from the air to produce a rubber like seal with properties similar to the two component systems. The material cures slowly and requires water spray or wet cooling to speed initial skin cure and prevent dirt pick-up.

Urethane sealants: Urethane sealants are based on polyurethane elastomers. The two part systems can be modified to provide a variety of physical properties. Most formulations provide good bond to all construction materials, possess high extensibility and a fair water resistance. These sealants cure to a tough, abrasion resistant, elastomeric rubber that facilitates their use in horizontal surfaces, such as joints in industrial floors. Other advantageous properties include high tensile strength, good recovery form elongation and compression cycles, outstanding resistance to ageing and thermal shock and nonstaining character. Although urethanes have a great tolerance to a wide range of chemicals than the other sealants, they have less resistance to concentrated chemicals, such as acids, alkalis, and some neutral salts. The one-part sealants cure by absorbing moisture from the air to produce a seal with properties similar to the

two-part systems. They can be used for general construction sealing, caulking and glazing application in which moderate movement be expected. Two of the main advantages of the material are:

- The good adhesion obtained without priming of the substrates under dry conditions
- They are very slow curing.

Silicone sealants: Silicone sealants are high performance, elastomeric sealants which have the widest service temperature and they are the only sealants which have some resistance to nitric acid. They exhibit excellent resistance to weathering and ageing and are suitable for sealing all types of construction joints. The sealants cure by absorbing moisture from the air to form a silicone rubber seal capable of a high degree of movement. The presence of moisture in the concrete can significantly reduce adhesion. Silicones are tougher and harder than polysulfide and possess good resistance to organic acids, salts and solvents. The high strength, low shrinkage, low toxicity, and excellent weathering qualities enable these sealants to be used indoors and outdoors without much concern. They are particularly suitable for a chemical environment that is also exposed to elevated temperature levels. The material is always packaged as a single component. Some limitations of the product are as follows:

- Generation of acetic acid in some formulations
- High price
- The need for a primer on some substrates for some formulations.

Epoxy sealants: Epoxy sealants are modified, flexibilized materials which possess high strength and adhesion in conjunction with their resistance to chemical attack. They can be formulated to produce a grade that is hard enough to resist wheeled traffic and keep the joint edges from fracturing. Such products will usually meet the requirements for an industrial floor joint sealant. Epoxies and modified epoxy sealants have the best chemical resistance to concentrated mineral acids, alkalis, high molecular weight organic solvents, and acids. The products are available mostly as two component systems consisting of a resin and hardening agent. Two types, self-leveling and nonsag grades are available. Limitations of epoxy materials include the service temperature limit of 60°C.

Kraton-based sealants: Kraton-based sealants are new styrene-butadiene styrene and ethylene- butylene's styrene thermoplastic elastometers, usually formulated with solids contents of approximately 90%. Therefore, there is little solvent release and low shrinkage of the applied sealant. Despite their ability to cure to a hard surface, they possess good elongation and can be formulated to provide clear cured rubber which finds use in architectural applications.

11.3.2 Sealant Properties

Some of the properties that must be evaluated when selecting sealants are adhering and cohesion, hardness, modulus, stress relaxation, compression set, and resistance to weather and application characteristics.

Adhesion and cohesion: These are two of the most important properties of a sealant being considered for use in a working joint. The sealant must adhere to the sides of the joint it is sealing, and withstand internal stress caused by joint movement.

Hardness: Hardness is measured by the resistance to penetration and is expressed as a reading on the durometer. The higher is the reading, the harder the sealant.

Modulus of a sealant: The lower modulus materials tend to dissipate the stress that results from strain induced by joint movement better than high modulus materials. In general, sealants with a high durometer hardness value tend to have high modulus, although low modulus sealants may have high modulus at lower temperature.

Stress relaxation: Stress relaxation in a sealant results from internal relaxation under conditions of strain. Low modulus sealants

exhibit the highest degree of relaxation by generating stress tending to pull an extended sealant back to its original shape. Consequently, a lower adhesive strength is required to keep it bonded to the faces of the joint.

Compression set: Compression set is the inability of a sealant to expand to its original shape after being compressed. A high degree of set is undesirable in a sealant for a working joint, because the result is a permanently reduced width of the sealant bead. When the sealant is so compressed, high internal stresses are produced as the joint widens to its former shape because of its reduced capability to accommodate movement. Subsequently, these stresses become large enough to cause adhesive failure at the bond line, or cohesive failure within the sealant mass.

11.3.3 Sealant Consistencies

Sealant products are available in four consistencies as brought out below:
- Putty-grade
- Gunnable or nonsag
- Pourable and self-leveling
- Extruded tapes.

The **putty grade** is used for glazing applications, and is of a consistency suitable for application by a putty knife.

Gunnable sealants are soft enough to be applied by a hand operated or air-powered caulking gun, yet stiff enough not to run or sag in vertical joints.

Pourable sealants are designed for joints and cracks in horizontal surfaces, such as plant floors. They are usually poured directly from their container into the joint, but some may be applied by gun.

Extruded tapes are soft tacky ribbons of sealant used in glazing applications.

11.3.4 Sealant Classes

Sealants can be grouped into three classes; low, medium and high range, according to their movement capability, service life, and cost.

Low range sealants: These materials are cheap and easy to use products, which have a rather short life due to their very limited movement capability. Oil-based and resin-based caulks, bituminous-based mastics, and polybutene products fail into this classification.

Medium-range sealants: This group can accommodate more movement than products in the low-range category; they are nevertheless limited in use to nonworking joints. Products included in this category are butyls, neoprene, and the higher quality latex materials. The latex used for sealant manufacture is the acrylic latex, rather than PVA latex. Medium-range products are generally used for interior work.

High range sealants: This group includes the silicones, urethanes, polysulfide, solvent release acrylics, and certain proprietary compounds. The products are designed for use in working or moving joints.

11.3.5 Sealant Selection

Movement, width and depth of the joint affect sealant selection. Job specification for repair should follow a three step procedure:

- First the anticipated joint movement is determined.
- Then the most appropriate sealant is selected to meet expected joint moment and other service requirements.
- Finally the depth of the sealant bead and the type of backer rod in the joint are specified.

The class of sealant is selected either by considering the type of joint in it will be placed or the type of substrate. Once the minimum sealant class has been determined, matching of sealant and geometry begins. In most joints, any sealant of the recommended class will probably be acceptable. Final product selection requires careful examination of product literature for information on recommended geometry and nature of primers that may require of various substrates. Joint depth affects the stress in a sealant. The base of the sealant bead should be free to move. Therefore, a bond breaker must be used to ensure that the sealant does not bond to the base. Of the many joint back up materials available, only closed cell, foamed, polyethylene rod can be used safely with all types of sealants.

11.3.6 Water Stops

In hydraulic structures, where the surface seal in a joint is subjected to water pressure or high rates of flow, special provisions to ensure water tightness should be provided. To ensure water tightness, a water stop must be incorporated.

Types: The **labyrinth type** is designed to overcome the need to split the concrete forms to install the water stop and is limited in its use to construction joints. The **flat, corrugated type** has numerous ribs running the length of the water stop. The **dumb-bell types** rely on the bulbs at the end of the section to provide anchorage in the concrete. The **metal types** can be of many forms, a typical type is one with a deep, crimped, V-groove in the center. The steel plate consists of a heavy gauge, flat plate, one half of which is embedded firmly in the concrete on the one side of the joint, the other half being coated with asphalt to allow it to slip in the concrete, whenever the movement occurs.

Properties: Water stops installed in concrete structure are not easily replaced. Considerable time is consumed in making the numerous joints and splices during water stop installation. Therefore, it is of paramount importance that a water stop should continue to perform well for the life of the structure. The durability characteristics of the materials used in these products thus plays a significant role in their selection for use. The properties that should be evaluated prior to selection are as follows:

- Mechanical fatigue
- Resistance to ageing when exposed to oxygen
- Resistance to ageing when exposed to liquid media and under stress
- Water tightness under hydrostatic conditions
- Tensile strength
- Flexibility at low temperatures

- Stiffness required to resist being folded over during placing of the concrete
- Ease of installation.

11.4 WATERPROOFING MATERIALS

Waterproofing materials are applied on concrete surfaces to form impervious coatings that prevent the passage of water in liquid form and may also retard vapour transmission in varying degrees, depending on the type of coating. Since the treatment prevents the ingress of water in liquid form, it also stops the transport of water soluble salts in either direction. Some of the marketed materials are given in the following sections.

11.4.1 Cement-based Coatings

Metallic waterproofing and cement/latex coatings provide the required adhesive bond to resist hydrostatic pressure. Metallic waterproofing consists of finely divided metallic aggregate and oxidizing agents, mixed with sand and Portland cement. The mixture is mixed with water to form a plaster coat which may be applied by brush or trowel depending on the consistency of the mix. After the coating is applied to the concrete surface, the metallic particle oxidize and expand to at least three times their original size, wedging themselves into pores and interstices, thus creating a relatively dense waterproof barrier. However, any subsequent cracking of the concrete wall will open cracks in the coating which would render it ineffective in preventing moisture entry.

11.4.2 Bentonite-based Products

Trowel or spray grades: Finely divided bentonite clay, formulated with fibers, is often used in thickness of about 5 mm in below grade locations. In contact with sufficient water, it swells 10 to 20 times its original volume forming a gel that prevents further penetration by water. The material has some ability to reseal itself over cracks that form subsequently in concrete.

Corrugated bentonite cardboard panels: Bentonite clay contained in biodegradable, corrugated, cardboard panels are suitable for applicable to exterior surfaces below grade. The panels are attached with tempered nails or roofing cement. Water borne bacterial decomposes the cardboard and water swells the bentonite to form the gel.

11.4.3 Mastic Asphalts

The products contain asphalt fillers, fibers, polymers, and solvents, and are usually reinforced with fiber glass mesh. These are usually laid by hand using wooden floats or squeegees. They are placed in two layers each of 10 mm thickness, as blow holes frequently form in the first layer and the second layer covers these holes. The main advantage of mastic asphalt is that a protective layer is unnecessary, it is however, prone to the problem of blowing and is labour intensive.

11.4.4 Polymer Resin-based Coatings

These are generally of two types:

- Resins blended with organic solvents
- Solvent free coatings.

Solvent based coatings are subdivided into single and two component coatings. The coatings on drying produce a smooth dense continuous film that provides a barrier to moisture and mild chemical attack of the concrete. Because of the resistance to moisture penetration, staining, and ease of cleaning, they are preferred for locations of high humidity and those in which a lot of swelling occurs.

Most products are low solid content materials which require multiple coats to produce a continuous film over concrete; since the materials are thermoplastic, and have a significant degree of extensibility they are capable of bridging minor cracks which may develop in the concrete surface if they are applied in sufficient thickness. The number of coats required depends on the surface texture, porosity, and the targeted dry film thickness. Although some of the newer products have some moisture tolerance, enabling them to be applied over damp surfaces; in normal usage they should be applied over dry surfaces. Due to their relative impermeability to water vapour, they could blister when applied to concrete surfaces with high moisture content or where the opposite surface of the concrete is in constant contact with moisture. Careful control of wet film thickness is therefore necessary during application.

Two component polymer coatings consist of a solution of a compounded polymer with or without solvent and a reactive chemical component called the curing agent hardener or catalyst. The materials are usually mixed just prior to use in accordance with the manufacturer's instructions. When using two components polymer based coatings, the following items are of importance to the application of the materials.

- Most produces are supplied as a kit containing the two components in the required proportions. Therefore, in order to realize the full potential of the product the correct mix ratio of the two components must be used.
- To ensure a complete reaction of the two components they must be mixed thoroughly.
- Some two-component materials require an induction period of 15 to 40 min after mixing. Therefore, such products cannot be used immediately after mixing.
- Viscosity reduction by the use of thinners should be resorted to only after the manufacturers are consulted.
- The storage temperature of solvent based coatings is not critical. They should be stored at a temperature 16 to 32°C just prior to use.

Some of the generic resin types used includes thin adhesive coatings based on epoxy, polyurethane, chlorinated rubber, and polyvinyl chloride, neoprene, and chloro-sulfonated polyethylene. To overcome some of the drawbacks of solvent-based coatings, a significant number of solvent-free coatings are on the market. These coatings are often specified for their high build, reduced risk of toxicity and fire hazard during applications where adequate ventilation of solvent-based materials would be uneconomical. Their major disadvantage is that they are relatively high viscosity materials that are difficult to apply in comparison to their solvent-based counter-parts. Suitable methods of application are rollers but in some cases it may be preferable to use squeegees. Their favored application is to large areas such as floors.

11.4.5 Membranes

Membrane waterproofing is applied to prevent water intrusion through openings resulting from construction defects, shrinkage cracking and structural movement. Vertical surfaces are waterproofed from the potentially wet side, the earth side of a foundation wall or the liquid side of a tank wall. Slab construction is treated similarly, except that the membrane is often encased between two courses of concrete to protect it from mechanical damage. More recently, traffic bearing membrane systems which combine an abrasion resistant top coat with an electrometric under layer have been used to protect bridge and parking decks. Membrane waterproofing systems can be grouped according to fabrication and application methods as follows:

- Built-up sheet membranes
- Prefabricated sheet
- Liquid applied materials.

Built-up Membranes

Hot-applied membranes built-up from multiple plies of saturated felt fabric or coated sheets, set in interplay applications of hot coal tar pitch or asphalt are the oldest type of membrane waterproofing. Applications of built-up membranes involve the following:

- Priming the concrete surface
- Reinforcing the corners with two or more plies of felt and adhesives
- Sealing the openings
- Applying a coating of hot waterproofing bitumen

- Brushing or rolling the sheet into hot bitumen
- Applying additional courses
- Applying protection board over the membrane.

The advantages generally associated with hot applied built-up waterproofing system are:

- Workmen are familiar with materials and methods
- Membrane is continuously bonded to concrete
- Allows the correction of filed errors
- Bridges existing joints and cracks in concrete

The disadvantages of the system are:

- Hot kettles of bitumen must be kept at correct temperature
- Hot bitumen causes burns and the fumes cause worker discomfort
- The method is labor intensive
- Organic felt and fabric deteriorate
- Conforming membrane to substrate contours and projections through concrete is difficult.

Prefabricated Sheet Membranes

Sheet materials are typically 1.25 to 2.4 mm thick, and are applied by three methods, viz. loosely laid, mechanically fastened and fully bonded. Loosely laid systems are usually used in the waterproofing of roof and require an overlay of smooth stones or earth for ballast. Mechanical fastening is usually required at perimeters and openings. Fully bonded systems use contact adhesives. Sheet membrane system includes:

- Vulcanized rubbers
- Thermoplastics
- Ethylene internally
- Impregnated asphalt composites

Advantages normally associated with prefabricates sheet waterproofing include the following:

- Workmen are familiar with application techniques.
- The membrane is more elastic than built-up membrane.

- Materials bridge existing joints and cracks in concrete.
- The membrane has uniform thickness controlled at the factory and not on site.

Disadvantages

- The seal at sheet laps and joints in some systems can be the weak link in membrane
- Conforming the membrane to irregular surfaces and projections through concrete is difficult.

Liquid Applied Membrane

A number of liquid applied membranes are available:

- Hot-applied rubberized asphalts
- Liquid applied neoprene
- Cold applied polyurethane

Many of these materials can be used in vertical surfaces as well as in horizontal slabs. Most must be covered with protection board or a traffic bearing slab or coating. A few formulations with sand sprinkled on the laid coating can accept pedestrian traffic. The performance of liquid applied waterproofing systems is related to membrane thickness, which should be checked during application using a wet film thickness gauge.

Advantages provided by the system are:

- Application is easy by several methods spray, roller, squeegee, etc.
- The continuous bond formed prevents lateral movement of water between membrane and concrete.
- No hot material is used.
- The membrane is more elastic than built-up membranes.

The disadvantages associated with the system include the following:

- The thickness depends on the skill of the applicator and may therefore vary.
- The material may require accurate mixing at job site.
- Compatible sheet material or fiberglass reinforcement material must be added to

bridge larger existing joints and cracks in concrete.

Multilayer Membrane Systems

Multilayer membrane systems are effective in reducing the ingress of moisture. They provide a watertight barrier that has elasticity to maintain integrity under the influence of temperature induced structural movement. Membranes have proven to be especially effective at preventing the onset of corrosion in new structure by preventing contamination of reinforced concrete by salt. There are two basic types of membrane systems.

- Liquid applied, traffic bearing elastomeric membrane system
- Asphalted systems.

Traffic bearing elastomeric membrane systems should have the following basic properties:

- Impermeability
- Adhesion
- Crack bridging characteristics
- Durability
- Ease of applications.

The asphaltic systems are much thicker than liquid applied elastomeric membranes and use asphalt cement based waterproofing membrane, which is either hot or cold applied. An asphalt paving wearing course 32 mm thick or more is applied on top of the membrane. The asphalt paving may be latex modified for added flexibility, or may be applied as a mastic system, which does not require compaction. Due to the softness of the bituminous membranes, the thicker the membrane layer the thicker the asphalt paving wearing course required to obtain a stable surface.

Final selection of membrane system is based on material properties or field experience and should involve the following items for consideration.

- Anticipated degree of structural movement
- Anticipated traffic loading, exterior or interior application
- Consideration of live load capacity and height limitations

- Nature and extent of deterioration of the concrete deck
- Installation procedure recommended by the manufacturer
- Compatibility with expansion and contraction joint seals
- Cost
- Long standing example of successful application available for inspection
- Good standing of the manufacturer and the length and conditions of the guarantee
- The case of repair of the system and the effectiveness of the bond between old and new membranes.

11.5 BONDING MATERIALS

Bonding materials are natural, compounded or synthetic materials used to joint individual members of a structure without mechanical fasteners. These products are often used in different repair applications, such as bonding of new concrete to old concrete. Two main types of bonding materials, namely latex emulsions and epoxies are frequently used in the building industry. Critical factors which govern bonding between new and old concretes, provided that sound concrete technology is followed, include:

- The strength and integrity of the old surface
- The cleanliness of the old surface.

Adequate bonding can be achieved by placing repair material directly against properly prepared substrate. There are special conditions when bonding agents are used. The surface condition plays an important role in bond development, although the magnitude of the bond varies depending upon other factors. Some of these factors are as follows:

- Although a good bond may be obtained with a grout bonding layer, generally a grout layer consisting of cement and sand slurry or cement/latex slurry increases bond strength.
- Proper compaction of the new concrete improves bond.

- The method of surface cleaning that produces a good bonding is determined by the density of the base concrete, e.g. for a sound base, concrete acid etching would suffice.
- Mechanical cleaning is essential if a weak or deteriorated surface layer existed on the old concrete.

11.5.1 Bonding Agents

Three main types of bonding agents are frequently used:
- Cement-based slurries
- Epoxies
- Latex emulsions

For **Portland cement-based** repairs and overlays, cement or sand cement slurry is used. After the substrate has been prepared and immediately before placing the repair materials, a thin coating of creamy grout must be vigorously and thoroughly broomed or brushed into the prepared surface.

In case of **latex-modified** repair materials or overlays, the bonding grout can be broomed in from the mix itself. For latex-modified materials, it can also be mixed separately. There are, now available, factory blended latex-modified cement slurry bonding agents which are mixed with water on site and applied to the prepared concrete surface. The following latex products are used as bonding agents:
- Styrene butadiene (SBR)
- Acrylic
- Polyvinyl acetate (PVA)

Latex Emulsions

There are a variety of different applications for latex emulsions as bonding agents. Some types have a greater degree of water resistance than others. The emulsions generally used in cementitious compositions and environments are of the oil-in-water type, and sometimes contain more than 50% of water. They are generally stable in the cement/water system. The following description of the advantages and limitations of the various types of emulsions used as bonding agents is intended to serve as a preliminary guide for the user and specifier.

A latex may be used as a bonding agent in three ways:
- Preparing a neat cement slurry, utilizing the latex as part of the gauging water
- The use of a 1:1 water-latex diluted materials
- The use of a re-emulsifiable latex which can be softened and retackified upon contact with water.

Styrene butadiene rubber (SBR): SBR lattices which are compatible with cementitious compounds are copolymers. They show good stability in the presence of multivalent cations, such as calcium and aluminium and are unaffected by the addition of relatively large amounts of electrolytes. The SBR lattice may coagulate if subjected to high temperatures, freezing temperatures, or severe mechanical action for prolonged periods of time.

Polyvinyl acetate (PVA) lattices: Commercial materials are copolymers manufactured by the emulsion polymerization process; two main types of polyvinyl acetate lattice are used in repair, i.e. nonre-emulsifiable PVA, and emulsifiable PVA.

Nonemulsifiable PVA: This PVA forms a film which offers good water resistance, good light stability, and good ageing characteristics. Because of its compatibility with cement, it is widely used as bonding agent and a binder for cementitious water-based paint and waterproofing coatings.

Emulsifiable PVA: These lattices produce a film that can be softened and retackified with water. Such latex permits application of a film to a surface long before the subsequent application of a water-based overlay. They are the most widely used as a bonding agent for plaster, and to bond finish or base coat gypsum, or Portland cement plaster to interior surfaces of cured cast-in-place concrete.

Acrylic lattices: Acrylic lattices are produced from acrylic resins which are polymers and copolymers of acrylic and methacrylic acids. These resinous materials range in physical properties from soft elastomers to hard plastics. The emulsion is used in cementitious compounds in the same manner as SBR lattices.

Epoxy lattices: Epoxy emulsions are produced from liquid epoxy resin, when mixed with the curing agent. The curing agent serves the additional functions of an emulsifying and wetting agent and as a surfactant. Emulsions are stable and water dilutable from the time of mixing until gellation occurs. Pot life may vary from 1 to 6 hours depending upon the curing agent selected, by adding high amounts of water.

100% solid epoxy bonding agents: Various epoxy products are available for bonding of freshly placed concrete to cured concrete as well as concrete to steel. Most products contain 100% solid resins which may or may not contain thixotropic fillers and other additives to enhance a particular property or reduce cost. Current products are available in a variety of consistencies, ranging from a highly filled paste to that of liquids with viscosities of 100 centipoises similar to that of water. It is a performance specification based on end use, there are no specific limits on chemical composition. Instead the materials must meet requirements related to physical properties, such as viscosity, bond strength, shrinkage and thermal compatibility.

The specification classifies the epoxy resin bonding system by type, grade, and class. The properties stipulated in the specifications are of great importance, namely bond strength, shrinkage, and thermal compatibility. The materials are placed in one of four categories:

- *General repairs:* These materials do not fit the other three categories, but are suitable for general repair work.
- *High early strength:* These materials have a compressive strength of 8 MPa at 4 hrs.
- *Self-leveling:* These materials are very flowable and will spread out to fill a cavity.

They are used on horizontal surfaces and are useful for hard to reach areas.

- *Slurry coats:* These are used for very thin repairs or for coatings.

11.5.2 Keys to Developing Bond

- Clean, sound substrate
- Roughened profile of substrate for mechanical interlock
- Open pore structure in substrate
- Repair materials/bonding agent with sufficient paste for absorption into substrate pores
- Repair materials are applied with sufficient pressure to facilitate contact between the repair materials and the substrate at the bond line.

Achieving an adequate bond between repair materials and existing concrete is a critical requirement for durable surface repairs. Various techniques are available to achieve the required bond. The bond at the interface between the repair materials and concrete substrate is likely to be subjected to considerable stress from volume changes, freeze-thaw, force of gravity, and sometimes impact and vibration. The stress states that develop at the bond lines will vary considerably, depending on the type and use of the structure. For example, the bond on the bridge deck overlay may be subject to shear stress in conjunction with tensile or compressive stress induced by shrinkage or thermal effects, and to compression and shear from service loads.

It is essential that the repair materials achieve a strong bond to the substrate and the subsequent stresses are not sufficiently great as to cause debonding. Repair which have bond lines in direct tension have the greatest dependency on bonding. Repairs that are subject to shear stresses at the bond line are capable of stress resistance not only by bonding mechanisms, but also by aggregate interlock mechanism, which add greatly to shear bond capacity. It should be remembered that high initial bond strength is generally not as important as bond durability.

12

Special Types of Repair Materials

12.1 INTRODUCTION

Concrete is no longer a simple mixture of cement, sand and aggregates, but it has matured into an engineered material containing a mixture of admixtures. Admixtures are materials other than water, aggregates and hydraulic cements. They are used as ingredients of concrete or mortar and added to the batch immediately before or during mixing. Almost every property of concrete can be modified to some extent. It can be used to improve concrete properties either in the handling process or consolidation of fresh concrete, in the performance of concrete both in fresh and hardened stages and even for economy in the cost of construction. However, the effectiveness of an admixture in a concrete mix depends on several factors, such as type and amount of cement, temperature of concrete, etc. It is important to note that no admixture of any type or amount should be considered as substitute for good concreting practices. Admixtures should be used judiciously and not for the purpose of covering up deficiencies in the mix.

In spite of several significant beneficial effects and extended applications provided by admixtures, it is important to note the serious problems presented by the user of a mixture of admixtures in concrete. Compatibility of the ingredients becomes an important issue and verification by trials with the actual materials to be used on a job site under the actual field conditions is therefore crucial. Admixtures can generally be classified as chemical and mineral admixtures. Chemical admixtures can be further classified as follows:

- Surface active chemicals, which are used for air entrainment or allowing reduction of water in concrete mixtures
- Set-controlling chemicals.

These may be either powders or in liquid form which are added in small quantities. Mineral admixtures are finely divided materials having pozzolanic characteristics, which can be added, to concrete in relatively large quantities (20 to 30%) by weight of cement. Power plants using coal or rice husk as fuel and metallurgical furnaces producing pig iron, silicon and ferrosilicon alloys are main sources of artificial mineral admixtures. Naturally available mineral admixtures are generally of volcanic origin. Improvements in impermeability, resistance to thermal cracking and chemicals, ultimate strength, durability, better finish, etc. are some of the benefits of using mineral admixtures.

12.2 CHEMICAL ADMIXTURES

Chemical admixtures, such as retarder and accelerators are helpful in avoiding the undesirable effects on rate of cement hydration due to ambient temperature conditions. Plasticizers and superplasticizers can be used to produce concretes of any required workability to suit the actual construction practices and any type of structural members, such as for pumping, members with highly congested reinforcement, self-leveling purposes, massive structures.

12.2.1 Superplasticizers (SP)

These are the most recognized category of water reducing admixtures having a remarkable plasticizing action on concrete and produce substantial enhancements in workability. Water reduction as high as 30% may be achieved. They may not affect the setting and hardening behavior of the system provided a proper compatibility with cement is established for the mix under consideration. This admixture was first introduced in the Japanese market in the 1960s followed by Germany. ASTM C494 classifies superplasticizers into two types:

- High range water reducer
- High range water reducer with retarder

Chemically, superplasticizers are organic polyelectrolytes, which belong to the category of polymeric dispersants. Some SPs are synthetic while others are derived from natural products and may be classified into the following four categories:

- Sulfonated melamine formaldehyde condensate (SMF)
- Sulfonated naphthalene formaldehyde condensate (SNF)
- Modified lignous sulphonates (LS)
- Hydroxy-carboxylic acid such as polyacrylate ester (PAE).

However, the performance of a superplasticizer in a cementitious system depends on its chemical composition, dispersion ability, thermal stability, molecular weight, and also on cement fineness and its composition. Table 12.1 gives typical results of the elemental analysis of superplasticizers.

Sulfonated melamine formaldehyde condensate: Though this product was originally developed in the 1950s as a dispersant for a variety of industries, it was not until some 10 years later that the possibilities for its use in concrete were recognized. The molecular weight is generally about 30,000. It has minimal effect on air entrainment and setting time.

Sulfonated naphthalene formaldehyde condensate: This was one of the first materials referred to in the literature as a water-reducing agent but only since 1970, it has found extensive application as admixture. The solid content of commercial products varies from 25 to 45%. The molecular weight is of the order of 2000.

Modified lignosulfonates: The crude lignosulfonates derived from wood pulp are commonly used as plasticizers, and in order to improve their effectiveness they can be refined and modified. The processing involves the removal of sugars and other impurities, selection of a higher molecular weight fraction and optimally, further sulfonation or partial polymerization.

Polyacrylate ester: Carboxylated synthetic polymers act as highly effective dispersant.

Mode of Action of Superplasticizers

The superplasticizers are adsorbed on to the initial hydrates of C_3A, C_3S and C_2S to form a complete monolayer around them. In case of SNF and SMF, the ionized sulfonate groups on the adsorbed superplasticizer molecules are strongly negatively charged and repel the

Table 12.1: Elemental chemical composition of some superplasticizers

	Carbon	Hydrogen	Oxygen	Nitrogen	Sulphur	Sodium
LS	42.7	4.5	35.4	0.0	7.8	7.1
SNF	48.3	2.6	22.6	0.0	13.8	12.7
SMF	22.0	3.5	26.7	25.5	12.4	9.3
PAE	47.8	8.0	37.8	0.0	0.9	4.3

cement particles thus breaking the weaker van der Waals forces of attraction, resulting in a well-dispersed system. The polyacrylate materials are similarly adsorbed and cause dispersion of the cement particles by the same electrostatic mechanism through the ionized carboxyl groups. The long flexible side chains attached to these polymers act as physical barriers, preventing the cement particles from coming within the range of the van der Waals forces. In doing so, water trapped within the original flocs is released and can contribute to the mobility of the cement paste and hence, to the workability. Superplasticizer does not cause much reduction in the surface tension of water; hence, there is little hydration but at 28 days, the products of C_3S hydration are the same as a superplasticized concrete.

Problems Associated with the Use of Superplasticizers

Some of the problems associated with the use of superplasticizers are discussed below:

Rapid loss of slump with time: This is especially true when a high early strength superplasticizer is used, along with the higher grade cement. It is therefore recommended that the superplasticizer be added only at the point of discharge of the concrete. One of the ways to deal the issue of slump loss is to use repeated dosages of superplasticizers. By the addition of second and third dosages, large increases in workability of superplasticizer concrete can be maintained for several hours, the amount of second and third dosage being the same as the initial dosage. Another method of extending the slump in superplasticizer concrete is to use blended admixtures, such as a blend of naphthalene based SP and lignosulphonate or a gluconate retarder.

Loss of slump with increase in ambient temperature: Loss of slump in concrete mix is accelerated when the concrete mixing temperature is high. When superplasticizers are mixed in such a situation, bleeding of the concrete mix might take place. This problem is true in the case of superplasticizer based on naphthalene sulfonate, when the mixing temperature is around 40°C. The use of melamine-based superplasticizer may be tried in such a case.

Prolonged retardation of concrete setting time: This is a typical problem when a retarding superplasticizer is used. This is mostly due to excessive dosage of the superplasticizer. It is always advisable to check the dosage of the superplasticizer to be used for given cement for its slump, loss of slump with time and air entrainment. When there is a change in the brand or type of cement, it pays to check the fresh concrete properties once again. Superplasticizers, because of their tendency to slightly increase air entrainment, may enhance the cohesion of the mix. Some category of SPs has little effect on air, while others reduce the natural air entrainment, thereby slightly reducing the cohesion. These factors should be borne in mind when selecting the superplasticizer category especially if changes in mix constituents are not feasible for producing a good cohesive mix.

Strength Characteristics of Superplasticizer Concrete

The dispersing capability of superplasticizer is also used to reduce the water requirements in comparison with control mix. In effect, the same slump as that of the control mix can be obtained with a lesser water cement ratio, leading to high strength of concrete. Increasing the cement content to achieve early high strength may result in excessive heat output, causing cracking as well as undesirable for a reduction in water/cement ratio to achieve high strength. It is clear, therefore, that any means, consistent with good concrete practice, of reduction of water–cement ratio closer to the theoretical level is desirable.

Superplasticizers are claimed to go a significant way towards satisfying this requirement, as water reduction in the region of 20 to 35% are obtainable by their proper use, although limited by grade of cement, type of aggregate and temperature. By comparison, water reductions of about 15% are obtainable with lignosulfonate based plasticizers. Concretes with low water content and containing superplasticizers are noted for their high early and ultimate strengths, excellent durability and waterproofing characteristics. The technique of producing low water/cement ratio concrete for the purpose of obtaining high strength or retaining workability at reduced water content is of considerable interest to concrete technologies.

The higher strength of the admixture concrete is achieved by a reduction in w/c ratio, the influence of which is more pronounced at early ages, but less effective at later ages. Superplasticizers strongly bond on to cement particles and make them disperse preventing flocculation. In India, with the advent of higher grades of cement like 43 and 53, there is a tremendous potential for SPs to produce high strength concrete. Superplasticizers, both high early strength type and retarding type, based on naphthalene and melamine are available. In comparison to other methods like a polymer impregnation, etc., water reduced concrete using superplasticizer offer the easiest and most economical way of producing high strength concrete.

12.2.2 Accelerators

Cement accelerators can be classified into two distinct types:

- Set accelerators, which reduce the period over which the cement remains workable, leading to early stiffening of the mix
- Strength accelerators, which give an increase in the early age strength of the cement.

An admixture, which falls into one of these categories, does not necessarily also meet the requirements of the other and may even have the reverse effect. Temperature is also an important factor and accelerators, which are very effective at 20°C, may be of little benefit at 5°C. The converse is also true, some accelerators only being effective at low temperatures. Accelerators are used for cold weather concreting, in repair applications where the concrete must be put into service at an early age and also in seawater concreting. They fall under three categories:

- Chloride based
- Chloride free
- Very rapid set accelerators.

Chloride Based Accelerators

Calcium chloride has been used as an accelerator for concrete since at least 1885. Both setting time and early strength are accelerated while at later ages strengths remain similar to or above the control mix. Calcium chloride, aluminium chloride, sodium chloride and ferric chloride are a few accelerators that can be used for cement concrete.

Chloride Free Accelerators

These can be divided into three categories:

- Materials containing active anions
- Materials containing active cations
- Organic chemicals.

Substances in the first category include calcium or sodium salts of formats, nitrates, nitrites, thiocyanates, thiosulfates, permanganates and chromates. Triethanolamine, 8-hydroxyquinoline and acetinst come under the category of organic accelerators. Triethanolamine though most popular, is generally used in combination with other formulation as its effect is very sensitive to dosage and the chemical composition of cement.

Very Rapid Set Accelerators

This type of accelerator can be used in applications where the desirable initial set is less than 3 minutes and the final set less than 12 minutes. These accelerators are essential ingredients of mixes for certain specialized applications, particularly in mining, tunneling and in sea defence works. The most active ingredients of these accelerators are alkali metal silicates, aluminates, hydroxides, sulfates, or chlorides.

12.3 MINERAL ADMIXTURES

Mineral admixtures are finely divided silicious materials, which can be added to concrete in relatively large amounts, generally in the range of 20 to 100% by weight of cement. They may be pozzolanic, or cementitious or both. A pozzolana is a siliceous or siliceous and aluminous material. Mineral admixtures can be divided into two groups, viz. natural and artificial. Naturally available mineral admixtures are of volcanic origin. However, they are not available in India. Power plants using coal or rice husk as fuel, and metallurgical furnaces producing pig iron, steel, copper, nickel, lead, silicon, and also ferrosilicon alloys are main sources of artificial mineral admixtures. Benefits of using mineral admixtures in concrete include improvements in impermeability, resistance to thermal cracking and chemicals, higher ultimate strength, better durability, etc. Fly ash (FA), ground granulated blast furnace slag (GGBS) and condensate silica fume (CSF) are the major industrial by-products suitable as mineral admixtures.

12.4 FLY ASH

Fly ash is the finely divided residue resulting from the combustion of ground or powdered coal which is transported from the fire box by the flue gases and is subsequently removed from the gas by electrostatic precipitators. During combustion of coal in thermal power plants, the volatile matter and carbon are burnt off, whereas mineral impurities in the coal, such as clays, quartz feldspar, etc. melt and are transported to lower temperature zones where it solidifies as spherical particles of glass. Some of this mineral matter agglomerates to form bottom ash, but most of it flies out with the flue gases called fly ash.

Classification

According to ASTM C618, mineral admixtures are classified into three:

- Class N: Low or calcined natural pozzolanas, such as diatomaceous earth, opaline charts, clays, and shale's tuffs and volcanic ashes or princes.
- Class F: Fly ash from anthracite or bituminous coal.
- Class C: Fly ash from lignite or sub-bituminous coal.

Influence of Physical Characteristics on Properties of Concrete

The shape, fitness, particle size distribution and density of fly ash particles influence the properties of fresh and hardened concrete. This is primarily due to the particle influence on the water demand of the concrete. The majority of fly ash is glassy, solid or hollow and spherical in shape. Clean, glassy, spherical particles of fly ash act as good void fillers in concretes and reduce water requirements. Rough textured and porous glass spheres in fly ash possess more reactivity than smooth spheres and hence, particles size distribution alone may not reflect the potential activity of a fly ash. Fineness of fly ash has an influence on the strength of concrete.

Strength Development

When fly ash is added as replacement of fine aggregates and not that of cement, the strength of concrete at given ages is enhanced. When used as a replacement of cement by fly ash, concretes exhibit lower gain in strengths at all ages. However, at ages beyond 28 days, they attain strengths equal to or greater than that of concretes without fly ash. Test results indicate that silica fume can be used to increase the early strength of fly ash concrete. Increases in impermeability and strength of fly ash concrete are due to:

- Presence of hydrated phases with low-density phase products and smaller voids in the hydrated paste
- Refinement of pore system
- Improvement in structure and strength of transition zone between the aggregate and hydrated cement.

12.5 GROUND GRANULATED BLAST FURNACE SLAG (GGBS)

Slags are by-products of the metallurgical industry. The most important slag from the standpoint of the quantity used as building material is iron blast furnace slag. Slags that have crystalline, stable and solid structure are used as aggregates while glassy structured material is used as a hydraulic binder. GGBS exhibits cementitious as well as pozzolanic characteristics. Hydrated GGBS gives the same hydrates as that of Portland cement. It reacts more slowly with water than Portland cement but can be activated with lime and sulfates either physically, by grinding or thermally. Typically silicon, calcium, aluminium, magnesium and oxygen constitute 95% or more of the blast furnace slag.

Specifications

ASTM C989 provides for three strength grades of GGBS, depending on their respective mortar strengths when blended with an equal mass of Portland cement. The classifications are made based on the slag-activity index and are classified into three grades as 80, 100 and 120. The slag activity index gives an indication for evaluating the relative cementitious potential of slags. The specification includes two chemical requirements: One limiting the sulfide sulfur to a maximum of 2.5% and the other limiting the sulfate content to a maximum of 4%.

Effect on Properties of Hardened Concrete

The extent to which GGBS affects strength is dependent on the slag activity index of the particular GGBS and the replacement level used in the mix. Other factors that affect the performance of GGBS in concrete are w/c ratio, physical and chemical characteristics of cement, curing conditions, etc. Concrete containing GGBS is more susceptible to poor curing conditions than concrete without GGBS. The rate of strength gain of grade 120 slag is greater in concrete mixtures which have high w/c ratio than in mixes with a low w/c ratio. Highest strengths at 28 days are obtained with 40 to 50% level of replacement. At least 70% GGBS may be needed to achieve significantly low heat of hydration. Comparatively, lower heat of hydration can be always expected when GGBS is used to replace equal mass of cement. Permeability is considerably reduced due to the pore filling with C-S-H and reduction in pore size.

12.6 CONDENSED SILICA FUME (CSF)

Condensed silica fume is a by-product from electric arc furnaces used in the manufacture of silicon metal or silicon alloys. This product is excellent for use as a supplement to Portland cement. Being composed of essentially non-crystalline SiO_2 in a finely divided state, CSF exhibits excellent pozzolanic characteristics. Unlike fly ash, the lime-silica reaction involving CSF is rapid and, therefore a long curing period is not necessary before the desired high strength and impermeability are realized. Due to very fine particle size, bulk density of CSF is generally between 200 and $250 \, kg/m^3$. This low bulk density causes great difficulties in handling and disposal of the materials. Silica fume is available in slurry, compacted and pellet forms.

12.7 ADMIXTURES FOR REPAIR/ REHABILITATION

Since the beginning of the use of concrete, strength has been regarded as the most significant and important property of concrete. However, it is now well recognized that this strength of concrete may deteriorated with time, is a result of a combination of various factors that may include moisture ingress leading to carbonation of concrete and/or reinforcement corrosion, chemical attack on concrete, poor quality of materials and workmanship. Concrete admixtures are vital in repair and rehabilitation works where rapid gains in strength, nonshrink characteristics, adequate bonding to parent concrete are a necessity. Although the use of admixtures for rehabilitation can encompass a fairly wide range of application in concrete from patch work of concrete in relatively small area to injection grouts for cracks, a few of the most common admixtures that are useful in rehabilitation are dealt with here.

12.7.1 Polymer Dispersions or Lattices

Various polymer systems may be used as an emulsion or dispersion of polymer droplets in water. One of the most common of these is styrene butadiene rubber latex solution, properly known as SBR latex. These are commonly used as an effective bonding agent when made into slurry with cement and water. Screeds toppings, mortars and concrete containing such admixtures are often described as latex modified concrete. SBR latex is a milky white liquid which contains microscopic particles of synthetic rubber dispersed in an aqueous surfactant solution produced from styrene and butadiene. When SBR latex is added to a standard concrete or a mortar mix, the following improvements in properties can be obtained:

- The w/c can be reduced from 0.5 to 0.3 due to the plasticizing effect of the latex.
- The adhesion of mixes to the apparent material can be increased by a factor.
- The tensile strength is normally doubled.
- The shrinkage during drying/curing is drastically reduced.
- The water penetration of latex modified mixes is greatly reduced.
- A slurry of SBR latex and Portland cement gives prolonged corrosion resistance to steel, because of adhesion, alkalinity, and low permeability.
- The greatest advantage of the SBR latex lies in the fact that it has similar properties like modulus of elasticity and thermal expansion.

12.7.2 Latex Modified System

Cement products are weak in tension because they contain numerous voids. As latex modified mortar dries out, the rubber particles coalesce together to form a continuous film and strands which stitch the opposite sides of the voids together. These films and strands also block up the voids thus increasing resistance to water penetration. They also adhere excellently well with old concrete, brick, steel asphalt, wood, etc. Recommendations for the use of SBR latex cementitious mixes are:

- Always use fresh cement-air set or warm cement leads to a rapid setting.
- Use washed sand complying with zone II; not more than 0.5% of the sand should be below 75 microns.
- Use the recommended dosage of the latex; normally 10 litres for every 50 kg of cement.
- Avoid excess mixing of water, for example when used as a floor screed, the reasonable w/c ratio is of the order of 0.3.
- Always apply a bond coat before applying the latex modified mortar/concrete to the old concrete.
- Allow a little time for the cement-latex slurry bond coat to be tacky.

12.8 EPOXY RESINS

Amongst the synthetic resins, such as epoxy, polyester, acrylic, polyurethane, etc. epoxy resins possess very high mechanical and adhesive strength properties most desirable for civil engineering applications. Epoxy resins when cured with different hardness offered a wide range of properties as brought out below:

- High adhesive strength to almost all materials
- Low shrinkage during curing
- Exceptional dimensional stability
- Natural gap filling properties
- Thermosetting
- Resistance to most chemicals and environments.

Epoxies for Crack Repair

The most appropriate method of crack repair depends on whether the crack is still actively moving or not. Active cracks may be due to inadequate provision of movement joint in the structure. The repair process involves converting the crack into a movement joint, by the use of a suitable sealant. The injection of a low-viscosity epoxy is a possible repair method for cracks between 0.02 mm to 6 mm in width. There are wide variations in the properties of different low-viscosity epoxy systems and it is therefore necessary to choose carefully to match the individual job requirements in relation to such matters as temperature of application, capability of bonding to moist concrete, shrinkage, thermal and elastic properties of the hardened resin. For epoxy injection to be effective the cracks must be free of dirt, grease, etc. Before injection can begin, the crack, where it appears on the surface of the member, must be sealed, so that the liquid resin does not leak and flow out of the crack before it begins to gel and cure.

Epoxy Mortars/Concretes

Epoxy resin systems are superior in almost all properties to concrete. The adhesive strength of the epoxy resin system is 50 times than that of cement concrete. However, the use of epoxy in mortar is recommended only in emergency repair, since they have a low modulus of elasticity and high creep than most Portland cement products. Fillers are a must when workability is to be improved and to minimize the use of resin. They are Portland cement, fly ash or silica flour.

Epoxy Coatings for Reinforcement

Extensive experimental investigations reveal that epoxy coatings when properly applied could be expected to adequately protect reinforcing steel bars from corrosion. However, powder based epoxies performed better than the liquid epoxies. Epoxy resins find use in grouting, bonding of old to new concrete, repairs of eroded concrete, coatings for reinforcements, etc. A judicious use of epoxy is called for in the light of their very high cost. Epoxies are also used for bonding of steel plates to the external faces of structural members like beams, etc. which have suffered damage revealed by cracks in the flexural, shear modes. The technique consists of *in situ* strengthening of failed concrete elements by boundary steel plates to concrete surface with epoxy resins. A large amount of experimental work done all over the world including India show that repaired and externally reinforced beam exhibit a 24% and 20% increase in strength over the original RCC beams in flexure and shear.

Other important admixtures include shrinkage compensating admixture that can be used for filling of grouts, etc. and accelerators for guniting/shotcreting works. A number of admixtures are now available for rehabilitation of concrete, but there are no cheap solutions to the problem of repair. Proper assessment of the problems and selection of the right combinations of repair materials are imperative in ensuring the success of the rehabilitation work.

12.9 POLYMERIC MATERIALS

Recent years have seen an increased emphasis on the repair and rehabilitation of all types of structure in preference to demolition and rebuilding. As a result, there is an increasing demand for repair materials. Durable repair can be obtained only by matching the properties of the base concrete with those of the repair materials intended for the use. The selection of repair materials has been identified and used for repair, rehabilitation and retrofitting of reinforced concrete structural elements. Polymeric materials, concrete polymer composites, fiber reinforced concrete/mortar, fiber reinforced polymer concrete/mortar-composites, fiber reinforced polymer composites, etc. are identified from extensive research work as high performance materials for repair, rehabilitation and retrofitting of damaged structural elements.

All matter in this world is composed of extremely small units called molecules. They are too small to be seen even under the most powerful microscope and are a complex association atom. Molecules come in different sizes and shapes. Molecules of plastics are much larger than the ordinary molecules. They are giant molecules in the form of long chain which is called polymers.

12.9.1 Types of Polymers

The distinction between types of polymers is based on their reaction to heating and cooling.

Thermoplastic Polymers

Thermoplastic polymers soften upon heating, and can be made to flow when a stress is applied. When cooled again, they reversibly regain their solid or rubbery nature. Continued heating of thermoplastics will lead ultimately to degradation, but they will generally soften at temperatures below their degradation points.

Thermosetting Polymers

Thermosetting polymers are materials which can be heated to the point where they would soften and made to flow under stress. However, they do not revert to the original solid state as the heating causes the material to undergo a curing reaction. Often, these polymers emerge from their synthesis reaction in a cured state. Further heating ultimately leads only to degradation and not softening and flow.

12.9.2 Applications of Polymers

In building construction, the applications of polymers can be classified in various ways, for example:

- Nonstructural polymers
- Structural and semistructural polymers
- Auxiliaries to other materials

The first group constitutes, by far the greatest volume and number of different uses. The second group includes patch repair, overlays, linings to concrete/steel products, injection to structural cracks, strengthening of structural elements, etc. Auxiliaries include adhesives, bonding agents, sealants, and decorative and protective coatings.

12.10 ORGANIC POLYMERS

Organic polymers are complex chemical compounds derived mainly from the petrochemical industry. These materials are often referred to as resins and the principal resins used in the construction industry are epoxide, polyurethane, polyester, acrylic, polyvinyl acetate and styrene-butadine.

12.10.1 Epoxide Resins

The formulators market epoxide resins and have the special properties required for the specific use to which they will be put. The basic characteristics of epoxide resins include the following:

- Outstanding adhesive qualities to such materials as concrete and steel
- Resistance to wide range of acids, alkalies and other chemicals
- Rather vulnerable to organic solvents
- Low shrinkage
- High coefficient of thermal movement compared with concrete
- High compressive, tensile and flexural strength
- Appreciable loss of strength at high temperatures
- High rate of gain of strength, which can be varied to suit the particular application
- Rather poor resistance to fire compared with concrete and clay bricks
- To obtain satisfactory results, the site conditions under which the resin will be applied must be stated to the formulator.

12.10.2 Polyurethanes

Polyurethanes, like epoxide resins, are products of the petrochemical industry. They can be obtained as elastomers, solid and rigid material and as flexible coatings. They are very durable in external conditions and retain their gloss well. Polyurethanes can be specially formulated to meet specific site requirements.

They can be combined with epoxies and will withstand relatively high temperatures as well as sudden changes in temperature, i.e. thermal shock.

12.10.3 Polyester Resins

Polyester resin can be divided into two classes:
- The so-called saturated polyester resins, such as polyethylene terephthalate or terylene
- The unsaturated polyester resins which can be cross-linked.

Commercially available unsaturated polyester can be divided into eight different classes or groups as follows:
- Gel coat/top coat resin
- General purpose resins
- Chemical resistant resins
- Reduced flammability resins
- Low styrene emission resins
- Low profile/low shrinkage resins
- Special purpose resins
- Casting resins.

All the strength properties decrease under-load at elevated temperatures. The resins had good electrical properties, such as loss factor, dielectric strength and are resistance chemicals. Polyester resins are used with Portland cement and selected aggregates to form a polymer cement aggregate mortar. Such mortars possess a number of desirable qualities, such as good resistance to wide range of chemicals, high resistance to abrasion, waterproofness under the range of heads likely to be met in liquid-retaining structures, and high bond strength with most common building materials.

12.10.4 Polyvinyl Acetate (PVA)

This material is used as a bonding agent and as an admixture in mortar to improve certain properties of the mortar. Manufacturers of proprietary compounds based on PVA claim increased tensile and flexural strength reduced drying shrinkage and reduced permeability.

12.10.5 Styrene-Butadiene and Acrylic Resins

These materials are also known as latexes and polymer emulsions. Some of the properties of a proprietary styrene-butadiene copolymer emulsion used with Portland cementing grout, mortar and concrete are:

pH	= 11.0
Total solid content	= 47% (nominal)
Specific gravity	= 1.1
Butadiene content	= 40%

Depending on the manner in which the polymeric materials are incorporated, polymer concrete composites can be classified under the following three major types, i.e. PMCC, PC and PIC.

Polymer-modified Cement Composites (PMCC)

In PMCCs, polymeric materials are incorporated into cement composites during the mixing stage. The composite is then cast to the required shape in the conventional manner and is cured in a manner similar to the curing of cement concrete. The hydrated cement and the polymer film, formed due to the curing of the polymeric material, form an interpenetrating network that binds the aggregates as shown in Fig. 12.1.

Resin Concrete (RC)/Polymer Concrete (PC)

In these, polymers are used as the binders of the aggregates, in lieu of the cement–water binder system adopted in cement composites.

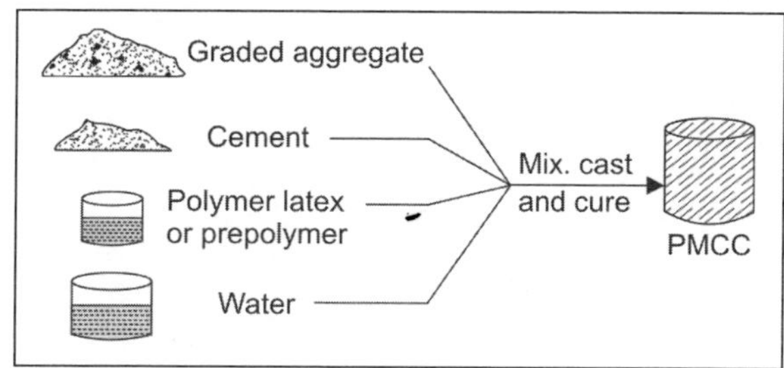

Fig. 12.1: Process technology for polymer modified cement composite

Monomers or prepolymers are mixed with the aggregates and the mix is cast to the required shape or form. The mix is then polymerized at the room temperature. The polymer phase binds the aggregates to give a strong composite as shown in Fig. 12.2.

Polymer Impregnated Concrete (PIC)

In PIC, monomers or prepolymers are impregnated into the pore system of hardened cement composites and are then polymerized. A very strong composite, viz. PIC, results in which cured polymer fills almost all the pores as shown in Fig. 12.3.

12.11 POLYMERIC REPAIR MATERIALS

Polymeric resins are complex chemical compounds derived from the petrochemical industry. The principal resins used for repair works are epoxies, polyesters, polyurethanes,

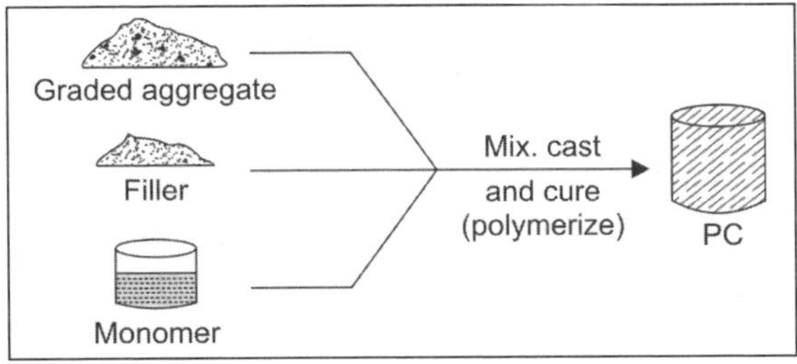

Fig. 12.2: Process technology for polymer concrete or resin concrete

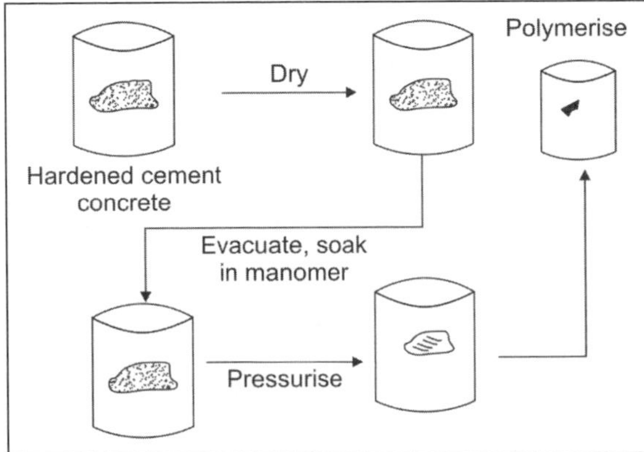

Fig. 12.3: Process technology for polymer impregnated concrete

acrylics and vinyl esters. They may be used in the form of latexes, such as styrene butadiene rubber or acrylics. While the range of use of these materials is now very wide, it is convenient and practical to divide into two categories; namely coatings and bonding aids in which the formulated compound is used on its own and polymer mortar/concrete composites in which the polymer is mixed with aggregate and sometimes cement.

12.11.1 Polymeric Coatings

Surface coatings and sealers rely upon the formation of a pinhole-free film of finite thickness over the concrete surface that acts as a protective barrier for concrete. They include epoxies, acrylics, polyurethanes, vinyl esters, etc. The hardened film may form by solvent evaporation, reaction with moisture or atmosphere, chemical reaction with a catalyst or hardener, or oxidation.

Acrylics

These form the base for a large proportion of coatings available in the market, viz. methacrylates, styrene methacrylate, styrene acrylics, and synthetic rubber latex. They have good wetting characteristics, satisfactory chemical resistance, and good colour retention characteristics. They also show good intercoat adhesion.

Urethanes

These can form an ideal coating for concrete. They can be easily applied by brush, spray or roller. Their main disadvantage is that they require careful surface preparation to ensure proper adhesion to the substrate and possess less intercoat adhesion.

Chlorinated Rubber

Coatings which were traditionally used for concrete were based on chlorinated rubber. While they have excellent diffusion resistance to carbon dioxide, they also do not permit the water vapour from inside the concrete to move out. This vapour barrier effect has resulted in many failures of this type of coating. Additionally, they are not permanently elastomeric and subsequently, crack and debond.

Epoxies

These can provide excellent protection to concrete in extremely aggressive environments where high chemical resistance is the main parameter. They have excellent alkali resistance also and bond well to concrete. The coatings are resilient and abrasion resistant but intercoat adhesion is poor. However, they may not be cost effective under other conditions. Coal tar epoxy systems are widely used as protective coatings for pipe lines and sewer appurtenances.

Silanes/Siloxanes

These are gaining popularity under the category of surface impregnates or pore liners. Hydrophobic materials like silicons line the pores of concrete and rappel water. Silicone resins are slow drying and pick up dirt. They should be applied on a perfectly dry concrete surface only. Silanes have high penetrating capabilities because of their smaller particle size. Their hydrophobic nature expels water from any pore of capillary into which they penetrate. This opens the surface of the substrate and allows gases to pass in more easily.

Latex Coatings

These are most widely used as interior coatings. Because they possess the desirable property of breathability, they are finding use in exterior applications also. Some of their advantages include ease of application, clean-up and wide range of available colours. Typical emulsions used are based on vinyl acetate, acrylic and styrene, and acrylic resins. Some of these resins have limitations, such as sensitivity to moisture and UV exposure.

12.11.2 Polymer Concrete/Mortar Composites

Polymer concrete/mortar (PC/PM) composites are being increasingly employed in repair and rehabilitation jobs owing to their very high strength and durability characteristics. There are essentially three basic types of PC composites:

- Polymer concretes (PC) formed by polymerizing a monomer binder and an aggregate mixture.
- Polymer modified concretes (PMC) in which a monomer or polymer is added to a fresh ordinary concrete mix in a liquid, powder or dispersed phase, and subsequently allowed to cure.
- Polymer impregnated concretes (PIC) in which ordinary hydrated Portland cement concrete (PCC) is impregnated with a monomer and is subsequently polymerized *in situ*.

Polymer Modified Concrete (PMC)

PMC has been of considerable interest to engineers because of its similarity in process technology to conventional concrete. Polymer latexes are usually copolymer systems, such as vinyl acetate, vinyl chloride, and butadiene, besides elastomeric systems like acrylonitrile butadiene (NBR), neoprene, and styrene-butadiene (SBR). PMC and PMM are increasingly used for rehabilitation because they are cement-based and therefore, give homogeneity to the system and the repair materials, and are more compatible with concrete compared to all other PC composites. An advantage of solvent-free PMC/PMM is the ease of adjusting the working rhythm as against the pot life and film-forming resistance of polymer solutions. They are economical while maintaining the technical value.

Polymer Concrete (PC)

Initial use of PC was almost exclusively for repair of ordinary PCC. The excellent bond of PC to PCC and very rapid cure time make PC an ideal repair material, especially in urban areas where fast, permanent repairs are essential. Added to these advantages are the possibility of tailoring its properties to suit any particular situation, excellent chemical resistance, and high bond strength. The most common types of monomers used to produce PC are methyl methacrylate, unsaturated polyester resin, and epoxies, besides furan, urethane, furfuryl alcohol, and vinyl ester which are also occasionally used. The materials used for patch repair range from unfilled resin binder to thick mortar sections. The PC repair can be carried out in two ways:

- Dry pack system in which the aggregates are prepacked and vibrated into the crater location and then infiltrated by a low-viscosity monomer like MMA.
- Premixed PC in which the aggregates and monomer are mixed together in a wheel barrow or a conventional concrete mixer and then directly applied to the surface and leveled.

Polymer concrete toppings consist of thermosetting resins with special sand fillers and are applied as a single 3–6 mm layer by troweling. They have the advantage of superior abrasion and chemical resistance, rapid cure, and tenacious bond to concrete base. Epoxies, polyesters and polyurethanes are more widely used while polyacrylics and phenolic are occasionally used for specific applications. PC overlays constitute the major field of PC application in civil engineering, MMA, UP resin, and epoxy-resin based PCs have been used for bridge deck overlays. Furan resins have been used for floor overlays where chemical resistance is important. PC overlays have an important role to play in situations requiring thin overlays which are impermeable, possess high bond strength with the substrate, besides high corrosion resistance, and low shrinkage. However, the most enticing aspect of PC overlays is the rapid turn-around-time from rehabilitation to reopening of bridges/pavements.

Polymer Impregnated Concrete (PIC)

Polymer impregnated concrete was developed in the 1950s and received wide publicity in the 60s and 70s. Recently, concrete sealing compounds like alkoxy silanes, alkoxy siloxanes, and metallic stearates have entered the market with claims of providing surface protection like surface impregnation. However, they do not provide the same extent of surface penetration and abrasion resistance, and their long-term durability and performance is suspect, due to possibility of removal from surface due to shallow depth of penetration. The process technology of PIC particularly useful for *in situ* applications.

12.12 FIBER REINFORCED CONCRETE

Fiber reinforced concrete can be defined as a composite material consisting of mixtures of cement, mortar or concrete and discontinuous, discrete, uniformly dispersed suitable fibers. Fibers used in concrete can be classified into two groups: Natural fiber materials like asbestos, cellulose, sisal, coir, jute, coconut and cane splits, and artificial fiber materials like steel, glass, carbon, polymer, etc. Fiber reinforced concrete is one of the most promising composites used in the construction. Natural fibers are not suitable for rehabilitation work but artificial fibers are suitable for rehabilitation work. The mechanical properties of fiber reinforced concrete (FRC) are influenced by the type of fiber, fiber length to diameter ratio (aspect ratio), the amount of fiber, strength of matrix, the size, shape and method of preparation of the specimen, and the size of the aggregate. Fibers influence the mechanical properties of concrete and mortar in all failure modes. The strengthening mechanism of the fibers involves transfer of stress from the matrix to the fiber by interfacial shear or by interlock between the fiber and matrix, if the fiber surface is deformed.

Besides the matrix itself, the most important variables governing the properties of FRC are the efficiency factor and the fiber content. Fiber efficiency is controlled by the resistance of the fibers to pull out, which in turn depends on the bond strength at the fiber matrix interface. Also, since pull out resistance is proportional to interfacial area, nonround fibers offer more pull out resistance per unit volume than larger diameter fibers. Therefore, for a given fiber length, higher aspect ratio is more beneficial. Most mixes used in practice employ fibers an aspect ratio less than 100, and failure of composites, therefore is primarily due to fiber pull out. However, increased resistance to pull out without increasing the aspect ratio is achieved in fibers with deformed surface or end anchorage; failure may involve fracture of some of the fibers, but it is still usually governed by pull out. An understanding of the properties of FRC is an important aspect of successful design.

12.13 BEHAVIOR OF STEEL FIBER REINFORCED CONCRETE

12.13.1 Compression

The effect of steel fibers on the compressive strength of steel fiber reinforced concrete (SFRC) concrete varies with fiber content. It is interesting to note that both increase and decrease of compressive strength with different fiber types have been experimentally observed. Even for the same material, there is mounting evidence to show that compressive strength may first rise, then drop with increasing fiber volume fraction. These observations suggest that the addition of fibers in a cement composite lead to a likely manifestation of increased resistance to microcrack sliding and extension, whereas strength degradation is a likely manifestation of increase in either pore or microcrack density, as a result of fiber addition.

12.13.2 Direct Tension

Because of the brittle nature of concrete, valid direct tensile testing of concrete in FRC is always difficult to carry out. Due to the importance of the tensile behavior of steel fiber reinforced concrete and concrete, many direct tensile tests of these materials have been attempted, using different designs of loading grips. Indirect methods of measuring the stress–strain curves have been attempted. SFRC has superior tensile properties,

particularly ductility, over plain concrete. Studies have indicated that the tensile stress–crack separation curve is the best alternative to characterize the tensile behavior of SFRC.

12.13.3 Flexure

The influence of steel fibers on the flexural strength of concrete and mortar greater than for direct tension and compression. Two flexural strength values are commonly reported. One corresponds to the first crack and the other corresponds to the maximum load. For large amounts of fibers, the two loads are quite distinct, but for very small fiber volumes, the first crack load may be the maximum load as well. Ultimate flexural strength generally increases in relation to the product of fiber volume concentration and the aspect ratio I/d. Concentrations less than 0.5 volume percent of low aspect ratio fibers have negligible effect on the static flexural strength properties.

12.13.4 Flexural Toughness

Toughness is an important characteristic for which SFRC is noted. Under static loading flexural toughness may be defined as the area under the load-deflection curve in flexure, which is the total energy absorbed prior to complete separation of the specimen.

12.13.5 Fatigue Strength

The behavior of SFRC in cyclic fatigue, despite its importance, has reckoned relatively little attention. FRC improves the dynamic properties like energy absorption, behavior under fatigue loading over the plain concrete.

12.13.6 Behavior under Cyclic Loading

The objective of subjecting the plain and SFRC specimens to cyclic loading is to investigate whether the specimens after subjecting them to cyclic loading would continuously possess their original integrity (i.e. without suffering damage). Since the peak strain for plain concrete is around 0.002, and these specimens fail suddenly, it is possible to subject them to cyclic loading only at very low strain levels. The performance of SFRC is found to be far superior to plain concrete even with 0.75% fiber volume fraction. The SFRC specimens, when loaded monotonically after cyclic loading at a strain of 0.003, reached almost the same peak load as was obtained under monotonic loading. All the SFRC specimens were able to sustain higher strain even after being loaded cyclically for fifteen cycles at a high strain of 0.007. This particular characteristic of SFRC could be beneficially used in the design of seismic resistant structures.

12.13.7 Shear and Torsion

Studies in the last few decades indicate that use of steel fibers as shear reinforcement in reinforced concrete beams helps in enhancing the tensile strength, resulting in increase in shear strength and possible prevention of shear failure. Studies carried out so far have shown that steel fibers up to about 1.5% by volume are effective as shear reinforcement either by themselves or in combination with vertical stirrups. The studies carried out on the torsional behavior of SFRC shown that there is an improvement in the torsional strength of concrete on addition of steel fibers of various types in varying volume fractions. Various researchers also studied the effect of reduction in longitudinal and transverse steel due to fiber addition.

12.13.8 Impact

Impact is a complex dynamic phenomenon involving crushing, shear failure and tensile fracturing. It is also associated with penetration, perforation and fragmentation end scaling of the target being hit. The addition of fibers improved the impact resistance of the plain concrete to a great extent. The improvement in the strength is dependent on the fiber types and fiber volume fractions. As there is no acceptable standard methods for determining impact resistance of SFRC, several types of tests have been used, namely weighted pendulum charpy-type impact test, drop-weight test, impact test, blast impact test, projectile impact test and instrumented impact test. The simplest of the impact test is the drop-weight test. This test yields the number of blows necessary to cause prescribed levels of distress in the test specimen. The test can be used to compare the relative merits of different fiber concrete mixes and to demonstrate the improved performance of FRC compared to conventional concrete.

12.13.9 Abrasion/Cavitation/Erosion

Both laboratory tests and full-scale trials have shown that SFRC has high resistance to cavitation force resulting from high-velocity water flow and the damage caused by the impact of large waterborne debris at high velocity. Several tests indicate that steel fiber addition do not improve the abrasion/erosion resistance of concrete caused by small particles at low water velocities. This is because adjustment in the mixture proportions to accommodate the fiber requirements reduce coarse aggregate content and increase paste content.

12.13.10 Creep and Shrinkage

There has been little work on the creep of steel fiber reinforced concrete. Fibers generally reduce the compressive and tensile creep. Tests have shown that steel fibers have little effect on free shrinkage of SFRC. However, when shrinkage is restrained, steel fibers can substantially reduce the amount of cracking and mean crack width.

12.13.11 Freeze–Thaw Resistance

Steel fibers do not significantly affect the freeze-thaw resistance of concrete, although they may reduce the sensitivity of the visible cracking and spalling as a result of freezing in concrete with inadequate air-void system. The freeze-thaw resistance of nonair-entrained concrete is similar for FRC and control concrete, where SFRC was found to be better in the case of air entrained concrete.

12.14 APPLICATION OF SFRC TO REPAIR OF DISTRESS STRUCTURES

The applications of SFRC fall in two categories: Repairs and new construction. Repairs are invariably to tackle problems of abrasion, cavitation or impact damage in various components of hydraulic structures, such as spillways, stilling basins, baffle blocks, outlet conduits, etc. Other area of applications of SFRC for repair purposes is in the field of shotcreting and overlays. Both steel and polypropylene fibers are used in shotcreting applications. Shotcreting with fibers improved the properties of the repaired concrete very much when compared to shotcreting with plain concrete. Similarly the behavior of overlays using SFRC is far superior to the conventional concrete.

12.15 BEHAVIOR OF FRC WITH OTHER FIBERS

12.15.1 Glass Fiber

The glass fibers are divided into three classes: E-glass, S-glass and C-glass. The E-glass is designated for electrical use and the S-glass for high strength. The C-glass is for high corrosion resistance, and it is uncommon for civil engineering application. Of the three fibers, the E-glass is the most common reinforcement material used in civil structures. Some typical properties of the glasses are shown in Table 12.2. It is produced from lime-alumina-borosilicate which can be easily obtained from abundance of raw materials like sand. The fibers are drawn into very fine filaments with diameters ranging from 2 to 13×10^{-6} m. The glass fiber strength and modulus can degrade with increasing temperature. Although the glass material creeps under a sustained load, it can be designed to perform satisfactorily. The fiber itself is regarded as an isotropic material and has a lower thermal expansion coefficient than that of steel.

Glass fibers possess high tensile strength and modulus of elasticity, but serious concern is expressed regarding their durability in an alkaline environment. For low w/c pastes, compressive strength is reduced by about 20% and for higher w/c ratio the decreases can be as high as 30%. Uniaxial tensile strength increases with age and amount of fiber. Aggregate grading does not influence the strength. In spite of the enhanced mechanical properties, question of the durability of alkaline resistant glass fiber concrete composite in alkaline environment remain unresolved.

Glass Fiber Reinforced Concrete

For the precast concrete industrial sector, the use of lightweight composite components (e.g. 60–80 kg/m^2 for main slab elements and 30–50 kg/m^2 for secondary slab elements), able to withstand the structural loads, would represent a significant improvement with respect to the current production. In fact the possibility to manufactured precast elements weighing about one-third of the conventional concrete products (e.g. 250–300 kg/m^2 for main and 100–150 kg/m^2 for secondary slab elements), would allow achieving some important objectives:

- A substantial reduction of overall costs
- A major improvement in the safety of the structure with respect to the seismic risk
- The possibility to construct buildings also on low bearing capacity soils.

The reduction in weight and carriage costs also allows to think about exportation of structural elements produced in an European factory to a foreign country. An opportunity has been recognized by the industrial proposes to realize these lightweight structural elements using glass fiber reinforced concrete (GFRC). Until now its main application has been in

Table 12.2: Typical properties of E- and S-glass

Typical properties	E-glass	S-glass
Density (g/cm^3)	2.60	2.50
Young's modulus (GPa)	72	87
Tensile strength (GPa)	1.72	2.53
Tensile elongation (%)	2.4	2.9

architectural cladding, in the form of panels 10–20 mm thick. Due to the tendency of GRC to age, or loose ductility with time, it never has been considered for structural applications. Although the estimated cost of such innovative, lightweight GRC components (i.e. 100–150 ECU/m^2 for main slab elements) is comparable or slightly higher than conventional ones, a reduction of overall costs of about 30% could be foreseen due to the following reasons:

- A slender carrying structure of the building (foundations, pillars, beams, etc.) is required.
- A complete automation of the whole production cycle is allowed.
- Carriage and assembly costs are minimal.
- The surface finishing (e.g. thermal insulation, waterproofing, painting, etc.) of precast elements can be carried out directly at the factory and no further operations are required on site.

Applications

Potential applications of structural GRC components are copious also limiting our consideration to the precast sector; some of these have already been identified by the proposers and represent an interesting market opportunity as brought out below:

- Main and secondary elements for slabs (e.g. in industrial shed or in building retrofitting).
- Façade panels, wall panels and roof elements (e.g. in stand alone technical shelter for mobile phone repeaters and in prefabricated houses).

The time-to-market for the above mentioned or similar applications is estimated to require at least one year after the end of the project, because it will be necessary to modify the current manufacturing plants for GRC production or to install new ones.

The lack of specific standards for designing and testing of these components could represent an obstacle to a wide acceptance of new products. To avoid this risk, one of the actions carried out by the consortium members in the exploitation phase will consist of a collaboration with national and international Standardization bodies in order to ensure the issuing of a specific normative in the shortest possible time.

12.15.2 Polypropylene

Polypropylene (PP) fibers have high tensile strength and low modulus of elasticity and they are dependent on the isotacticity molecular weight and other structural features of PP. The elastic properties would be influenced by the extent to which calcium hydroxide interacts with fiber at the interface and therefore dependent on the fiber type. The brittleness of the composite is probably also affected by the amount and size of the calcium hydroxide crystals present. Further, crystallization of the calcium hydroxide in the contact zone may actually result in a weakening of the bond between the fiber and matrix. Surface modification of PP can result in improving interfacial bond. Reinforcement of cement matrices with continuous fibers, fibrillated filaments film; tape or woven fabric generally results in increased flexural and tensile strength.

12.15.3 Natural Fibers

Some investigation have already been carried out on the use of natural fibers from coconut husk, sisal, sugarcane bagasse, bamboo, akara, plantain and musamba in cement paste, mortar, and concrete. These investigations have shown encouraging results. Flexural strength increases with fiber addition to a maximum and then decreases. The decrease in the strength for longer fibers was mainly due to bundling effect of the fibers. Impact strength depends on curing period and fiber volume fraction for both jute and their fiber reinforced concretes. Durability is of major importance in evaluating the suitability of natural fibers for inclusion in cement matrix. Coir fiber exhibits ductile failure characteristics while most of the other fibers exhibit brittle failure.

Coir also shows the greater resistance to alkali attacks. Resin coatings provide a reasonable measure of protection against alkali attack.

12.15.4 Aramid Fibers

These are synthetic organic fibers consisting of aromatic polyamides. The aramid fibers have excellent fatigue and creep resistance. Although there are several commercial grades of aramid fibers available, the two most common ones used in structural applications are Kevlar® 29 and Kevlar® 49. Typical properties of aramid fibers are shown in Table 12.3. The Young's modulus curve for Kevlar® 29 is linear to a value of 83 GPa but then becomes slightly concave upward to a value of 100 GPa at rupture, whereas for Kevlar® 49 the curve is linear to a value of 124 GPa at rupture. As an anisotropic material, its transverse and shear modulus are in order of magnitude less than those in the longitudinal direction. The fibers can have difficulty achieving a chemical or mechanical bond with the resin.

Experimentation was conducted on the use of a new aramid fiber for the reinforcement of Portland cement-based concrete and slurry. The study showed that synthetic fibers manufactured by cutting an epoxy-impregnated, braided bundle made of aramid filaments act similarly to steel fibers in reinforcing both Portland cement-based concrete and slurry matrices. The advantage of epoxy-coated aramid over steel is in the lack of corrosion problems, and over polypropylene in the higher performance.

12.15.5 Carbon Fibers

The graphite or carbon fiber is made from three types of polymer precursors, viz. polyacrylonitrile (PAN) fiber, rayon fiber, and pitch. The tensile stress–strain curve is linear at the point of rupture. Although there are many carbon fibers available on the open market, they can be arbitrarily divided into three grades as shown in Table 12.4. They have lower thermal expansion coefficients than both the glass and aramid fibers. The carbon fiber is an anisotropic material, and its transverse modulus is in order of magnitude less than its longitudinal modulus. The material has a very high fatigue and creep resistance.

Since its tensile strength decreases with increasing modulus, its strain at rupture will also are much lower. Because of the material brittleness at higher modulus, it becomes critical in joint and connection details, which can have high stress concentrations. As a result of this phenomenon, carbon composite laminates are more effective with adhesive bonding that eliminates mechanical fasteners. Widespread use of carbon fiber in cement has been limited due to cost consideration. Initially it was used in the pipe manufacturing only. Alternative uses are now being exploited as a result of the development of less expensive discontinuous fiber. Although discontinuous randomly distributed carbon fibers are less efficient than continuous aligned fibers, the properties of composites containing these carbon fibers are significantly improved. Tensile and flexural strengths increase with fiber content and they are generally less than for those with continuous fibers. At low water–cement ratios, the strengths are similar. Compressive strength of carbon fiber reinforced cements generally decreases with fiber addition.

Table 12.3: Typical properties of aramid fibers

Typical properties	Kevlar 29	Kevlar 49
Density (g/cm³)	1.44	1.44
Young's modulus (GPa)	83/100	124
Tensile strength (GPa)	2.27	2.27
Tensile elongation (%)	2.8	1.8

Table 12.4: Typical properties of carbon fibers

Typical properties	High strength	High modulus	Ultra-high modulus
Density (g/cm³)	1.8	1.9	2.0–2.1
Young's modulus (GPa)	230	370	520–620
Tensile strength (GPa)	2.48	1.79	1.03–1.31
Tensile elongation (%)	1.1	0.5	0.2

12.16 FIBER REINFORCED POLYMER COMPOSITES

Although the concept of fiber reinforced materials can be traced back to the use of straw as reinforcement in bricks manufactured by the Israelites in 800 BC and in more recent times to the use of short glass fiber reinforcement in cement in the United States in the early 1930s; fiber reinforced resin matrix materials or fiber reinforced composites is developed only in 1940s as advanced fiber reinforced polymers (FRPs). Because of the highly desirable lightweight, corrosion resistance, and high strength characteristics in composites, research emphasis went into improving the material science and manufacturing process. That effort led to the development of two new manufacturing techniques known as filament winding and pultrusion, which helped advance the composite technology into new markets. With the increasing demand for composites, new and improved manufacturing process, such as pultrusion, resin transfer molding, and filament winding were developed and implemented in the early 1990s. With this enhancement in place, the current focus is to rebuild the infrastructure using FRP composites for maintenance and rehabilitation of existing bridges as well as new construction.

There are three basic manufacturing techniques in producing composite structural products, with many variations and patented processes: (1) The pultrusion process involves a continuous pulling of the fiber rovings and mats through a resin bath and then into a heated die. The elevated temperature inside the die cures the composite matrix into a constant cross-section structural shape. (2) The filament winding process can be automated to wrap resin wetted fibers around a mandrel to produce circular or polygonal shape. (3) The lay-up process engages a hand or machine build-up of mats of fibers that are held together permanently by a resin system. This method enables numerous layers of different fiber orientations to be built up to a desired sheet thickness and product shape.

Characteristics of Composites

The mechanical properties of composites depend on many variables, such as fiber types, orientations, and architecture. The fiber architecture refers to the preformed textile configurations by braiding, knitting or weaving. Composites are anisotropic materials with their strength being different in any direction. Their stress–strain curves are linearly elastic to the point of failure by rupture. The polymeric resin in a composite material, which consists of viscous fluid and elastic solids, responds viscoelastically to applied loads. Although the viscoelastic material will creep and relax many excellent structural qualities and some examples are high strength, material toughness, fatigue endurance, and light-weight. Other highly desirable qualities are high resistance to elevated temperature, abrasion, corrosion, and chemical attack.

Some of the advantages in the use of composite structural members include the ease of manufacturing, fabrication, handling and erection. Project delivery time can be short. It took the Russell country engineer one day to install the deck panels in the first vehicular composite bridge. Composites can be formulated and designed for high performance, durability and extended service life. They have excellent strength-to-weight ratios. If durability can be proven to last 75 years, composites can be economically justified using the life cycle cost method.

Some of the disadvantages in the use of composites in bridges are high first cost, creep and shrinkage. The design and construction require highly trained specialists from many engineering and material science disciplines. The composites have a potential for environmental degradation, for example, alkali attack and ultraviolet radiation exposure. There are very little or nonexistent designing guidance and/or standards. There is a lack of joining and/or fastening technology. Because of the use of thin sections, there are concerns in global and local buckling. Although the lightweight feature may be an advantage in the response to earthquake loading, it could render the structure aerodynamically unstable. In manufacturing with the hand lay-up process, there is a concern about the consistency of the material properties.

As the name implies, advance fiber reinforced polymer composites is made of fiber reinforcements, resin, and additives. The fibers provide increased stiffness and tensile capacity. The resin offers high compressive strength and binds the fibers into a firm matrix. The fillers serve to reduce cost and shrinkage. The additives help to improve not only the mechanical and physical properties of the composites but also workability. The discussions that follow immediately will explain the basic functions and behaviors of the constituents.

12.16.1 Fiber Reinforcements

The fiber is an important constituent in composites. A great deal of research and development has been done with the fibers on the effects in the types, volume fraction, architecture, and orientations. The fiber generally occupies 30–70% of the matrix volume in the composites. The fibers can be chopped, woven, stitched and/or braided. They are usually treated with sizings, such as starch, gelatin, oil or wax to improve the bond as well as binders to improve the handling. The most common types of fibers used in advanced composites for structural applications are the fiberglass, aramid, and carbon. The fiberglass is the least expensive and carbon being the most expensive. The cost of aramid fibers is about the same as the lower grades of the carbon fiber. Other high-strength high-modulus fibers, such as boron are at the present time considered to be economically prohibitive.

12.16.2 Resin Systems

The resin is another important constituent in composites. The two classes of resins are the thermoplastics and thermosets. A thermoplastic resin remains a solid at room temperature. It melts when heated and solidifies when cooled. The long-chain polymers do not chemically cross link. Because they do not cure permanently, they are undesirable for structural application. Conversely, a thermosetting resin will cure permanently by irreversible cross linking at elevated temperatures. This characteristic makes the thermoset resin composites very desirable polyesters, epoxies, and vinyl esters; the least common ones are the polyurethanes and phenolic.

12.16.3 Fillers

Since resins are very expensive, it will not be cost effective to fill up the voids in a composite matrix purely with resins. Fillers are added to the resin matrix for controlling material cost and improving its mechanical and chemical properties. Some composites that are rich in resins can be subject to high shrinkage and creep and low tensile strength. Although these properties may be undesirable for structural applications, there may be a place for their use.

12.16.4 Additives

A variety of additives are used in the composites to improve the material properties, aesthetics, manufacturing process, and performance. The additives can be divided into three groups—catalyst, promoters, and inhibitors; coloring dyes; and releasing agents.

12.16.5 Fiber Reinforced Polymer (FRP) Composite Laminates

Fiber reinforced polymer (FRP) composite materials are produced from their main fiber types, namely carbon, glass and aramid. Each fiber has different engineering properties and therefore selection must be made to suit the requirements of a particular application. Carbon fibers have great strength. Stresses at failure can be in excess of 3000 N/mm^2. However, they are very expensive. Conversely, glass fibers are relatively inexpensive but have less strength and greater elasticity. Aramid fibers, renowned for their high impact resistance, tread the middle ground and may be yet to realize their true potential in structural engineering applications.

12.16.6 Fiber Reinforced Polymer Composites for Repair and Strengthening

Perhaps the most commercially successful use of FRP composite construction materials has been in the strengthening existing structures. In the lifetime of building, modifications are common. Changes in use often involve increases in loading or necessitate the introduction of alternative load paths. The solution in many cases has been to install FRP laminates or wraps as externally bonded reinforcement. The use of FRP materials for structural repair presents several advantages and has been recently investigated all over the world. An extensive experimental work has been carried out on strengthening of RC elements/beams by external plate bonding technique using carbon fiber reinforced polymer laminates and glass fiber reinforced polymer laminates.

12.16.7 FRP Laminates Technology and Benefits

The pultruded FRP laminate reinforcing consists of bonding the strip with the concrete structure using a high-strength epoxy resin as the adhesive. The strips are manufactured using a pultrusion process. The pultrusion principle is comparable with a continuous press. Normally 24,000 parallel filaments are pulled through the impregnated bath, formed into strips under heat, and hardened. These strips are unidirectional; the fibers are oriented only in the longitudinal direction. Correspondingly, the strip strength in this direction is proportional to the fiber strength and, thus very high. Strips are produced with strengths of approximately 3,000 MPa in the longitudinal direction, and with a thickness of up to 1.5 mm and widths of up to 150 mm.

In order to achieve an optimum composite action, the preparation of the bonding surfaces of the strip and concrete is critical. The strips must have the outermost layer of their bonding face, normally matrix-rich, removed to expose the fibers. Just before the bonding, the bonding surface is carefully cleaned with acetone. The concrete surface is treated by sand blasting, high-pressure water jets, stoking or grinding. Shortly before the bonding, it is cleaned with a vacuum cleaner. Concrete must be at least 6 weeks old, and have a minimum tensile strength of 1.5 MPa. Highly filled epoxy resin adhesive is used for the bonding.

FRP laminate reinforcing technology provides a solution for strengthening problems of concrete structures. It provides great strength, high modulus of elasticity and outstanding fatigue resistance it is a very lightweight noncorrosive material, that requires minimal preparation of laminates, and it is alkali resistant. It is an economic method that requires very short contract times.

12.17 FERROCEMENT

As per ferrocement modal code, ferrocement is a type of reinforced concrete commonly constructed of hydraulic cement mortar reinforced with closely spaced layers of relatively small wire diameter mesh. The mesh may be made of metallic or other suitable materials. The fitness of the mortar matrix and its composition should be compatible with the opening and tightness of the reinforcing system it is meant to encapsulate. The matrix may contain discontinuous fibers.

12.17.1 Materials Used

Cement

The cement should comply with ASTM C150–859, ASTM C595–85, or an equivalent standard. The cement should be fresh, of uniform consistency and free of lumps and foreign matter. It should be stored under dry conditions for as short duration as possible.

Fine Aggregate

Normal weight fine aggregate used in ferrocement, should comply with ASTM C33–86 requirements or an equivalent standard. It should be clear, inert, free of organic matter and deleterious substances, and relatively free of silt and clay.

Mixing Water

The mixing water should be fresh, clean and potable. The water should be relatively free from organic matter, silt, oil, sugar, chloride and acidic material. Salt water is not acceptable, but chlorinated drinking water can be used.

Admixtures

Conventional and high-range water reducing admixtures should be used.

Reinforcement

Wire mesh with closely spaced wires is the most commonly used reinforcement in ferrocement. Expanded metal mesh, welded wire fabrics, square steel wire mesh, hexagonal or chicken wire mesh are used. The various types of reinforcing are shown in Fig. 12.4. Properties and types of constituent materials used in ferrocement construction are shown in Table 12.5.

12.17.2 Ferrocement as a Repair Material

The major objective of repair or renovation work for concrete is to return the concrete to a satisfactory level of performance at a reasonable cost. For many repairs and renovation programs of civil engineering structures cite the suitability of ferrocement because of the following:

- Better cracking behavior

Table 12.5: Guidelines for material selection

Materials	Range
Wire mesh	
Diameter of wire Φ	$0.5 \leq \Phi \leq 1.5$ mm
Types of mesh	Chicken wire or square woven or welded galvanized mesh or expanded metal
Size of mesh opening (S)	$6 \leq S \leq 25$ mm
Volume fraction (V_r) of reinforcement in both direction	$2\% \leq V_r \leq 8\%$
Specific surface (S_r) of reinforcement in both direction	$0.1 \leq S_r \leq 0.4$ mm^2/mm^3
Elastic modulus (E_r)	140–200 kN/mm^2
Yield strength (F_y)	250–460 N/mm^2
Ultimate tensile strength (F_u)	400–600 N/mm^2
Skeletal steel	
Type	Welded mesh, steel bars, strands
Diameter (d)	3 mm $\leq d \leq 10$ mm
Grid size (G)	50 mm $\leq G \leq 200$ mm
Yield strength	250–460 N/mm^2
Ultimate tensile strength	400–600 N/mm^2
Mortar composition	
Cement	Any type of Portland cement (depending on application)
Sand/cement ratio (s/c)	$1 \leq s/c \leq 3$ by weight
Water/cement ratio (w/c)	$0.35 \leq w/c \leq 0.65$ by weight
Gradation of sand	5 mm to dust with not more than 10%* passing 150 µm BS test sieve
Compressive strength (cube)	30–50 N/mm^2

*For crushed stone, sand may be increased to 20%.

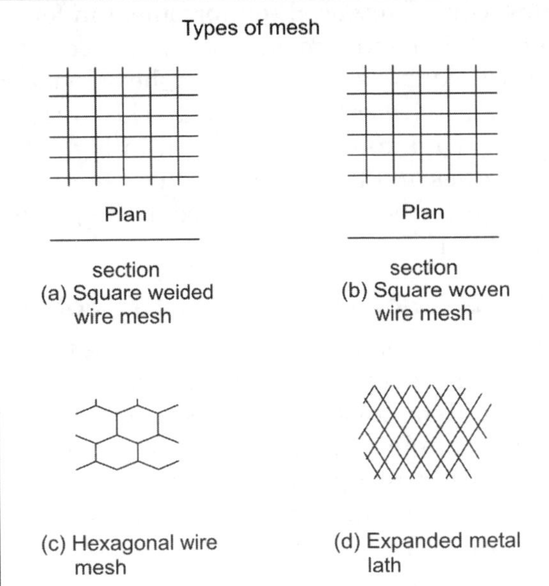

Fig. 12.4: Typical sections of wire mesh

- Capability of improving some of the mechanical properties of the treated structures
- Further modification or repair of ferrocement treatment is not difficult
- Imposition of little additional dead load requiring no adjustment of the supporting structures
- Ability to withstand thermal changes very efficiently
- Ability of achieving water proofing property without providing any surface treatment
- Readily available constituent materials
- No need for special equipment
- Ability to be used in repair program with no distortion or downgrading of the architectural concept of the structures
- Flexibility of further modification

Ferrocement has been used extensively in the area of home application for the rehabilitation of distressed elements due to overloading, fire damaged beams and columns and general repair in the deteriorating structure. The ferrocement material is ideally suitable due to its ability to arrest crack and high tensile strength-to-weight ratio.

Many infrastructure deterioration problems and failures can be traced back to the cracking and brittle nature of concrete. It is no wonder that significant research efforts have gone into attempt at enhancing the ductility of concrete is by means of fiber reinforcement. Most fiber reinforced concrete results in enhanced post peak tension-softening behavior under tensile load. This improved toughness is useful in resisting the propagation of cracks into major fractures by energy absorption in the bridging actions of fibers in the fracture process. While the fracture toughness of concrete can increase by an order of magnitude by fiber reinforcement, the tensile strain capacity usually remains little changed. In recent years, efforts to convert this quasi brittle behavior of fiber reinforced concrete (FRC) to ductile strain-hardening behavior resembling ductile metal have met with varying degrees of success. In most instances, the approach is to increase the fiber content as much as possible and processing techniques are adapted to overcome the resulting workability problem. Cement-based composites, such as slurry infiltrated fiber concrete (SIFCON) and slurry infiltrated mat concrete (SIMCON) are good examples of this approach, which utilizes from 5 to 20% volume content of steel fibers.

12.18.1 Sifcon

Slurry infiltrated fiber concrete is a combination of cement, aggregate, fiber and water with some admixture. SIFCON is a recently developed construction materials using steel fibers with cement matrix. The matrix consists of cement mortar slurry. The fiber bed is infiltrated with cement based slurry by gravity flow. Hence, the optimization of the flow of matrix of cement slurry is as important step in the design of its composition. Slurry infiltrated fibrous concrete (SIFCON) can be considered as a special type of fiber concrete with high fiber content. The matrix usually consists of cement slurry or flowing mortar. SIFCON has excellent potential for application in areas where high ductility and resistance to impact are needed. Only very limited information is available about its behavior under different types of loading. Tests on 22 mm thick SIFCON specimens were carried out to study their behavior in flexure and under subjection to abrasion and impact loads. Toughness characteristics of SIFCON were also evaluated by testing another set of specimens 100 mm × 100 mm × 500 mm per ASTM C1018. Both strength and deformation characteristics of the specimens were studied. The results obtained from these tests are compared with those carried out on companion plain mortar and conventional fiber reinforced mortar (FRM) specimens.

The investigations confirm the superior characteristics of SIFCON as compared with plain mortar and normal FRM. SIFCON (slurry infiltrated fiber concrete) have number of outstanding properties, such as strength (compression, tension, bending, and shear), ductility, toughness, durability, stiffness and energy absorption capacity under monotonic and cyclic loads. The properties of SIFCON are achieved through an optimized combination of matrix properties, fiber reinforcing parameters, fiber content and interface characteristics between fiber and matrix. It is observed that the presence of SIFCON in over reinforced concrete beams leads to ductility indices exceeding three times those obtained without it. Crack widths and spacing are more than an order of magnitude smaller than in conventional reinforced concrete. Experimental results also suggest that there is no need for stirrups in flexural members with a SIFCON matrix.

12.18.2 Simcon

The cost of civil infrastructure constitutes a major portion of the national wealth. Its rapid deterioration has thus created an urgent need for the development of novel, long-lasting and cost-effective methods for repair, retrofit and new construction. A promising new way of resolving this problem is to selectively use advanced composites, such as high performance fiber reinforced cementitious composites (HPFRCCs). With such materials, novel repair, retrofit and new construction approaches can be developed that would lead to substantially higher strengths, seismic resistance, ductility, durability, while also being faster and more cost-effective to construct than conventional methods. SIMCON which stands for slurry infiltrated mat concrete is well suited for the development of novel repair, retrofit and new construction solutions that lead to economical and improved structural performance.

SIMCON (slurry infiltrated mat concrete) is a nonwoven steel fiber mat that is infiltrated with concrete slurry. It is shown that the SIMCON concept of reinforcement represents a significant improvement in the reinforcement efficiency in high density, high strength fiber reinforced concrete. Comparable levels of flexural strength and flexural energy absorption capacity can be achieved at greatly reduced fiber loadings relative to the short, discrete fiber reinforcement approach. SIMCON uses a manufactured continuous mat of interlocking discontinuous steel fibers, placed in a form, and then infiltrated with flowable cement-based slurry. The use of continuous mats, typically made with stainless steel to control corrosion in very thin members, permits development of high flexural strengths and very high ductility with a record volume of fibers.

12.19 MISCELLANEOUS MATERIALS

It is imperative that materials be tested and evaluated, prior to use in the repair of abrasion-erosion damaged hydraulic structures. Abrasion-resistant concrete should include the maximum amount of the hardest coarse aggregate that is available and the lowest practical w/c. In some cases where hard aggregate was not available, high-range water-reducing admixtures (HRWRA) and condensed silica fume have been used to develop high compressive strength concrete 97 MPa to overcome problems of unsatisfactory aggregate. Apparently, at these high compressive strengths the hardened cement paste assumes a greater role in resisting abrasion-erosion damage, and as such, the aggregate quality becomes correspondingly less important. The abrasion-erosion resistance of vacuum treated concrete, polymer concrete, polymer impregnated concrete and polymer Portland cement concrete is significantly superior to that of comparable conventional concrete that can also be attributed to a stronger cement matrix.

The increased costs associated with materials, production and placing of these and any other special concretes in comparison with conventional concrete should be considered during the evaluation process. While the addition of steel fibers would be expected to increase the impact resistance of concrete, fiber reinforced concrete is consistently less resistant to abrasion-erosion than conventional concrete. Therefore, fiber reinforced concrete should not be used for repair of stilling basins or other hydraulic structures where abrasion-erosion is of major concern. Several types of surface coatings have exhibited good abrasion erosion resistance during laboratory tests. These include polyurethanes, epoxy resin mortar, furan resin mortar, acrylic mortar, and iron aggregate toppings. However, some difficulties have been reported in field applications of surface coatings, primarily the result of improper surface preparation and thermal incompatibility between coatings and concrete.

Conventional materials: While proper material selection can increase the cavitation resistance of concrete, the only totally effective solution is to reduce or eliminate the causes of cavitations. However, it is recognized that in the case of existing structures in need of repair, the reduction or elimination of cavitations may be difficult and costly. The next best solution is to replace the damaged concrete with more cavitation resistant materials. Cavitation resistance of concrete can be increased by use of a properly designed low w/c, high-strength concrete. Vital to increased cavitation resistance are the use of hard, dense aggregate particles and a good aggregate to mortar bond. Typically, cement-based materials exhibit significantly lower resistance to cavitations compared to polymer-based materials.

Other cavitations resistant materials: Cavitations damaged areas have been successfully repaired with steel fiber concrete and polymer concrete. Some coatings, such as neoprene and polyurethane, have reduced cavitation damage to concrete, but since near perfect adhesion to the concrete are critical, the use of the coatings is not common. Once a tear or a chip in the coating occurs, the entire coating is likely to be peeled off.

12.20 REQUIREMENTS FOR THE REPAIR MATERIALS

The repair materials should fulfill the following requirements:
- It must be thoroughly bonded to concrete.
- Shrinkage of the material should be small enough not to jeopardize the bond.
- The material and old concrete should respond to changes in temperature, moisture and load similarly enough to avoid gross differences in movement.
- The permeability of the material should be sufficiently low.
- The material should be resistant to weather action.
- In addition to above requirements, it is sometimes necessary to match the colour and texture of the old concrete.

13

Selection and Evaluation of Materials

13.1 INTRODUCTION

Deterioration of any concrete structure is the natural phenomenon of gradual degradation of constituent materials brought about by physical, chemical or mechanical processes. Occasionally, natural/man-made calamities and accidents accentuate this process. As a consequence, the ability of a structure to perform the intended function is impaired and there arises a need to repair or upgrade a structure to restore the designed level of serviceability or even enhance it. Experience over the last 40–50 years has shown that the current design specifications, even if properly implemented, cannot ensure adequate protection of reinforced concrete in aggressive environments. It is everybody's experience that premature deterioration of concrete is still unacceptively high and corrosion of reinforcement is one of the most dreaded causes of failure.

Repair and rehabilitation have therefore become established facets of construction industry. They have become the preferred alternative compared to demolition and rebuilding of structures due to social, economic and psychological factors. The cost of materials constitutes a substantial portion of the repair and refurbishment cost. Therefore, there is an ever increasing surge of new products formulated to provide a variety of properties in order to meet the market demand. Since a great majority of these products are proprietary in nature with very little information on their composition and performance, the user is confounded with the problem of selecting the appropriate repair material for a particular job.

13.1.1 Development in Repairs

Repair programme is a job of an expert from the beginning and it should be carried out in a planned way. At present there are no widely accepted guidelines for design, specifications and execution of repair projects. Unfortunately in India whilst a large number of structures are heading for repairs, there is no coordination between diagnosis of repairs, understanding and specifications of consultant and manufacturer's product specifications. To top it all, the repair person is not knowledgeable about different chemicals available. This is resulting into incorrect repairs and a time has come to start repairs to re-repairs.

13.2 SELECTION OF REPAIR MATERIALS

Selecting repair materials is an important and complex process involving an understanding of requirement of the repair by the owner, engineer, service and exposure conditions, and also installation technique. After requirements are established and material properties are defined, the selection of specific materials can be made. Final selection of material is made based on the relationship between cost, performance and risk. Selecting repair materials requires an understanding of material behavior in the cured and uncured states in the anticipated service and exposure conditions.

One of the greatest challenges facing the successful performance of repair materials is their dimensional behavior relative to the substrate. Relative dimensional changes cause internal stresses within the repair material and within the substrate. High internal stresses may result in tension cracks, loss of load carrying capability, delamination or deterioration.

Another challenge is selecting repair materials for structural applications. There are two obstacles to achieve 100% repair efficiency in structural applications. First, initial loading on repair material and removal of load from the structure during repair. Next, the dimensional behavior of repair materials on substrate. These are based on load subjected, temperature and moisture change. For the selection of materials of repair it is necessary to understand the properties of various materials and their limitations. The selection of material also depends upon the basic nature of the structure to be repaired. Selection of the repair material thus depends on the following:

- Type of concrete to be repaired and its age
- Magnitude and thickness of repairs
- Site conditions
- Temperature
- Environmental influence.

13.2.1 Criteria for Selection of Repair Materials

A careful selection of repair material is necessary for the following reasons:

- Almost every repair job has unique conditions and special requirements.

- The composition and properties of repair materials have a profound effect on the performance and durability of a repair.
- The repair materials perform adequately only if they are prepared, applied and cured as per the specified procedures, which may necessiate the use of appropriate tools and considerable skill.
- Repair materials, being generally proprietary in nature, are very costly.

The following factors have to be considered in selecting a repair material:

- Function of repair material
- Physical and mechanical properties
  - Dimensional stability
  - Coefficient of thermal expansion
  - Strength in compression, tension and flexure
  - Stiffness, elastic modulus and Poisson's ratio
- Durability related properties
  - Permeability to liquids, vapours and gases
  - Diffusion resistance
  - Water absorption, sorptivity and porosity
  - Weathering resistance
  - Resistance to chemicals
- Chemical computability with other repair materials
- Identification of anticipated service conditions
- Conditions prevailing at the time of repair.

13.2.2 Methodology for the Selection of Repair Materials

The basic emphasis of the selection methodology is to maximize the performance of a repair and ensure durability. The following procedure is adopted for selection of repair materials.

- *Definition of service condition:* Identification of major and minor chemicals, traces, spillages, cleaning chemicals, slurries and abrasives, their characteristics and their interaction with the environment under stagnant conditions, operating temperature range, dilute conditions, alternate wetting and drying effect.

- *Determination of appropriate application condition:* Viscosity, flow characteristics, pot life, curing requirements, coverage, film/layer thickness, and size of repair.
- *Tools and equipment required:* Pump, sprayer, injecting/grouting machine, dryer, mixer, batching plant, heater, etc.
- *Repair, rehabilitation, retrofitting and maintenance schedule:* Anticipated durability of repair, maintenance requirements, replacement/renewal/strengthening requirements.
- *Product performance record/history:* Durability, functionality, environment friendliness.
- *Material testing and assessment for quality assurance and quality control.*
- *Selection of applicator/contractor.*
- *Material and job specifications.*

13.3 CLASSIFICATION OF REPAIR MATERIALS

The materials used for concrete repairs are divided into following categories on the basis of the type of application:

Patch repair materials
- Cementitious mortar/concrete
- Polymer modified cementitious mortar/concrete
- Polymer mortar/concrete
- Quick setting compounds
  - High alumina cement based
  - Calcium sulfate based
  - Magnesium phosphates
  - Sulfur concrete.

Injection grouts
- Cementitious grouts (with or without fibers)
- Gas forming grouts
- Sulfoaluminate grouts
- Polymer grouts.

Bonding aids
- Polymer emulsion type
- Polymer resin type.

Resurfacing materials
- Protective coatings and membranes
- Impregnates and hydrophobic sealers
- Toppings/screeds
- Overlays
- Gunite/shotcrete.

Other repair materials
- Corrosion inhibitors
- Rebar protective coatings
- Cathodic protection
- Realkalization.

13.4 EVALUATION TESTS FOR REPAIR MATERIALS

Presently the construction industry is flooded with large number of materials with claims which often need to be verified. If not properly evaluated, the repair materials may fail during service life due to severity of exposure or incompatibility with substrate or detrimental interaction between the constituents. The best approach to judge the suitability of a repair material for a particular situation is to evaluate the performance of the material under verifiable conditions, which are identical to the expected service history. Generally, the evaluation tests involve the determination of one or more measurable parameters, which bear a fundamental relationship with the resistance and performance of the repair material under given exposure conditions. The tests may be destructive or nondestructive and many times partly destructive in nature. It may often become necessary to resort to accelerated testing to get a rapid assessment of the suitability of repair materials when several varieties of materials are available.

This may require the intensification/magnification of the magnitude of the most potentially destructive parameter. However, such tests may not yield a quantitative measure of the prospective service life of a material under actual service conditions. But, they are useful tools for the selection of materials and deciding on the construction features, such as curing requirements, techniques for finishing, curing and protection of surface treatment. There are a large number of field/laboratory tests developed for cementitious, polymeric and polymer-cement composites. The choice of a particular set of tests for material selection and quality control depends mainly on the magnitude of work and its importance, besides the availability of test facilities and time constraints. Therefore, no generalization can be made as regard to the choice of a particular set of tests for quality control purposes.

13.4.1 Test for Compressive Strength and Elastic Modulus

This is the single most important property for quick quality control checks for all repair materials based on cementitious composites and is important for polymer bound systems. The usual practice is to specify the compressive strengths at the end of 1, 3, 7 and 28 days of curing period for cementitious systems. This is determined by static destructive tests on 50 mm or 75 mm size cube specimens in case of mortars and 100 mm or 150 mm cube or cylinder specimens in case of concretes. For finding the compressive strength of resin systems, BS:2728-1970 and BS:6319-1983, Part-2 recommend the use of 12.5 mm cubes or 9.5 mm diameter × 9.5 mm height cylinder specimens while ASTM D695 specifies 12.5 mm × 25 mm cylinders. For polymer mortar, ASTM C579-91 and IS:9612-1979 suggests the use of 50 mm cubes or 25 mm diameter × 25 mm height cylinder specimens. In case of polymer concrete, cylinder specimens of minimum diameter equal to three times the maximum size of aggregate and height of two times the diameter are to be used.

The resin specimens are to be tested so as to complete the test in about 1 to 2 minutes. ASTM C579-91 specifies that the specimens have to be tested at a load rate of 42 MPa per minute or at a cross head speed of 2.5–3.2 times

$h/25$, where h is the height of the specimen in mm. Before testing, the specimens are to be conditioned at temperature and humidity conditioned at specified temperature conditions (25–27°C) for specific duration (up to 7 days). JIS A6024-1992 suggests the use of 13.8 mm × 13.8 mm × 40 mm prisms for determining the compressive systems and resin mortars/concretes, the procedure recommended by IS:516 or BS:1881 or ASTM C39/C109/C579 may be followed. JIS A6203-1986 recommends the use of broken test piece obtained from flexural tests on 40 mm × 40 mm × 160 mm prisms for determination of compressive strength of polymer modified mortars. For determining the elastic modulus in compression, cylindrical specimens with aspect ratio of 2 may be preferred. Generally the specimens are loaded to 3–40% of their compressive strength and initial tangent/chord or secant modulus obtained from the stress–strain curve.

13.4.2 Tensile Strength and Elongation

This is a very important property for all repair materials as patch repairs and overlays are often required on the tension faces of concrete beams. For testing pure resin systems, ASTM D638 and BS:2782 recommends dumb-bell shape specimens, 150 mm long × 20 mm wide and 3 mm thick. The straight portion of the specimen should be 60 mm long × 10 mm wide, with a recommended gauge length of 50 mm for strain measurement. The type of material and temperature dictates the speed of testing. For thermosetting resins like epoxies, BS: 2728, BS:6319, Part 7-1981 recommend test duration of 0.5 to 1.5 minutes. IS:9162 recommends 44.5 mm × 76 mm × 25 mm briquettes with a central neck of 25 mm cross-section in case of epoxy resin compositions. For tensile modulus, either a secant or chord modulus obtained from the stress–strain plot is used.

For cementitious systems and resin mortars/concretes ASTM D190 recommends, direct tensile strength for briquette type specimens 25 mm × 25 mm in cross-section or split tensile strength recommended by ASTM C496, IS:516, BS:1881, determined from cylindrical specimens may be used. Tensile elongation at failure is often specified for pure resin systems as one of the requirements and can be measured as per ASTM D638. ASTM D882 specifies a minimum elongation of 1% and 30% respectively for bonding materials used for general and skid resistant overlay applications.

13.4.3 Flexural Strength and Flexural Modulus

This is another set of useful measure of relative strength and stiffness of repair materials. For pure resin systems, BS:2728 and ASTM D790 recommend prisms of size 12.5 mm × 9.5 mm × 112.5 mm to be tested over a span of 100 mm under three point bending. IS:9162 and ASTM C580 recommend 25 mm × 25 mm × 250 mm prisms to be tested for flexural strength of resin mortar under center-point loading at the rate of 4.2 MPa/minute over a span of 229 mm. For polymer concrete, specimens of size 50 mm × 50 mm × 300–400 mm are recommended by ASTM C580 for small size aggregates. The flexural modulus is determined from the initial straight portion of a load-deflection curve obtained by three point bending test on specimens loaded at the rate of mm/minute. For cementitious systems and resin bound systems, JIS A6203, JIS A6024 and ASTM C348 suggest the use of 40 mm × 40 mm × 160 mm specimens. IS:516 and BS:1881 give the procedure for determination of flexural modulus. ASTM C580 defines two types of flexural modulus, which are obtained from the load deflection characteristics. The initial tangent modulus is obtained from the slope of the curve at zero deflection whereas secant modulus is obtained from the slope of the chord connecting the origin to 50% of the maximum deflection.

13.4.4 Properties of Raw Materials and Green Mixes

For cementitious systems and polymer modified cementitious systems, general sampling procedures laid down for fresh concrete (ASTM C172) may be used. The minimum numbers of specimens to be tested per batch are generally laid down in most of the specifications and ranges from three to five in most of the cases. The important tests, which are often required to be conducted on fresh cementitious mixes are:

- Slump or flow test for workability
- Unit weight and air content
- Sieve analysis
- Specific gravity and water absorption for coarse aggregates and fine aggregates
- Organic impurities in fine aggregates.

The important properties to be tested in case of polymeric latexes and resin systems include the following:

Specific solids content, drying time, viscosity, consistency, thixotrophic index, gel time, dry film thickness, pH, glass transition temperature (T_g) coverage, minimum film forming temperature (MFFT), slump and sag.

Specific gravity: Specific gravity of powders and resins is easily obtained by specific gravity bottle/pycnometer method or density gradient method.

Solids content: The nonvolatile solids content of a resin solution is determined by finding the loss of weight on heating the resin solution, taken in a small petri-dish, in an oven and drying off the volatile compounds. The recommended drying time ranges from one hour for phenolics, two hours for emulsions to three hours for unmodified epoxies. This is an important test for emulsions and superplasticizers. For rubber latex, determination of dry rubber content is often necessary. This requires the coagulation of latexes by adding acetic acid, washing the coagulum with water, drying it at a temperature of about $70 \pm 2°C$, cooling in a desiccator and taking the dry weight. Tests are also specified for epoxy and amine content of epoxy resin compositions.

Drying time: This is an important test for coatings and surface treatments. This is usually determined by the touch dry test in which the coating is applied on a glass plate kept inclined at 45° to the vertical and shielded from air currents and direct sunlight. After the completion of specified time, the coated panel is placed horizontally, sprinkled with about 0.5 g of sand from a height of about 50–150 mm and brushed in an inclined position. After 10 seconds, if all the sand can be brushed without damaging the surface, the coating is surface dry. It is considered to have dried hard when the coating film does not move when pressed with a finger and no impression is noticeable on the coated surface after wiping with a soft cloth.

Viscosity: Viscosity of resins and latexes is a very important property as it is related to the molecular weight and to its solvent strength. The techniques commonly employed for viscosity measurements are:

- Bubble tubes for clear liquids and curing agents
- Ostwald (U tube) viscometer for liquid resins
- Falling sphere viscometer (for high viscosity resins).

Consistency test: Consistency test is specified by ASTM C881 for epoxy systems. It is defined as the flow of the lower edge of a semicylindrical resin bead along its length measured from its original position to the nearest 3 mm. Maximum consistency specified for epoxies is 6.4 mm at the temperature corresponding to the service condition.

Thixotropic index: Thixotropic is the typical property of viscous liquids to show an apparent increase in viscosity in stationary condition.

Gel time/working time: Gel time is the interval of time, in connection with the use of synthetic thermosetting resins, which extends from the introduction of a catalyst/harden into a liquid resin until the start of formation of a soft, gelatinous mass. At this time viscosity increases to a point where it barely moves when probed with a sharp instrument. A shop iron wire terminating into a loop may be used

for agitating the resin taken in a test tube. The end point is indicated by the cessation of the motion of test tube relative to the wire agitator. In this collection, the wire agitator can lift the tube filled with gelatinous resin. The minimum specified gel time is generally 30 minutes.

Dry-film thickness (DFT): This is another important property for coatings. A coat-meter based on magnetic flux principle or electromagnetic induction can be used for this purpose. For low thickness coatings, adhesive tape method is recommended. In this approach, an adhesive tape is placed on the substrate and coated. It is subsequently removed and DFT is obtained by weighing or thickness measurement. Graduated metal wedge scales, which can measure the wet film thickness, are also available.

pH: The pH of latexes measured using a pH electrode is a convenient test for latex quality.

Glass transition temperature (T_g): Glass transition temperature corresponds to the temperature at which a polymeric material changes from relatively hard, elastic, glass-like substance to a relatively viscous rubbery material. Classical methods for determining the T_g include methods based on dimensional change (thermal expansion), optical methods (based on refractive index), electrical methods (based on dielectric properties like loss factor), differential thermal analysis (DTA) and nuclear magnetic resonance (NMR) techniques. However, for civil engineering purposes, a more simple measure of temperature susceptibility of resin systems called heat deflection/distortion temperature (HDT) is commonly used.

Coverage or wetting power: A checker board is often conveniently used for the test. The number of square meters of a checker board having alternate black and white squares which can be satisfactorily obliterated to remove contrast between black and white squares by a fixed quantity of the coating is determined as a measure of coverage or wetting power. Alternatively, a calibrated metal wedge called (Pfund cryptometer) can be used to check for wet film thickness, which in turn can be expressed in terms of coverage.

Minimum film forming temperature (MFFT): For polymer dispersions the procedure specified in ASTM D2354 is used. The ambient temperature at the site should not be less than MMFT to ensure film formation in case of polymer dispersions.

Sag flow test: Sag test is intended for evaluation of the tendency of a coating to sag when applied and allowed to dry on a vertical nonabsorbing substrate at a specified film thickness. An antisag meter in conjunction with draw down charts is required for this test.

13.4.5 Coefficient of Thermal Expansion

This is an important measure of the thermal susceptibility of a repair material. ASTM D696 and BS:4618 specify the measurement of coefficient of thermal expansion by a quartz tube dilatometer 500 mm long. The coefficient of expansion is obtained by measuring the fractional change in length per degree temperature change when the specimen temperature is raised from $-30°C$ to $+30°C$ using a dial guage of least count 0.0025 mm. For volume expansion, a pyrex glass tube dilatometer is used to test a bar specimen of size 6 mm × 6 mm × 150 mm with rounded edges. Frequently, direct measurement techniques using a travelling microscope or optical methods are used. In these cases, the relative movement of a mark ruled on the specimen in the form of a strip is observed against an index mark on a steel reference bar, the specimen being placed in a temperature controlled oven/water bath. ASTM C531-91 recommends the use of 25 mm × 25 mm × 250 mm specimens for resin mortars.

13.4.6 Shrinkage

Plastic shrinkage of resin binders is a very important property, which determines the performance of a repair system. ASTM D955 and BS:2728 require the measurement of diameter of the cold moulded specimen using 50–125 mm discs of 3 mm thickness. ACI committee 548 recommends two methods, viz. du Pont method and Ohama-Demura method for measuring the plastic/curing shrinkage of PC overlay materials, which can yield a continuous record of shrinkage strain with time. In the former case, a teflon mould with inside dimension of 75 mm × 300 mm is used and the strain is measured with a displacement transducer by sensing the displacement of a movable prong with reference to a fixed prong. In the latter method, the material is cast in a similar teflon mould but the displacement is measured by sensing the movement of 25 mm diameter, brass expansion plugs placed at the ends of the mould using direct current displacement transducers. A very simple test procedure often used for evaluating the shrinkage of repair mortars is the coutino ring test. In this test, the repair mortar is cast to a uniform thickness of 50 mm around the circumference of a 112.5 mm diameter × 65 mm wide steel pipe and exposing it to 50% RH and recording the time of cracking.

13.4.7 Adhesive/Bond strength

ASTM C882-91 specifies a slant shear test for determining the bond strength. In this method, the bond strength is obtained by testing a cylinder 75 mm diameter × 150 mm height formed by bonding together two replicate halves of Portland cement mortar, each of which has a diagonally cast bonding area at 30° inclination to the vertical. After suitable curing of the bonded specimen, the integrated cylindrical specimen is tested in compression and the failure load divided by the bonded area gives the bond strength.

For latex modified cement concrete overlays, ACI committee 548 recommends the use of a pull out test as laid down in ASTM C900. This test involves embedding an insert with a head of 25–30 mm diameter at the time of casting to a depth equal to the diameter. The shaft of the insert is about 15 mm in diameter. A pull out apparatus bearing around the inserts and secured firmly to the insert is used to rupture the surface within 2 to 2.5 minutes. The pull out load divided by the vertical component of the fractured area gives the pull out strength.

JIS A6203 requires the coring of a 40 mm × 40 mm area through the polymer mortar overlay so as to reach the concrete substrate. A hexagonal head is fixed to the core surface using a high strength epoxy adhesive. The bolt head is pulled out at the rate of about 2.7 kN/min. The pull out load is the measure of the bond strength. ASTM C327-94 and IS:9162 suggest a cross brick push off test for resin mortars. The bricks should have a flexural strength of at least 6.9 MPa. An instrument called elcometer adhesion tester is used for evaluating the adhesion of coatings to concrete substrates. For epoxy injection grouts, JIS A6024 recommends a flexural bond test in which two replicate prisms of dimensions 40 mm × 40 mm × 80 mm are bonded together and then tested under two pint loading with a shear span of 40 mm at a speed of 50 kN/m.

13.4.8 Durability Related Tests

Thermal compatibility test: This test specified in ASTM C884-92 consists of applying the repair material to a 300 mm × 300 mm × 75 mm concrete block to a uniform thickness of 12.5 mm. The repair mortar is cured for 7 days at 23 ± 1°C. At the end of 7 days, the specimen is placed at 0°C for 24 hours in a deep freezer followed by 24 hours storage at room temperature. Four such cycles are completed. If any delamination of overlay or horizontal cracking of concrete at the interface is observed, it is interrupted as failure of the epoxy system.

Water absorption: For resin binders ASTM D570 specifies 50 mm dia. × 3 mm thick disk specimens to be used for water absorption tests. The specimens are conditioned at 110°C for 24 hours and then cooled in a desiccator and immediately weighed. The specimen is immersed on edge in distilled water at 23 ± 1°C for 24 hours and then weighed after surface drying. The change in mass gives the water absorption of the resin system. This procedure is also recommended ASTM C884 for epoxy resin adhesives. IS: 9162 and ASTM C413-94 suggest that 25 mm dia. × 25 mm long cylindrical specimens may be dried at 105–110°C, cooled and then soaked in boiling water for 2 hours to determine water absorption of polymer mortar/concrete compositions.

For polymer modified mortars, JIS A5203-1980 specifies the use of 40 mm × 40 mm × 160 mm specimens. After casting, the specimens are subjected to a 28-day curing cycle involving 2 days moist curing, five days water curing. The specimen is then dried to attain constant mass at 80 ± 2°C for 48 hours. The change in mass gives the water absorption. For purely cementitious systems, BS:1881-Part 5 1970 requires a 75 mm dia. specimen with thickness less than 150 mm and greater than 32 mm to be oven dried at 105 ± 5°C for 3 days, cooled for 24 hours, weighed and then soaked in water for 25 minutes. The change in mass gives the water absorption. Soaking in boiling water for 5 hours is recommended for determination of void content. For chemically resistant resin mortars/concretes, ASTM C413-94 requires 25 mm × 25 mm cylinder specimens to be kept in boiling water for 2 hours, cooling the cold water and observing the change in mass for obtaining the water absorption.

Water permeation: This type of test is suitable for polymer modified system, coated specimens and cementitious systems. It may be also used for polymer concrete specimens. As applied to polymer modified mortars, JIS A6203 specifies cylindrical disks of 150 mm dia. × 40 mm height prepared and cured as specified for water absorption test. The specimens are then dried to constant mass at a temperature of 80 ± 2°C, brushed lightly at the centre, cleaned and weighed. The specimen is then subjected to a hydraulic pressure of 1 kg/cm^2 for one hour. The change in weight of the specimen gives the amount of water permeation.

Carbonation resistance: In this test, 40 mm × 40 mm × 160 mm prisms coated with epoxy resin on the top and bottom faces, are immersed in 2.5% NaCl solution and 20°C for 7 days. At the end of this period, the specimens are split and the exposed faces sprayed with 1% phenolphthalein alcohol. The depth of colourless portion of the rim measured with a slide calipers gives the depth of carbonation.

Chloride penetration and chloride ion diffusion: In this test, 40 mm × 40 mm × 160 mm prisms coated with epoxy resin on the top and bottom faces, are immersed in 2.5% NaCl solution and 20°C for 7 days. At the end of this period, the specimens are split open and sprayed with 0.1% sodium fluorescein and 0.1 N silver nitrate solutions. The depth of colourless rim is measured using a slide calipers to obtain the chloride ion penetration depth. By periodic observation of chloride penetration depth, the diffusion coefficient can be computed.

ASTM C1202-94 suggests an accelerated chloride diffusion test. This test consists of monitoring the amount of electrical charge passed through a vacuum saturated cylindrical disk of 100 mm diameter × 50 mm thickness, over a six hour period under a potential difference of 60 V DC. One end of the specimen is immersed in 3% NaCl solution while the other end is immersed 0.3 N NaOH solution. The total charge passed in six hours is a measure of chloride diffusion resistance. A total charge exceeding 4000 coulombs indicates high permeability.

Penetration depth of impregnates: To assess the depth of penetration of hydrophobic treatments like silane-siloxanes and high molecular weight methacrylates, cube samples treated with them may be split and dipped in water. The impregnated/penetrated rim portion will stay dry making it possible to determine the depth of penetration.

Freeze-thaw resistance: ASTM C666-92 suggests a procedure for determining the freeze-thaw resistance of samples. Cylindrical specimens/prisms of minimum dimension 5–125 mm and length 275–400 mm are tested after the end of curing period. The specimen is kept in a freeze-thaw apparatus and cooled to a temperature of about –1.1°C to +2.2°C. At this temperature, it is tested for fundamental transition frequency, weighed and its initial length determined with a length comparator. The specimen is then subjected to alternate freezing and thawing cycles consisting of alternatively lowering the temperature from 4.4 to –17.8°C and raising it from –17.8 to 4.4°C in not less than 2 hours and not more than 5 hours. The specimens are removed at intervals not exceeding 36 cycles and tested for fundamental frequency, weight and length changes. The test is continued for 300 cycles or until the relative dynamic modulus of elasticity reduces to 60% of the initial value whichever occurs first. The durability factor is expressed as $DF = PN/M$, where P is the relative dynamic modulus, N = number of cycles at which P reaches the minimum value, M = specified number of cycles at which exposure is terminated.

13.4.9 Miscellaneous Tests

Cracks Bridging Action

This is a special requirement for coatings and is intended to ascertain its ability to accommodate the movements in the substrate and allow the opening and closing of cracks without fracturing. Thin slate or ferrocement plates are coated on one face and then tested under monotonic or cyclic tension to induce specified crack width in the substrate. This enables the maximum crack width that a coating can withstand without fracture to be obtained for different thicknesses.

Weathering Resistance

The weathering of polymeric repair materials is mainly attributed to the breakdown of polymer chains under concerted action of atmospheric agents, such as ozone and UV radiation. To assess the weathering resistance a thin ferrocement or slate slab precracked in tension may be provided with the given repair material and exposed to ozone concentration of 7500 ppm at an elevated air temperature of about 40°C. To test for UV resistance the coated specimens are exposed to UV radiation for duration of about 2000 hours. To simulate the action of rain, water is sprayed on the specimens at fixed time intervals for specified duration. The coating is periodically examined for cracks, blisters, peeling and breakages. 2000 hours duration is supposed to be equivalent to about 5 years of service life under severe exposure conditions. Occasionally, the specimens are held at the specified crack width in a tensioning device during the period of exposure to accelerate the tests.

Chemical Resistance

Most of the polymeric materials are resistant to alkalies, gasoline and oils, dilute acids, ammonia, salts and some solvents. However, strong oxidizing acids attack the polymeric repair materials. To evaluate the chemical resistance of resin mortars IS: 9162 recommends that 25 mm × 25 mm cylinder specimens be immersed in the specified chemical for duration of 56 days. It is taken out periodically, cleaned under running water and dried wiped and a blotting paper and dried for about 30 minutes and again reimmersed in the chemical solution. The change in mass and physical signs of deterioration, such as cracking, spalling, softening, etc. is periodically recorded.

13.4.10 Test for Surface Quality

Inadequate preparation of concrete surface to be repaired is often the source of failure of patch repairs and overlays. The tests generally recommended by the American Concrete Institute for evaluation of surface preparation are discussed below:

Cleanliness of the Surface

The following surface conditions are considered:

- *Dust condition:* The surface to be repaired is wiped with a clean dark cloth. If any white powder is noticed on the cloth, the surface is considered to be too dusty and therefore unsatisfactory.
- *Oily condition:* If water sprinkled on the surface immediately spreads out, without forming droplets, the surface may be assumed to be free from oils or dust.
- *Acid condition:* Using a pH paper, if the pH of the concrete surface is found to be less than 4, it is considered unacceptable.
- *Presence of laitance:* This may be detected by scrapping the surface with a putty knife and observing for presence of any loose, powdery material.

Dryness of the Surface

Many polymeric bonding agents, repair materials and barrier systems do not perform satisfactorily, if moisture is present at the surface of concrete substrate. A simple test for evaluating the dryness of the surface consists of tapping a 1.2 mm × 1.2 mm size polythene sheet to the substrate surface and determining the time required for the moisture to collect underneath the sheet, when the ambient conditions are as near to the likely conditions during repair as possible. If the observed time period is less than the cure period of repair material, it may be concluded that concrete is adequately dry. However, decision should be taken giving due consideration to the nature of repair material. Certain cementitious and water-based repair materials are to be applied to damp concrete surface. In those cases, it is enough if care is taken to prevent any water stagnation on the concrete surface.

Test for Surface Soundness

This test is recommended to ascertain the quality of surface preparation at the rate of approximately one test for every 10 sq m of area to repair. The test procedure consists of bonding the flange portion of an aluminium T-section using a small quantity of epoxy compound to the cleaned, dried, and prepared surface, curing it for 24 hours and then measuring the tensile load required to pull out the T-section, when pulled at a uniform rate.

Strength of less than 1.2 MPa indicates that the concrete substrate may be suspiciously weak and there is a need for investigation of the quality of concrete before undertaking full scale repairs.

Patch Test for Adhesion of Repair Material

This is another simple field test to check whether adequate bonding has been obtained on properly repaired substrates. Although the test procedure is recommended for epoxies, it may be used for other polymeric materials and barrier systems also. The test procedure consists of drilling through the repaired patch down to the substrate surface to produce a 50 mm diameter cored disk. A standard 37.5 mm diameter pipe cap with 50 mm diameter bottom flange is bonded to the cored disk using strong epoxy glue. After the curing of the glue, the pipe stub is subjected to tensile pull out and the failure load and mode of

failure are observed. Failure in concrete substrate at stress levels below 1.25 MPa indicates weakness of the substrate. If a cohesive type of failure occurs in the patch or the adhesive fails, the test is repeated. Repeated failures indicate inadequate surface preparation, defect in the repair material or in its application.

13.4.11 Other Tests

There are many other important tests to assess the properties of repair materials and repaired structures, such as impact resistance, abrasion resistance, skid resistance, low temperature flexibility test, thermal stability and degradation, flammability, fire resistance, chemical resistance, climatic cycling resistance/weathering resistance, ozone and UV ray resistance, salt spray resistance, etc., which are especially important for polymeric materials used for overlays and floorings.

SHORT ANSWER QUESTIONS

1. What is meant by binder?
2. What is the benefit of binder?
3. Define fine aggregate as a repair material.
4. Write the benefit of fine aggregate.
5. Define coarse aggregate as a repair material.
6. What is the benefit of coarse aggregate?
7. What is meant by special fillers?
8. What are the benefits of polymer modifier?
9. Define polymer modifiers.
10. Write the benefits of polymer modifier.
11. What are the uses of chemical modifier?
12. Write the benefit of chemical modifier.
13. What is meant by repair materials?
14. What are the basic needs for selection of repair materials?
15. What is the material properties considered when selecting a repair material?
16. What is PMCC?
17. What is PC and RC?
18. What is PIC?
19. How are the polymers used in repair system?
20. What is the action of polymer latex?
21. What are the properties of polymer based repair materials?
22. Explain about polymer bonding aids.
23. What are the requirements for the repair materials?
24. Write short notes on bentonite.
25. What is the use of bituminous coat as a repair material?
26. Write short notes on epoxies.
27. Write about polymer.
28. Name the materials available for repair techniques.
29. Define SBR latex.
30. What are the properties obtained from SBR latex with cementitious mixer?
31. How does the latex works in cementitious materials?
32. Write down the recommendations for the use of SBR latex cementitious mixes.
33. Name the epoxy resins for structural repairs and rehabilitation.

34. What is the necessity of careful selection of repair materials?
35. What are the factors considered in selecting a repair material?
36. Define the methodology for the selection of repair materials.
37. Define the classification of repair materials.
38. How do you evaluate the repair materials?
39. What is meant by thixotropic index?
40. What are the properties involved in green mixes?
41. What is the limitation of compressive strength and elastic modulus of repair materials?
42. In which type of repair tensile strength and its elongation is required for repair materials?
43. What are the properties to be tested in case of polymeric latexes and resin systems?
44. Name the durability related tests.
45. Name the tests for surface of repair materials.
46. Write about patch test for adhesion of repair material.
47. What are the mechanical properties of fiber resin forced concerted as a repair material?
48. What is meant by aspect ratio in FRC?
49. Name some of the natural fibers used as a repair material.
50. Expand the terms: FRC, GFRP, CFRP and SFRC.
51. What are the types of polymers used as repair materials?
52. Write in short the applications of polymers.
53. What are the characteristics of epoxide resins as a repair material?
54. What is the use of polyurethanes as repair materials?
55. Classify polyester resins as a repair material.
56. What is the function of polyvinyl acetate as a repair material?
57. What are the properties of SBR and AR with Portland cement mortar?
58. Define PMCC, RC/PC and PSC.
59. Differentiate polymer coatings and latex coatings.
60. Compare silages/siloxanes.
61. Define admixtures and classify it.
62. Write short notes on chemical admixtures.
63. Write short notes on mineral admixtures.
64. Define SIFCON.
65. Define SIMCON.
66. What is meant by ferrocement?
67. What are the materials used in fiber-cement?
68. Write about the reinforcement used in fiber-cement?

THEORY QUESTIONS

1. Explain the classification of repair materials with selection criteria.
2. Discuss the methodology used in the selection of repair materials.
3. Explain about the various tests for surface quality of repair materials.
4. Discuss about various miscellaneous test/chemical test for repair materials.
5. Explain about the durability related tests for repair materials.
6. Write about test for physical and mechanical properties of repair materials.
7. Explain the behavior of steel fiber reinforced concrete as a repair material.
8. Explain the behavior of glass fiber as a repair material.
9. Explain the application of carbon fibers as a repair material.
10. Write about organic polymer as repair materials.
11. Explain the application of fiber reinforced polymer composite for repair and strengthening of concrete structures.

12. Explain the types of polymer concrete composites with neat sketches.

13. Discuss the process technology of polymer composites.

14. Explain about the fiber reinforced polymer composite laminates as a repair and rehabilitation material.

15. Explain about the polymeric repair materials.

16. Discuss about chemical admixtures as a repair material.

17. Discuss about mineral admixtures as a repair material.

18. Explain about ferrocement as rehabilitation material.

19. Define SIFCON and explain the characters and application in construction industry.

20. Explain briefly about SIMCON and its application in construction industry.

21. Write about the role of epoxies and latexes as repair materials.

Part 4

Repair and Rehabilitation of Concrete Structures

14. Planning and Designing of Concrete Repair
15. Repair of Cracks
16. Concrete Removal and Preparation for Repair
17. Strategies and Techniques for Repair/Rehabilitation/
Retrofitting of Structures

14

Planning and Designing of Concrete Repair

Sometimes concrete needs to be repaired. A successful repair relies on many factors. One must consider the best repair strategy, material and procedure. The best attempt at a repair is likely to fail if the wrong material is used for the repair.

When we are planning a concrete repair, we must consider various options and evaluate the best type of material to implement in the repair or rehabilitation of concrete structures. One may find the choices for repair to be controversial in some venues. Appropriate experts may need to be consulted for ensured success. There are however many facts that are proven and can be trusted. We will deal with them in this chapter.

14.1.1 Compressive Strength

How much compressive strength is needed in material for concrete repairs? If it is determined that the existing concrete structure is of adequate compressive strength, then the repair material should be of a similar compressive strength. There are few instances where beefing up the compressive strength in a repair is beneficial. An exception is the repair of concrete that is damaged by erosion. When erosion is at fault for a defect, using a higher compressive strength is a valid decision.

T-beam construction requires that the flange and web must be built integrally or otherwise effectively bonded together.

14.1.2 Modulus of Elasticity

Modulus of elasticity is a measure of stiffness with higher modulus materials. When carrying out a repair, the modulus of elasticity should be similar to that of the concrete substrate. This allows for uniform load transfer across a repaired section. Using materials with a lower modulus of elasticity will exhibit lower internal stresses. This reduces the potential for cracking and delamination of a repair. Clear spacing between ribs in joint construction cannot be more than 750 mm.

14.1.3 Drying Shrinkage

Drying shrinkage is a concern in concrete repairs. Existing concrete that is in need of repair is unlikely to shrink. However, patches and repairs that are made fresh are subject to shrinking and this can compromise the repair. To avoid shrinkage, the material used for repairing old concrete should be made

14.1.4 Bonding

The bonding between repair material and concrete is a key element in a successful repair material and concrete is a key element in a successful repair. For concrete to accept a good bond with a repair, the concrete should be properly prepared. Polymer adhesive provide a better bond of plastic concrete to hardened concrete than can be obtained with cement slurry or the plastic concrete alone. This creates some controversy. Many experienced people indicate that polymer bonds are less than 25 percent better than properly prepared concrete surfaces without adhesives. Slab thickness in joint construction must not be less than 5 cm in depth.

14.1.5 Thermal Expansion

Thermal expansion is going to happen with concrete. If a polymer is used as a repair material, the result will often be cracking, spalling or delamination of the repair. The coefficient of thermal expansion must be considered for suitable repair materials. When we compare a polymer to concrete, the coefficient of thermal expansion for a polymer is likely to be 14 times greater than that of concrete. Large repairs and overlays are especially vulnerable to cracking due to thermal expansion.

There are repair products available that offer minimum shrinkage. The goal is to use a material that will provide minimum shrinkage. When estimating the thickness of concrete to be applied as the minimum amount required to cover material, the floor finish may be counted for nonstructural consideration.

14.1.6 Creep

Repair materials should have a creep factor consistent with the material being repaired. Stress relaxation through tensile creep reduces the potential for cracking. One should refer to the manufacturer's documents when selecting an appropriate creep rate for various repairs.

14.1.7 Permeability

Good concrete is relatively impermeable to liquids. However, moisture evaporates at surface and replacement liquid is pulled to the evaporating surface by diffusion. This has to be considered when making a repair, any large patch or overlay that is made with an impermeable material can trap moisture between the existing concrete surface and the seal made by the repair. If this happens, the repair is likely to fail. We should use a repair material that has low water absorption and high water vapour transmission characteristics.

The minimum number of longitudinal bars in compression members is four within rectangular or circular ties.

Planning a concrete repair requires consideration of many factors including the following:

- Application conditions
- Geometry
- Temperature
- Moisture
- Location
- Service conditions
- Downtime
- Traffic
- Temperature
- Chemical attack
- Appearance
- Service life.

Depth and orientation of a concrete repair are important considerations. Thick sections have heat generated during curing of some repair materials. Thermal stress can occur that is beyond an acceptable level. Shrinkage is another concern. Thin layers of concrete used as a repair are subject to spalling. An advantage to polymer materials is that they can be used in thin layers. Aggregate size is determined by the thickness of a repair. Repairs made overhead must be made in such a way that they will not sag. The minimum number of longitudinal bars in compression members is three bars within triangular ties.

Portland cement hydration stops at or near freezing (32°F). Latex emulsions do not coalesce to form films at temperature below about 45°F. Materials that can be used in colder temperatures generally require longer setting times. In contrast, high temperatures may make a repair material set faster and result in a decrease in the working life of the material.

14.3 CAUSES AND REPAIR APPROACHES FOR SPALLING AND DISINTEGRATION

The causes and repair approaches for spalling and disintegration are brought out in Table 14.1.

Having water come into contact with fresh concrete is not acceptable in most repairs. Grouting external waterproofing or diversion systems are commonly used to prevent interference from moving water while working with concrete. If we are working with polymers, some of them will not adhere in moist conditions. On the other hand, some polymers are not affected by moisture.

The minimum number of longitudinal bars in compression members is six bars enclosed by spirals.

14.4 REPAIR METHODS FOR SPALLING AND DISINTEGRATION

The repair methods for spalling and disintegration are brought out in Table 14.2.

The location of a needed repair can have an impact on both materials and procedures. Some locations limit the type of equipment that can be used. Some repair materials are odorous, toxic, or combustible. All of these factors must be considered when planning a repair.

Sometimes materials that will set quickly are needed to reduce downtime. In the case of heavy vehicular traffic, repair materials need to have a high strength rating and good abrasion and skid resistance. We also have to consider concrete deterioration.

High-service temperatures can affect the performance of some polymers. While polymers can be sensitive to solvents, most polymers resist most acids and sulfates. Soft water can damage Portland cement products. Matching patches and repairs can be very

Table 14.1: Causes and repair approaches for spalling and disintegration

Cause	Deliberation likely to continue		Repair approach
1. Erosion (abrasion, cavitation)	×		Partial replacement surface coatings
2. Accidental loading (impact, earthquake)	×		Partial replacement
3. Chemical reactions			
Internal	×		No action total replacement
External	×	×	Partial replacement surface coatings
4. Construction errors (compaction curing finishing)	×		Partial replacement surface coatings No action
5. Corrosions			Partial replacement
6. Design errors			Partial or total replacement based on future activity
7. Temperature changes (excessive expansion caused by elevated temperature and inadequate expansion joints)			Redesign to include adequate joints and partial replacement
8. Freezing and thawing	×		Partial replacement no action

Note: This table is intended to serve as a general guide only.
Courtesy: United States Army Corps of Engineers.

Table 14.2: Repair methods for spalling and disintegration

Repair approach	*Repair method*
1. No action	Judicious neglect
2. Partial replacement (replacement of only damaged concrete)	Conventional concrete placement
	Dry packing
	Jacketing
	Preplaced aggregate concrete
	Polymer impregnation
	Overlay
	Shotcrete
	Underwater placement
	High strength concrete
3. Surface coating	Coatings
	Overlays
4. Total replacement of structure	Remove and replace

Note: Individual repair methods are discussed in other chapters.

difficult. Its appearance is a factor; we will have to do our research to come up with a close, visual match. We must also keep in mind how long will the repair be expected to last when we are choosing the type of repair material to use.

Manufacturer's Data

Most manufacturers offer information on the qualities of their products. We should be able to find information on: (1) the compressive strength, (2) the tensile strength, (3) the slant-shear bond and (4) the modulus of elasticity.

Even with this information, we may have more questions that are not answered in the manufacturer's documents, for example drying shrinkage, tensile bond strength, creep, absorption and water vapour transmission. If we are not provided with this information in the planning phase, one has to request it from the manufacturers.

The planning phase is critical to a successful outcome. You may have to spend a few hours on research, but the result should be a better repair. Do not guess. Make decisions based on facts.

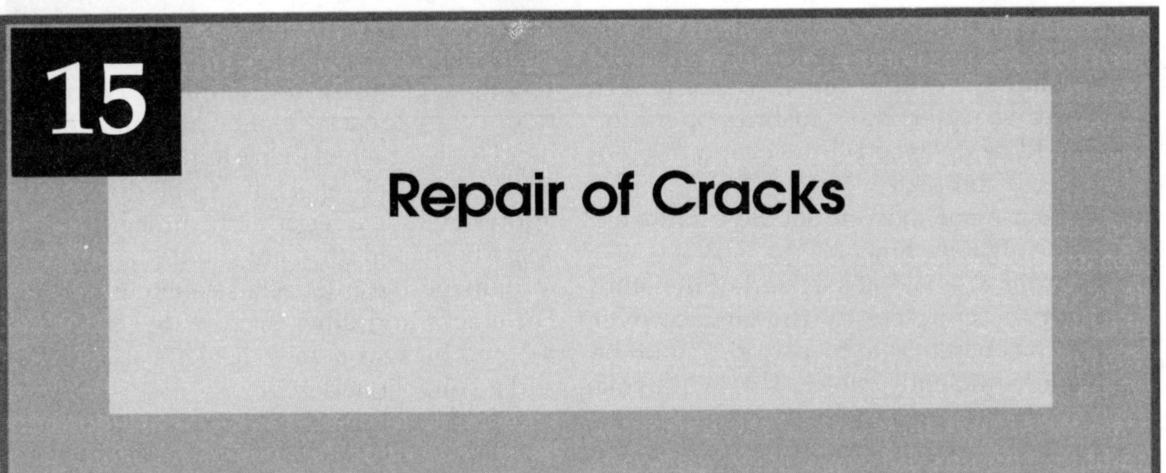

Repair of Cracks

15

Cracks are almost common in a most of the buildings resulting in ugly appearance and a feeling of instability to the occupants. But all cracks are not harmful to the life of a building. Nonstructural cracks do not affect the main structure if they are treated approximately in time whereas structural cracks can be eliminated by proper design and through supervision.

Cracks may be differentiated into following kinds:

- With regard to depth—surface, shallow, deep, through cracks
- With regard to direction—map cracks or pattern cracks roughly hexagonal and single continuous cracks
- With regard to width—thin (less than 1 mm), medium (1–2 mm), and wide (more than 2 mm) cracks
- With regard to load transfer—structural and nonstructural.

15.1.1 Factors that Affect Cracking

Before going into details it is necessary to know the factors that affect cracking. The various factors are given below:

Water: The more is the water the greater is the cracking tendency. Water increases shrinkage and reduces strength.

Cement: Richer concrete cracks more. Increasing the volume of aggregates by 10% can be expected to reduce shrinkage by about 50%.

Aggregate: Presence of clay binders in aggregates causes high shrinkage and cracking. The smaller the maximum size of well graded aggregates the greater the shrinkage of concrete at the same strength.

Bleeding: Excessive bleeding induces cracks.

Improper curing: When rate of evaporation is very high so that fresh concrete dries rapidly, early curing and appropriate screening is essential.

Exposure: The southern and western exposures of structure are usually more severe than northern and eastern exposures.

Cover: Clear cover less than the specified value expedites corrosion of reinforcement.

15.1.2 Measures to Rectify Cracking

Following measures may be adopted to rectify cracking:

- The fresh concrete should have low water/cement ratio to reduce shrinkage and well compacted to prevent settlement due to shrinkage.
- The aggregates used should be well graded to give high quality concrete.

- In order to avoid formation of hair cracks, dry cement should not be allowed to be sprinkled over the wet surface to absorb bleeding water, but finishing operation should be delayed till the evaporation of water is complete.
- The curing of concrete should be allowed for a sufficient time.
- In order to avoid absorption of moisture from the concrete by the surface over which concrete is to be placed, should be made sufficiently moist. This will avoid drying shrinkage.
- Formwork used should be sufficiently strong.

The repair of concrete structures may vary between a cosmetic treatment and a total replacement. By a proper investigation and by using well designed equipment, tools and materials can be reinstated economically. An appropriate repair method can be selected depending upon the cause and extent of damage, importance of the structural element, and its location. The choice of the method will determine its success. Repair of concrete structures is decided upon the factors, such as:

- The cause of damage
- Type, shape and function of the structure
- The type and extent of damage
- The capabilities and facilities available with builders
- The availability of repair materials.

15.1.3 Stages of Concrete Repair

Repair of concrete structures is carried out in the following stages:

- Removal of damaged concrete
- Pretreatment of surfaces and reinforcement
- Application of repair material
- Resorting the integrity of individual sections and strengthening of structure as a whole.

Removal of Defective Concrete

Prior to the extension of any repair, one essential and common requirement is that all deteriorated or damaged concrete be removed.

Removal of defective concrete can be carried out using tools and equipment, the types of which depend on the damage. Normally, removal of concrete can be accomplished by hand tools, or when that is impractical because of the extent of repair, it can be done with a light or medium weight air hammer fitted with a spade shaped bit. Care should be taken not to damage the unaffected concrete portions. For cracks and other narrow defects, a saw-toothed bit will help to achieve sharp edges and a suitable undercut.

Pretreatment of Surfaces and Reinforcement

The preparation of a surface/pretreatment for repair involved the following steps:

- Complete removal of unsound material
- Undercutting along with the formation of smooth edges
- Removal of the cracks from the surface
- Formation of a well-defined cavity geometry with rounded inside corners
- Providing, rough but uniform surface for repair.

The cleaning of all loose particles and oil and dirt out of the cavity should be carried out shortly before the repair. This cleaning can be achieved by blowing with compressed air, hosing with water, acid etching, wire brushing, scarifying or a combination. Brooms or brushes will also help to remove loose material.

Application of Repair Material

After the concrete surface has been prepared, a bonding coat should be applied to all the cleaned exposed surfaces. It should be done with minimum delay. The bonding coat may consist of bonding agents such as cement slurry, cement sand mortar, epoxy, epoxy mortar, resin material, etc. Adequate preparation of surface and good workmanship are the ingredients of efficient and economical repairs.

15.2 REPAIR PROCEDURE

The repair of cracked or damaged structures is discussed under two distinct categories, namely, ordinary or conventional procedures, and special procedures using the latest techniques and newer materials, such as polymers, epoxy resins, etc. A repair procedure may be selected to accomplish one or more of the following objectives:

- To increase strength or restore load carrying capacity
- To restore or increase stiffness
- To improve functional performance
- To provide water tightness
- To improve appearance of concrete surface
- To improve durability
- To prevent access of corrosive materials to reinforcement

15.3 DURABLE REPAIR DESIGN

15.3.1 Modulus of Elasticity and Strength of Repair Material

The modulus of elasticity of the repair material not only affects the resultant flexural stiffness of the repaired members, but also the tensile stress present within the repair material and the debonding stress at the interface when differential movement occurs between the repair material and substrate. A higher difference in modulus of elasticity between the repair material and substrate may adversely affect the stress distribution within the repaired composite material's cross-section and may lead to considerable stress concentrations. Therefore, the repair material selected should have as similar a modulus of elasticity as the substrate as possible. Thus, considering the strength of repair material alone seems less important. In fact an over emphasis on strength may cause repairs to experience cracking arising from excessive drying shrinkage, creep and heat of hydration.

15.3.2 Coefficient of Thermal Expansion of Repair Material

Tensile stresses in the repair material caused by changes in the temperature of the surrounding environment are proportional to the difference in the coefficient of thermal expansion and the change in temperature. Therefore, the repair material selected should have as similar a coefficient of thermal expansion as the substrate as possible.

15.3.3 Thickness of Repair

The internal stresses within the repair material and substrate are affected not only by the differential movements, but also by the relative thickness. A thinner repair layer is more easily cracked or denoted by the higher tensile stress which occurs in the repair material. For most available repair materials, there seemed to be an optimum thickness of repair material which results in the lowest tensile stress occurring within the repair material for a given amount of differential movement between the repair material and substrate. This optimum value is affected by the ratio of the modulus of elasticity of the repair material to substrate and thickness of repair material to substrate.

15.3.4 Shrinkage and Creep of Repair Materials

Differential shrinkage of the repair material and substrate is another important consideration for a durable repair. The most common damages in concrete is between substrate and repair materials and it is proportional to the differential shrinkage. Therefore, the repair material selected should have free shrinkage properties that are as low as possible. For the repair material under tension, creep may mitigate against the tensile stresses caused by differential shrinkage. However, for the repair material under compression, creep may decrease the compressive stress within the repair material and aggravate the compressive stress in the substrate caused by the differential shrinkage. The creep of the repair material should be controlled based on the state of stress that the repair material will be subjected to in-service.

15.4 DURABLE REPAIR APPLICATION

15.4.1 Preparation of the Repaired Surface

The most important surface characteristics of the receiving substrate are its roughness, soundness, cleanliness and moisture condition prior to application of the repair materials. The first step in a repair to be carried out is the removal of the damaged concrete. It is very important to select a method most appropriate for the specific *in situ* condition. Any method that weakens the sound concrete and creates microcracking should be avoided. Otherwise, the durability and bond will be decreased by these defects. Commonly used methods *in situ* include sand blasting, chipping with jack hammers, and hydro demolition among which the last is highly recommended. A sound surface of adequate roughness can be created by this method.

Higher plastic shrinkage of the repair material near the interface should be avoided. This requires that the substrate be prewetted for at least 7 hours prior to the application of the repair material in order to decrease the absorption by and expansion of substrate caused by the uptake of moisture from the repair material. Otherwise, the higher uptake of moisture by the substrate after the repair material is cast may lead to higher plastic shrinkage of the repair material near the interface and higher expansion of the substrate, and thereby resulting in the possible debonding of the repair material at an early age.

15.5 REPAIR METHOD

The application method and surface preparation are equally important considerations with regards to the performance of the repaired structures. The repair method adopted not only affects the resultant quality of the repair material, but also the quality of the interfacial transition zone. Shotcrete seems to be an ideal method because it has good compatibility with the substrate concrete. Furthermore, good compaction with a relatively lower water/cement ratio of the repair material can be achieved using the shotcreting process. This ensures good/high mechanical properties of the component parts and durability of the repair structure.

Bonding Agents

Use of polymer bonding agents was not recommended earlier as their modulus of elasticity is substantially different from that of the substrate. However, use of a cementitious bonding agent with a low water/cement ratio was used. This type of bonding agents not only has good compatibility with the substrate and repair material, but can also alleviate the effects of differential shrinkage and thermal movement between the repair material and concrete substrate thus enhancing the bond strength and durability. However, subsequently polymeric bonding agent was found to be more affective.

Curing of Repair Materials

Excessive loss of water may result in higher shrinkage (plastic and drying) and cause debonding failure of the repair material at an earlier age. Therefore, specification of proper curing after completion of the repair is very important. Curing time should be at least the same as that adopted for usual concrete practice or in accordance to manufacturer's recommendation if a commercially available material is used due to the restraint afforded by the substrate.

Workmanship

It is customary to specify that the specialized contractor should provide satisfactory proof of experience in concrete repair works. Carelessness in any of the above aspects may affect the workmanship and quality of repair work.

15.6 METHODS OF REPAIR

In dealing with cracks, the following considerations are to be taken care of and observed:

- Determination of extent, location and width of cracks
- Classification of cracks as structural and nonstructural.

Dormant cracks: Are caused by some event in the part (e.g. drying shrinkage), which is not expected to recur. They remain constant in width, and may be repaired by filling them with a rigid material.

Active cracks: Do not remain constant in width, but open and close as the structure is loaded, or due to thermal and hydral changes in the concrete.

Growing cracks: Increase in width become the original reason for their occurrence persists, e.g. continuing foundation settlement or reinforcement corrosion.

The repair of cracks can be achieved with the following techniques:

- Resin injection
- Routing and sealing
- Stitching
- External stressing
- Bonding
- Blanketing
- Overlays
- Dry pack
- Vacuum impregnation
- Polymer impregnation
- Autogenous healing
- Flexible sealing
- Drilling and plugging
- Bandaging
- Coating
- Grinding
- Sand blasting
- Resurfacing
- Acid etching
- Caulking, etc.

Bonding cracks of width less than 0.075 mm with epoxy compounds by pressure injection make a grove 12 mm by 12 mm by chiseling; wash crack with water under pressure to remove any debris and drying by hot blown air; drilling 30 to 40 mm deep, 15 mm diameter holes at 100 mm spacing along the length of the crack; clean holes by compressed air, inserting 12 mm diameter aluminium nipples to the holes, sealing the crack lengthwise with epoxy putty and pressure grouting the holes using epoxy grout mix until the crack fully filled; after curing projecting part of nipples are cut and surface finished smooth.

15.6.1 Resin Injection

Epoxy resins are usually selected for crack injection because of their high mechanical strength and resistance to most chemical environments encountered by concrete. However, epoxies are relatively rigid and not suitable for active cracks. The method is used to restore structural soundness of members where cracks are dormant or can be prevented from further movements.

The polymer/resin injection under pressure will ensure that the sealing material or sealant penetrates to the full depth of the crack. The injection holes are drilled at close intervals along the length of the crack and the epoxy is injected under pressure in each hole in turn till the injection material starts to flow out of the next hole. The hole which is injected is then sealed off and the next hole is treated. Before injecting the epoxy, it should be ensured that the crack at surface is sealed between the holes with rapid curing resin. Usually pressure injection is resorted to in sealing the cracks. Figures 15.1, 15.2 and 15.3 show schematic crack injection procedures generally adopted for concrete structures.

15.6.2 Routing and Sealing

This is the simplest and most common method of crack repair. It can be executed with relatively unskilled personnel and can be used to seal both fine pattern cracks and larger isolated cracks. The system can be used to repair dormant cracks that are of no structural

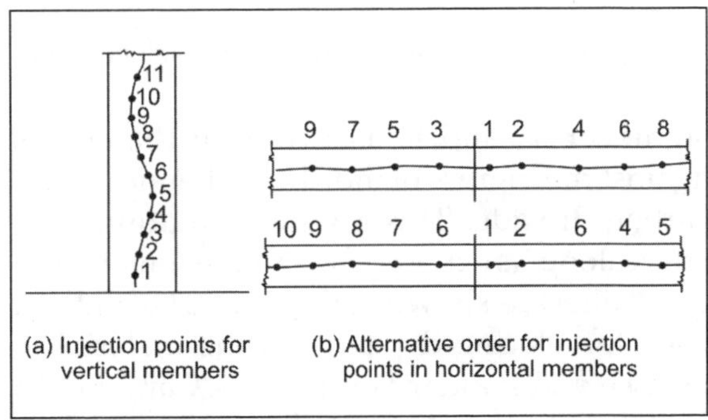

Fig. 15.1: Diagram of crack injection

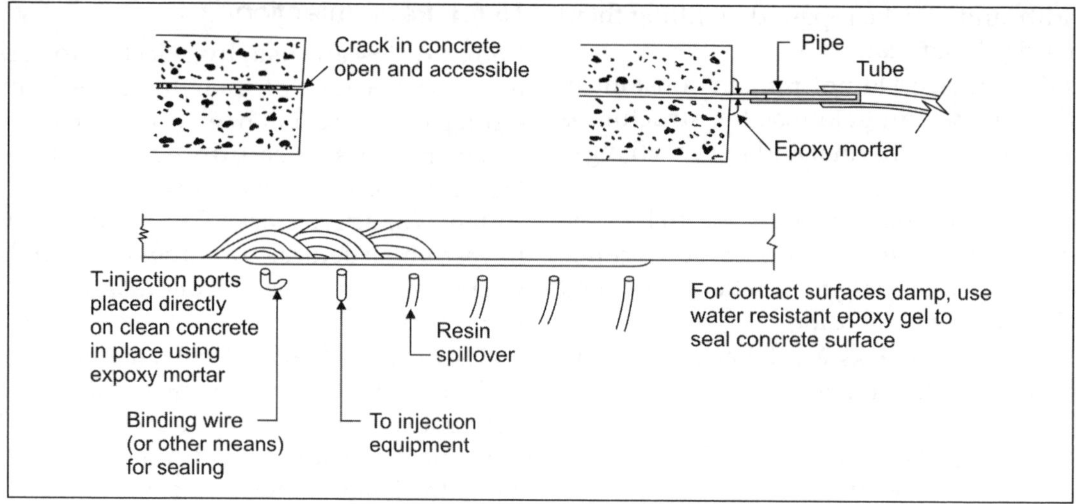

Fig. 15.2: Use of injection ports in drilled holes

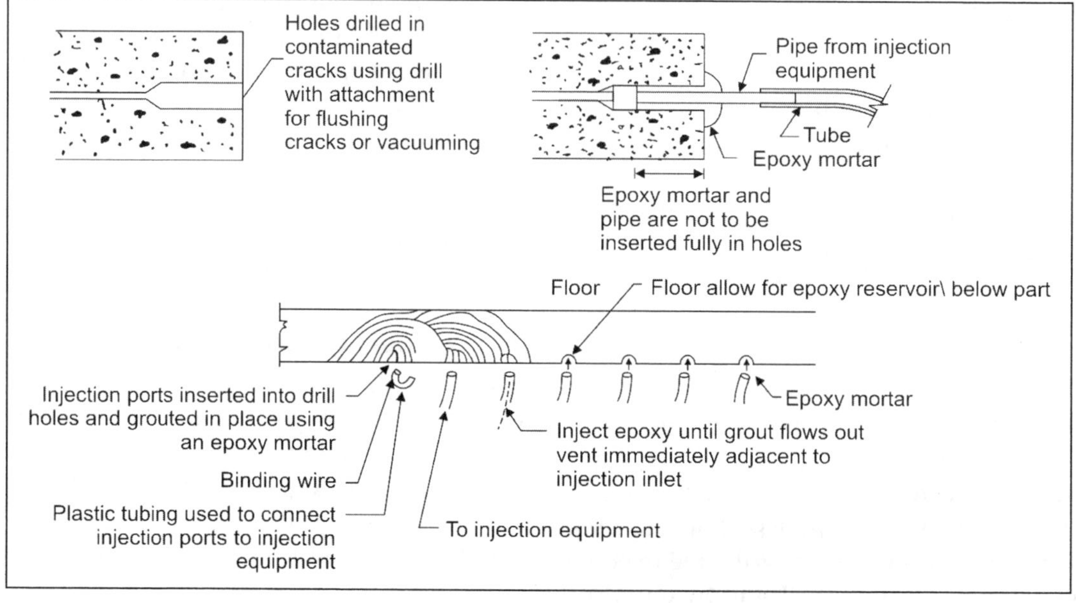

Fig. 15.3: Use of T-injection ports placed flush on concrete surface

significance, and is used to seal the cracks against the ingress of moisture, chemicals and carbon dioxide. This involves enlarging the crack along its exposed face and sealing it with crack fillers as shown in Fig. 15.4. The method used for placing depends on the material to be used and standard techniques. Care should be taken to ensure that the entire crack is routed and sealed.

15.6.3 Stitching

In this technique, the crack is bridged with U-shaped metal units called stitching dogs before being repaired with a rigid resin material. This can establish restoration of the strength and integrity of cracked section; due care is to be given to make an analysis check to ensure that this will perform well under applied load as shown in Fig. 15.5.

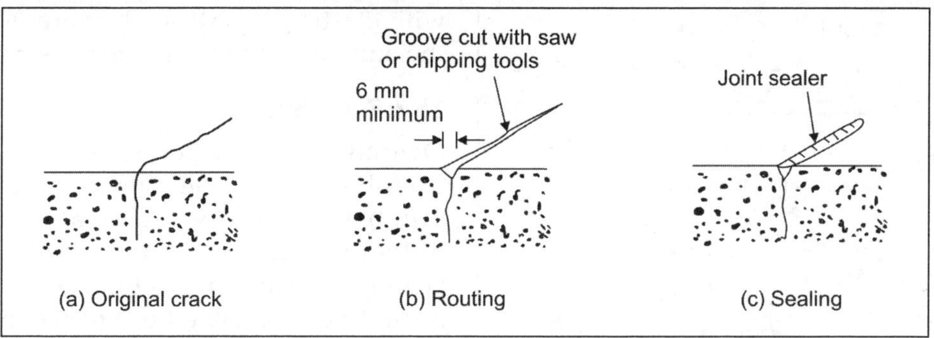

Fig. 15.4: Repair of crack by routing and sealing

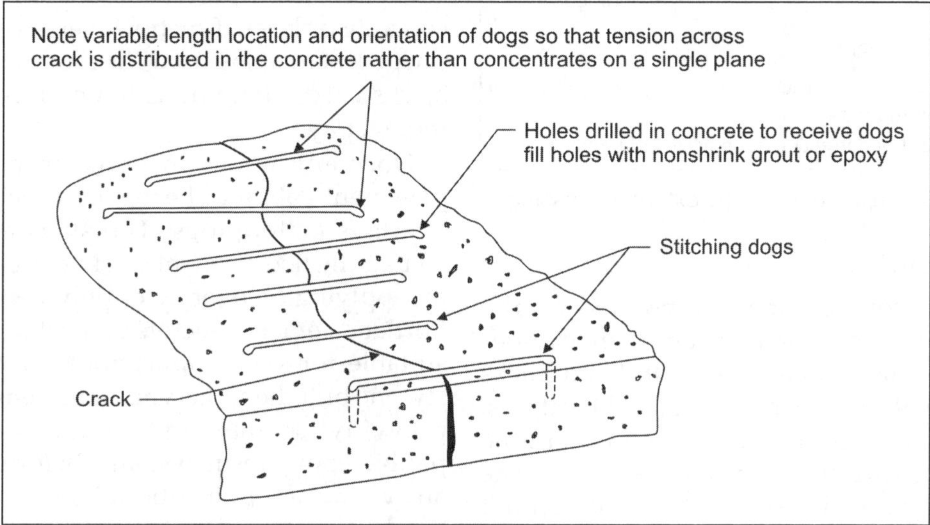

Fig. 15.5: Repair of crack by stitching

A nonshrink grout or an epoxy resin based adhesive should be used to anchor the legs of the dogs. Stitching is suitable when tensile strength must be re-established across major cracks, although stitching will not close the crack, and it is a way of stopping the movement of an active crack and thereby preventing it from spreading. Stitching dogs should be of variable length and/or orientation and so located that the tension transmitted across the crack is not applied to a single plane within the section but is spread over an area. Since stitching may accentuate restraints causing cracking, strengthening of adjacent areas of the structure to take the additional stress may be required.

15.6.4 External Stressing

Development of cracking in concrete is due to tensile stress and can be arrested by removing these stresses. Cracks can be closed by external stressing by inducting compression force sufficient to overcome the tension, which has caused the cracking; the compression force can be applied by use of the usual prestressing wires or rods; some form of abutment is needed for anchorage as shown in Fig. 15.6. The compressive force also is applied by wedging, i.e. opening the crack and filling it with an expanding mortar by jacking and grouting or by actually driving wedges.

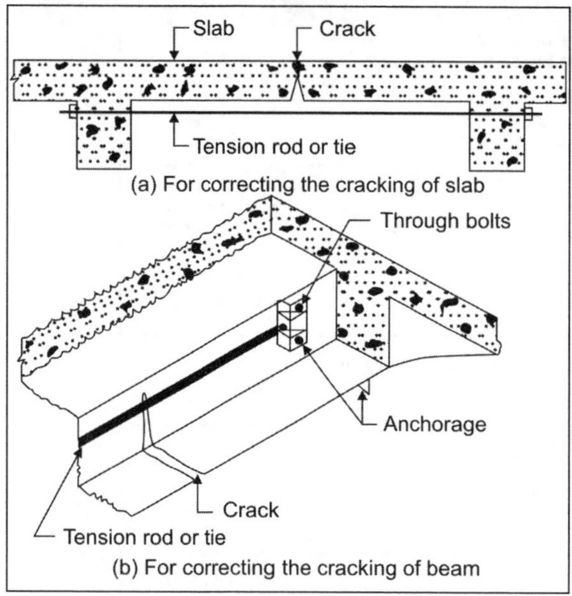

Fig. 15.6: Repair of crack by external stressing

15.6.5 Bonding

Cracks in concrete may be bonded by the injection of epoxy bonding compounds under pressure. A usual practice is to drill into cracks from face of the concrete at several locations. Water or a solvent is injected to flush out the defect. The surface is then allowed to dry. The epoxy is injected into the drilled holes until it flows out through the other holes.

Bonding with epoxies' cracks as narrow as 0.075 mm can be sealed with epoxy compounds, usually pressure injection is resorted to in sealing the cracks.

15.6.6 Blanketing

It is similar to routing and sealing but used on a large scale and applicable for sealing both active and dormant cracks and joints.

Type I: Where an elastic sealant is used the sealant material is one which returns to its original shape when the externally induced stress is removed.

Type II: This is similar to the sealant chase of an elastic sealant except that the bond breaker is omitted and the sealant is bonded to the bottom as well as to the sides of the chase. The sealant is mastic rather than a compound

with elastic properties. They are used where the anticipated movements are small.

15.6.7 Overlays

Overlays are used to seal cracks. They are useful and desirable, where there are large numbers of cracks and treatment of each individual defect would be too expensive. Sealing of active cracks by use of an overlay should be extensible but not flexible. A two or three ply membranes of roofing felt lay in a mop coat of tar with tar between the plies covered with a protective course of gravel, concrete or brick, functions very well for this purpose. Gravel is used for roofs, concrete or bricks used where fill is to be placed against the overlay.

Dormant cracks in both structural and pavement slabs may be repaired using bonded overlays or toppings. The slabs and decks containing fine dormant cracks can be repaired by applying an overlay of polymer-modified Portland cement concrete or mortar. Polymers suitable for such applications are latexes of styrene butadiene, acrylic, nonre-emulsifiable polyvinyl acetate, and certain water-compatible epoxy resin systems. Before applying an overlay or topping the old floor slab should be thoroughly shot-blasted or scrabbled, cleaned and saturated with clear water.

15.6.8 Dry Packing

Dry packing is the hand placement of a low w/c ratio mortar which is subsequently rammed into place to produce a dense mortar plug having tight contact to the existing concrete. Because of the low w/c ratio, there is little shrinkage and the patch remains tight, with good durability, strength, and water tightness. Dry pack is used for the repair of dormant cracks but is not recommended for filling or repairing active cracks. Prior to commencing a dry pack repair, the crack is routed at the surface to a slot 25 mm wide and 25 mm deep and cleaned of any debris. The cavity surfaces are then coated with a bond coat consisting of a thin layer of grout or mortar, just before placing the dry pack material. Dry pack mortar must be mixed just before it is to be placed. Curing of the placed material is carried out by applying a curing compound or supporting a strip of wet burlap along the length of the crack.

Relatively narrow cracks and deep holes can be rectified by dry packing; the crack is slightly undercut, the surface prepared, and mortar with just enough water for sticking is applied. Dry pack is a mortar mixture with a very low water–cement ratio, applicable for small area of repair. It is normally placed by hand. Materials commonly used in dry packs are Portland cement, sand and water.

15.6.9 Polymer Impregnation

This technique consists of flooding the cleaned, dried cracked concrete surface with a monomer which is then polymerized in place, thus filling and structurally repairing the crack. A monomer system is a liquid that consists of small organic molecules capable of combining to form a solid plastic. The monomer systems used for impregnation contain a catalyst or an initiator and the basic monomer. They also contain a cross-linking agent. When heated the monomers join together or polymerized becoming a tough, strong and durable plastic that greatly enhances a number of concrete properties. For repairing fractured elements, polymer impregnation method may be used. In this case the facture is dried first, and then it is temporarily encased in a water tight sheet metal. Now the fracture is soaked with monomer and polymerized. On polymerization of the monomer the fractured element becomes a solid mass with enhanced strength and properties.

15.6.10 Vacuum Impregnation

For treating the concrete surfaces that contain a large number of cracks, vacuum impregnation may be used. The part of the structure to be repaired is enclosed within an air tight plastic cover and then the air from all cracks within the cover is sucked by applying vacuum. After exhausting the air from all cracks, the monomer of resin grout is forced under one atmosphere pressure in cracks and pores of the concrete surface. On completion of impregnation, the cover is removed before the impregnate hardens. The process is extensively used to reduce the permeability of weak concrete or masonry. This process may also be used to improve the abrasive resistance of industrial concrete floor slabs.

15.6.11 Autogenous Healing

The natural process of crack repair known as autogenous healing has a practical application in closing dormant cracks in a moist environment. Healing occurs through carbonation, calcium hydroxide in cement paste by carbon dioxide, which is present in surrounding air and water. Calcium carbonate and calcium hydroxide crystals precipitate, accumulate, and grow within the cracks. The crystals interlace and twine, producing a mechanical bonding effect. As a result, some of the tensile strength of the concrete is restored across the cracked section, and the crack may become sealed.

15.6.12 Flexible Sealing

It is often appropriate to convert active cracks into movement joints. A recess is cut along the line of the crack and filled with a suitable flexible material. The width of recess and strain capacity of the sealing material, determine the movement the joint will accommodate. Therefore, the sealant bond should have a suitable width and shape factor for the expected movement. A bond breaker should be used to prevent the sealant from bonding to the bottom of the recess. This allows the sealant to change shape without concentration of stress at the bottom of the recess. Figure 15.7 shows the repair technique of an active crack by flexible sealing.

15.6.13 Drilling and Plugging

It is particularly useful for vertical cracks in retaining walls, and consists of drilling down

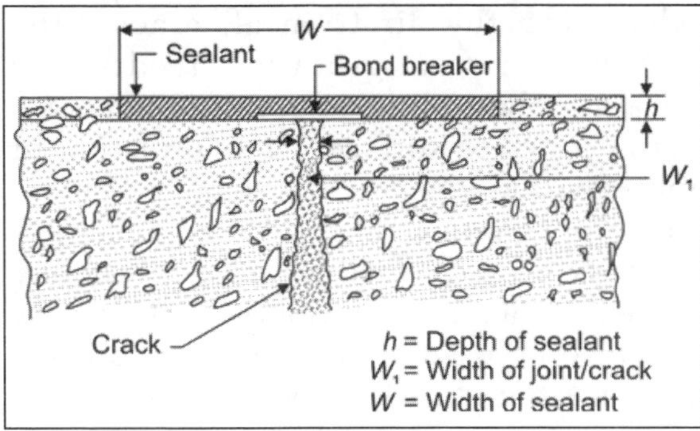

Fig. 15.7: Repair of an active crack by flexible sealing

the length of the cracks and grouting it to form a key. The method is only applicable for the cracks running in reasonably straight lines and accessible at one end. It consists in drilling a hole down the length of crack and grouting it to form a key. A hole 50 to 60 mm in diameter should be drilled, centered on and following the crack as shown in Fig. 15.8.

The hole must be large enough to intersect the crack along its full length and provide sufficient repair material to structurally take the loads exerted on the key. The drilled hole is then cleaned and filled with grout. For problems concerning water tightness, the drilled hole should be filled with a resilient material of low modulus in lieu of grout. In case where both load transfer and water tightness is desired, two holes are drilled, one

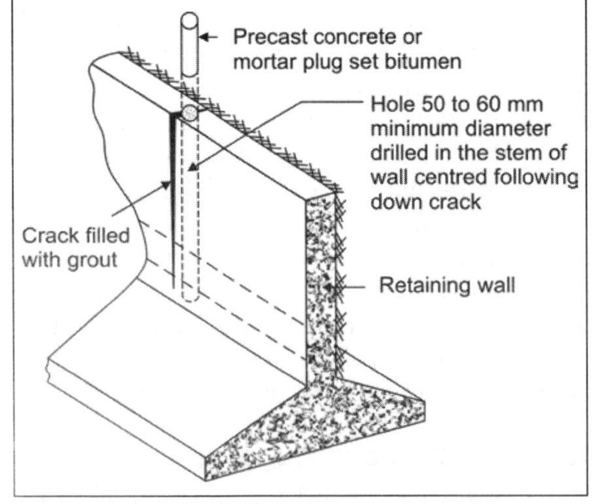

Fig. 15.8: Repair of crack by drilling and pluging

filled with resilient material and the second being grouted.

15.6.14 Bandaging

A flexible strip is fixed over the crack with only the edges of the strip bonded. Where movement is not all in one plane, where there is excessive movement beyond that which can be accommodated by a recess of convenient size, or if there are factors which prohibit the cutting of recesses, a surface bandage can be used. In areas which are subject to traffic, the flexible bondage will be coated over with a wearing course.

15.6.15 Surface Coatings

A wide range of surface penetrating sealers and coatings ranges from thin, purely cosmetic treatments to thick membranes can be applied to cracked concrete. If cracking has reached a stable condition, then a coating can usually be applied successfully. Low viscosity, low solids resin solutions, such as epoxies, urethanes, and acrylics have been used to seal the surface of concrete in areas that are not subjected to wear. The method is effective in sealing dormant, random, multiple cracks. Bridge and parking decks and interior slabs may be effectively coated using a heavy coat of epoxy, urethane or solvent based acrylic resin. The uncured coating is usually seeded with a hard wearing aggregate. The epoxy-based compounds are useful in following conditions:

- Where an adhesive is needed to bond plastic concrete to hardened concrete or to bond rigid materials to each other
- For patching
- When a coating over concrete.

A layer of material applied to a surface, which forms a continuous membrane is termed coating. Application of suitable coatings afford the necessary protection while still allowing free passage of water vapor, is required for the protection of the structure. The coatings generally adhere to the concrete and form films after application.

Anticarbonation Coatings

The coatings applied to concrete surface to arrest the carbonation process are known as anticarbonation coatings. They are normally based on chlorinated rubber, polyurethane resin or acrylic emulsions. The cracks which are very fine and not considered to have structural significance may be protected by applying the coating locally over cracks. These are termed conventional coatings. The coatings should have flexibility as well as ability to bridge cracks. High-build polyurethane and epoxy-polyurethane formulations have been successful.

15.6.16 Grinding

Grinding is a slow and expansive repair method, but often is a most logical solution to certain problems with concrete. Among the most common applications of grinding are the leveling of slab surface that have irregularities either from casting or wear or the removal of unsound concrete from the surface in preparation for some other treatment. Grinding is also used to remove concrete that is stained, pitted or marred with shallow cracks. Grinding is also used in repairing concrete to get an even surface.

Surface Grinding

Surface grinding can be used to repair some bulges, offsets, and other irregularities that exceed the desired surface, exposure of easily removed aggregate particles, or unsightly appearance. For these reasons, grinding should be performed with certain limitations. Grinding of surfaces exposed to public view should be limited in depth so that no aggregate particles more than 6 mm in cross-section are exposed at the finished surface. In no event should surface grinding result in exposure of aggregate of more than half the diameter of the maximum-size aggregate. Where surface grinding has caused or will cause exposure of aggregate particles greater than the fore-mentioned limits, the concrete must then be repaired by excavating and replacing the concrete.

15.6.17 Sand Blasting

Sand blasting may be a satisfactory method by itself to repair some type of damage, but it is usually used in conjunction with other techniques. Sand blasting is frequently used to remove foreign materials that could impair the bond of the patching materials to the concrete. In repairing for architectural treatment, sand blasting is quite useful.

15.6.18 Resurfacing

Protective coatings can greatly reduce the effect of such adverse conditions, and significantly improve the durability characteristics of the concrete. A variety of coatings are available, and some are tailored for greater chemical resistance, where others are formulated to resist wear and erosion. Resurfacing of a concrete surface requires thorough cleaning or removal of the existing coating. If the old coating is in good condition, then compatibility between the old and the new coating should be determined. General resurfacing materials are:

- Bituminous coatings and mastics
- Polyesters and vinyl esters
- Urethanes
- Epoxies
- Neoprene
- Coal tar epoxy
- Acrylics

Intumescent Coatings

These are specially formulated fore resistant paints applied on structural elements for the purposes of increasing their fire rating. These products are finding increased use in the repair of fire damaged buildings where they are used for fire protection of steel work.

15.6.19 Acid Etching

Acid etching is a particularly efficient method of surface preparation, resulting in a sharp even finish, ideal for paint application. Although sulphonic acid is recommended where stainless steel is cast into the concrete, a 10% solution of hydrochloric acid is more generally used. The acid reacts readily with the alkaline constituents of the concrete. This exposes particles of fine aggregate within the mortar layer which form a first-class key. The acid should be sluiced off with fresh water as soon as the reaction finishes, preventing the formation of salts, which might subsequently cause blistering of the paint film.

Several applications are normally necessary to achieve a suitable finish, particularly if the surface laitance is relatively intact. Where large areas are involved, there may be problems with disposal of the spent acid solution, which is a potentially harmful substance. Care should also be taken when handling the acid, because fumes are given off, and splashes will cause burns. Appropriate protection and proper ventilation is essential when using acid etching technique. Etching is specially suitable for areas demanding clean conditions, because no dust is produced.

15.6.20 Caulking

Caulking compounds are flexibilized polymeric materials which are used for two purposes:

- To plug irregular gaps between two surfaces
- To provide a dynamic bridge between two surfaces that are more relative to one another.

In this capacity, they perform the following functions:

- Accommodate joint movement
- Prevent the infiltration of the joint by solid matter such as dust and sand
- Prevent the ingress of water into the structure.

The generic types of sealants used in industrial and domestic building, etc. include acrylic latexes, butyls, polysulfides, urethanes, silicones, etc.

16

Concrete Removal and Preparation for Repair

It is very common for some concrete to be removed before making a repair to a concrete structure. Many jobs require existing concrete to be prepared properly to receive a repair. When the damage done to concrete will probably not have to be removed.

Environmental impact of concrete removal is often an issue. This is especially true when the work is done near a waterway. In some cases, concrete debris can be allowed to enter a waterway because some of the aggregate used in concrete is natural river gravel anyway. Therefore, allowing it to fall into the water is essentially returning the gravel from its place of origin. Another possibility, when allowed, is to dispose of large piece of concrete in water so it can create an artificial reef for fish. When this is an acceptable means of disposal it reduces the cost of deconstruction considerably.

When concrete is removed, it is done in such a way that sound concrete can be reached to base a repair on. It is not uncommon for multiple methods to be used on a single structure to remove concrete. For example, a presplitting method might be used to weaken concrete so an impacting method can be used to complete the concrete removal. Another example could be the cutting of a section of concrete to delineate an area in which an impacting method will be used for concrete removal.

When demolishing concrete, we must be aware of any utilities or other embedded items in the existing concrete. A review of as built drawings can render assistance in projecting obstacle locations.

16.2 GENERAL CLASSIFICATION OF CONCRETE REMOVAL METHODS

A general classification of concrete removal methods applicable for concrete repair is given in Table 16.1.

16.3 REMOVAL METHODS

Several methods that can be used to remove concrete are as follows:

- Blasting
- Crushing
- Cutting
- Impacting
- Milling
- Presplitting

The category, method, features and considerations to be made are brought out in tabular form in Table 16.2.

Table 16.1: A general classification of concrete removal methods applicable for concrete repair

Category	Description	Specific methods
Blasting	Blasting methods employ rapidly expanding gas confined within a series of boreholes to produce controlled fracture and removal of concrete	Explosive blasting
Crushing	Crushing methods employ hydraulically powered jaws to crush and remove the concrete	Mechanical crushing boom mounted, mechanical crushing portable
Cutting	Cutting methods employ full depth perimeter cuts to disjoint concrete for removal as a unit or units	Abrasive water jet cutting, diamond blade cutting, diamond wire cutting, stitch during thermal cutting
Impacting	Impacting methods employ repeated striking of the surface with a mass to fracture and spall the concrete	Mechanical impacting hand-held, mechanical impacting boom mounted, mechanical impacting spring action
Milling	Milling methods generally abrasion or cavitation erosion techniques to remove concrete from surface	Hydro milling, rotary head milling, presplitting, chemical expansive
Presplitting	Presplitting methods employ wedging forces in a designed pattern of presplitting boreholes to produce a controlled cracking of the concrete to facilitate removal of concrete by other means	Agents presplitting piston jack splitter, presplitting plug and feather splitter

Table 16.2: Category, method, features and considerations

Category	Method	Features	Considerations
Blasting	Explosive blasting	Method applicable for removal from mass concrete structures, method is most expedient and in many cases the most cost effective means of removing large volunteers when 230 mm (10 in) or more of face is to be removed, produce reasonably small size debris that is easily handled	Requires highly skilled personnel for design and execution of blasting plan, stringent safety regulations must be complied with regard to the transportation storage and use of explosive because of their inherent dangers. Sequential blasting techniques must be employed to reduce peak blast energies and thereby limit damage to surrounding property resulting from air blast pressure ground vibration and fly rock control blasting techniques should be employed to limit damage to concrete that remains
		Method applicable for removing	
Crushing	Mechanical crushing boom mounted	Concrete from decks, walls, columns and other concrete members where shearing plane depth is 1.8 m (6 ft) or less, boom allows removal from vertical and overhead members, steel reinforcing can be cut, limited noise and vibration is produced. Pulverizing jaw attachment can deboned the concrete from the reinforcement for purpose of recycling both. Method produces relatively small debris that is easily handled	Method is more applicable for total demolition of a concrete member than for removal to rehabilitate or repair it, boundaries must be saw cut to limit over breakage, removal must be started from a free edge or a hole cut in member, exposed reinforcing steel is damaged beyond reuse, production rates vary depending on condition of concrete and amount of reinforcement

Contd.

Table 16.2: Category, method, features and considerations *(Contd.)*

Category	Method	Features	Considerations
	Mechanical crushing portable	Method applicable for removal from decks walls and other members where shearing plane depth is 300 mm (12 in) or less method can be used to remove concrete in areas of limited work space, limited noise and vibration is produced, produce small size debris that is easily handled	Requires two men to handle (weighs approximately 45 kg (1001b)), reinforcing steel is damaged beyond reuse, crushing must be started from a free edge or a hole cut in member, boundaries must be saw cut to limit over breakage, production rates are low.
		Method applicable for making	
Cutting	Abrasive-water-jet cutting	Cutout through slabs, walls and other concrete members where access to only one face is feasible and depth of cut is 300 mm (20 in) or less. Abrasive enable jet to cut steel reinforcing and land aggregates irregular and curved cutouts can be made. Cutout can be made without overcutting corners Cuts can be made flush with adjoining members No heat vibration or dust is produced Handling of debris is more efficient as bulk of concrete is removed as units	Cutting is typically slower and more Controlling flow of waste water may be required. Personnel must wear water may be required Personnel must wear hearing protection because of the high levels of noise produced. Additional safety precautions are required because of the high levels of noise produced. Additional safety precautions are required because of high water pressure (200–340 MPa (30,000–50,000 psi)) produced by system.
	Diamond blade cutting	Method applicable for making cutouts through slabs, walls and other concrete members where access to only one face is feasible and depth of cut is 600 mm or less, precision cuts can be made. No dust or vibration is produced, handling of debris is more efficient as bulk of concrete is removed as units.	Selection of the type diamonds and metal bond used in blade segments based on the type (hardness) and percent of coarse aggregate and on the percent of steel reinforcing in cut. The higher the percent of steel reinforcement in cuts, the slower and more costly the cutting. The harder the aggregate the slower and more costly the cutting, controlling flow of waste water may be required. Special blades with flush-cut arbors are required to make cuts flush with adjoining members.

Contd.

Table 16.2: Category, method, features and considerations *(Contd.)*

Category	Method	Features	Considerations
	Diamond-wire cutting	Method applicable for making cutouts through concretes where depth of cut is greater than can be economically cut with the diamond blade saw cuts can be made through mass concrete and in areas of difficult access. Overcutting of corner can be avoided if cut started from drilled hole at corner. No dust or vibration is produced. Handling of debris is more efficient as bulk of concrete is removed as units.	The wire saw is a special tool that for many jobs will not be as cost effective as other techniques, such as blasting impacting and presplitting selection of type diamonds and metal bond used in beads is based on type (hardness) and percent of coarse aggregate and percent of steel reinforcing in cut. The higher the percent of steel reinforcement in cuts, the slower and more costly the cutting. The harder the aggregate, the slower and more costly the cutting bends with embedded diamonds last longer, but are more expensive than beads with electroplated diamonds (single layer). Wires with beads having embedded diamonds should be of sufficient length to complete cut as replacement will not fit into cut (wear reduce wire diameter and thereby cut opening as cutting proceeds). Deep cutouts that are formed by three or more boundary cuts may require tapering to avoid binding during removal, controlling flow of waste water may be required.
	Stitch drilling	Method applicable for making cutouts through concrete members where access to only one face is feasible and depth of cut is greater than can be economically cut by diamond blade saw. Handling of debris is more efficient as bulk of concrete is removed as units.	Rotary percussion drilling is significantly more expedient and economical than diamond core for reinforced concrete. Diamond core drilling is more applicable than rotary percussion drilling for reinforced concrete. The greater the percentage of steel reinforcement contained within a cut, the slower and more costly the cutting. Depth of cuts is dependent on accuracy of drilling equipment in maintaining overlap between holes with depth and on the diameter of boreholes drilled. The deeper the cut, the greater bore hole diameter required to maintain overlap between adjacent holes and the greater the cost. Uncut portions between adjacent boreholes will prevent removal.

Contd.

Table 16.2: Category, method, features and considerations (*Contd.*)

Category	Method	Features	Considerations
			Concrete toughness for percussion drilling and aggregate hardness for diamond coring will affect cutting rate and cost. Personnel must wear hearing protection because of the high levels of noise produced.
	Thermal cutting	Method applicable for making cutouts through heavily reinforced decks, beams, walls, and other reinforced members where site conditions allows efficient flow of molten concrete from cuts. Method is an effective means of cutting pre-stressed members, irregular shapes can be cut, minimal vibration and dust produced Handling of debris is more efficient as bulk of concrete is removed as units.	Method is of limited commercial availability and is costly Remaining concrete has thermal damage with more extensive damage occurring around steel reinforcement. Noise, smoke and fumes are produced personnel must be protected from heat and hot flying rock produced by cutting operation Additional safety precautions are required because of hazards associated with storage handling and use of compressed and flammable gases
Impacting	Mechanical impacting boom mounted breaker	Method is applicable for both full and partial depth removals where required, production rates are greater than can be economically achieved by the use of hand-held breakers Boom allows concrete to be removed from vertical and overhead members Boom mounted breakers are widely available commercially. Method produce easily handled debris.	The blow energy delivered to the concrete should be limited to protect the structure being repaired and surrounding structure from damage resulting from the high cyclic energy generated. Performance is a function of concrete soundness and toughness. Productivity is significantly reduced when boom is operated from top of wall because of the operator's limited view of the removal operation. Care must be taken to avoid damage to supporting members, concrete that remains may be damaged (microcracking) along with reinforcing steel Saw cuts at boundaries should be employed to reduce the occurrence of feathered edges. Dust is produced. Personnel must wear hearing protection because of the high levels of noise produced.
	Hand-held breakers		

Contd.

Table 16.2: Category, method, features and considerations (*Contd.*)

Category	Method	Features	Considerations
	Mechanical impacting hand-held breaker	Method is applicable for work involving limited volumes of concrete removal and for removal in areas of limited access. Hand-held breakers are widely available commercially. Breakers can be operated by unskilled labor. Method produces relatively small debris that is easily handled	Not applicable for large volume of removal, except where blow energy must be limited. Performance is a function of correct soundness and toughness, significant loss in productivity occurs when breaking action is other than downward. Removal boundaries will likely require 23 min (1-in) deep or greater saw cut to reduce the occurrence of feathered edges. Concrete that remains may be damaged (microcracking). Size of breakers for bridge decks is typically limited to 14 kg (30 lb) class for removal above reinforcement and 7 kg (23 lb) class from around reinforcement Dust is produced, personnel must wear hearing protection because of the high levels of noise produced Method is more applicable for total demolition of a concrete member than for removal to rehabilitate
	Blasting	Steel reinforcing is left undamaged for reuse, method produce easily landed aggregate size debris	Additional material and formwork thereby increasing repair time and cost. Method requires large sources of potable water (the water demand for some units exceeds 4,0001/hr. (1,000 gal/hr.) Laitance coating that is deposited on remaining surface during removal should be washed from surface before coating dries. Flow of waste water may have to be controlled. An environmental impact statement will be required if waste water is to enter a waterway. Personnel must wear hearing protection because of the high level of noise produced. Fly rock is produced. Additional safety requirements are required because of the high pressure (100–300 MPa (16,000–40,000 psi)) range produced by the system.

Contd.

Table 16.2: Category, method, features and considerations (*Contd.*)

Category	Method	Features	Considerations
		Method is applicable for removing deteriorated concrete from mass structures	Removal is limited to concrete, outside structural steel reinforcement
	Rotary head milling	Method is applicable for removing deteriorates concrete cover from reinforced members, such as pavements and decks, where it is unlikely that the reinforcement will be contacted. Boom allows removal from vertical and overhead surfaces. Concrete containing wire mesh can be cut without significant losses in productivity. Method produces relatively small debris that is easily handled.	Significant loss of productivity occurs in sound concrete. Productivity is significantly reduced when boom is operated from top of wall as operator view of cutting is very limited. Concrete that remains may be damaged (microcracking). Skid loader units typically mill more uniform removal profile than other rotary head and water jet units noise, vibration and dust are produced. Personnel must be restricted from presplitting area during early hours of product hydration as material has the potential to blow out.
Presplitting	Chemical, presplitting expansive agents	Method is applicable for presplitting concrete members where depth of boreholes is 10 time borehole diameter or greater. Expansive products can be used to produce vertical presplitting plagues of significant depth, some products form a clay type material when mixed with water that allow the material to be packed into horizontal holes. No vibration, noise or flying rock is produced other than that produced by the drilling of bore-holes and the secondary breakage method.	Boreholes can cause injury Presplitting with expansive agents is typically costly. Expansive products that are prills or become slurries when water is added are best used in gravity-filled vertical or near vertical holes. A liner may be required to contain the expansive material in holes drilled into concrete with extensive cracks Products are limited to a specific temperature range. Rotary head milling or mechanical impacting methods will be required to complete removal. Development of presplitting plane is significant, decreased by pressure of reinforcing steel, normal to plane. Loss of control of presplitting plane can result, if boreholes are too far apart or holes are located in severely deteriorated concrete.
	Mechanical presplitting piston jack	Method is applicable for presplitting more massive concrete structures where 230 mm or more of face is to be removed and presplitting requires	Large-diameter 90 mm boreholes are required that increase cost. Splitters are typically used in pairs to control presplitting plane.

Contd.

Table 16.2: Category, method, features and considerations *(Contd.)*

Category	Method	Features	Considerations
	Splitter	Boreholes of depth greater than can be used by plug and feather splitters Splitter can be reinserted into boreholes to continue removal for full depth of holes. Splitter can be used in areas of difficult access. No vibration, noise or flying rock is produced other than that produced by the drilling of boreholes and the secondary breakage method	Hand-held breakers and pry bars are typically required to complete removal Development of presplitting plane is significant, decreased by presence of reinforcing steel normal to presplit plane Loss of control of presplitting plane can result if boreholes are too far apart or holes are located in severely deteriorated concrete Availability of splitters is limited in the US.
	Mechanical presplitting plug-and feather splitter	Method applicable for presplitting slabs, walls and other concrete members where presplitting depth is 120 cm or less. Method typically less costly than cutting methods. Limitation of direction of presplitting can be controlled by orientation of plug and feathers, splitters can be used in areas of limited access Limited skills required by operator No vibration, noise, or flying rock is produced other than that produced by the drilling of boreholes and the secondary breakage method.	Splitter cannot be reinserted into boreholes to continue presplitting after presplit section has been removed, as the body of the tool is wider than the borehole Development of presplitting plane in direction of borehole depth is limited Development of presplitting plane is significantly decreased by presence of reinforcing steel normal to plane Secondary means of breakage will typically be required to complete removal. Loss of control of presplitting plane can result if boreholes are too far apart or holes are located in severely deteriorated concrete.

16.3.1 Blasting

Blasting is done with boreholes that use rapidly expanding gas to create controlled fracture and concrete removal. The use of explosive blasting is usually not considered for jobs that require the removal of less than 250 cm. Blasting is a fast, and often cost effective method of removing concrete but there is a downside. When blasting is used there is a risk of damaging sound concrete that is not meant to be disturbed. A solution to this is a method known as smooth blasting. Whenever there is a possibility of lateral forces causing a transfer of moment on columns, the shear resulting from this action must be prohibited by the use of reinforcement.

Smooth blasting uses detonating cord to distribute blast energy throughout a borehole. This can avoid energy concentrations that might damage surrounding structural elements. An extended version of smooth blasting is known as cushion blasting, which is essentially the same as smooth blasting with the exception that cushion blasting requires the filling of boreholes with wet sand. Cushion blasting is seldom used. Cutting is often used in conjunction with blasting. Selected cuts are made at prime locations to control the breakaway of concrete, when blasting is employed. Sequential blasting allows for more delays to be used with each firing. This optimizes the amount of explosive detonated with each firing. Air blast pressure is maintained at acceptable levels, as are ground vibrations and fly rock.

16.3.2 Crushing

Crushing is done with hydraulically powered jaws. Boom mounted mechanical crushers and portable mechanical crushers are both used. When total demolition is desired, a boom-mounted crusher is normally used. This type of equipment is most effective when not used for partial removal. Portable crushers work best when the shearing plane depth is 250 cm or less.

16.3.3 Cutting

Cutting is used to remove full sections of concrete. The following methods can be used for cutting:

- Abrasive water jets
- Diamond saws
- Diamond wire
- Stitch drilling
- Thermal tools

Abrasive water jets can be used to make cutouts through slabs, walls, and other concrete members. This type of cutting is capable of cutting through steel reinforcing and hard aggregate. A drawback to abrasive water cutting is that it tends to be slow and costly. Other concerns are controlling the waste water and maintaining personal safety when working with very noisy, high pressure systems.

Diamond blades are used when the cutting depth is less than 2 feet. Blade selection is dependent on the type of aggregate and reinforcing that must be cut. The time required for blade cutting depends on the hardness of the material being cut, but the process can be slow.

Diamond wire cutting is used when the cutting depth is greater than a diamond blade can handle. This type of cutting is well suited for cuts where access is limited or difficult. Wire cutting is specialized and tends to be expensive, but sometimes it is the best alternative available.

Shear head arms must not be interrupted within a column section.

Stitch Cutting

Stitch cutting is used when the cut depth is greater than what can be reached with a diamond saw blade. This is most often the case when only one face of the concrete is accessible. If two faces are accessible, wire cutting is more likely to be used.

Thermal Cutting

Thermal cutting is used for cutting through heavily reinforced concrete structures where site conditions will allow efficient flow of molten concrete from cuts. Flame tools are favored for this type of cutting for depths up to 60 cm, cutting lances are used when the cutting depth is greater. Thermal cutting is expensive and is limited in places where it can be used. The dangers associated with thermal cutting include working with compressed and flammable gases, hot flying rock, and high temperatures.

Negative movement reinforcement in a continuous member must be anchored in or through the supporting member by embedment length, hooks, or mechanical anchorage.

16.3.4 Impacting Methods

Impacting methods involve the use of some sort of mass to hit concrete repeatedly. A single example of this would be banging concrete with a sledge hammer. The impact is meant to fracture and spall concrete. Reinforcement must be cut out when impacting methods are used for concrete removal.

Boom Mounted Concrete Breakers

Boom mounted concrete breakers can be found attached to back holes. This type of breaker is fast and cost effective. The breaker may be operated by compressed air or hydraulic pressure. Saw cuts are typically used to outline the breakout area. This reduces feathered edges. Concrete that remains after this type of breaking is done can suffer from micro-cracking, so one must check for that. If it is found, a high pressure water jet might remove microfractured concrete. The water pressure should be set a minimum of 20,000 pounds per square inch (psi).

Continuous reinforcement is required at interior supports of deep flexural members to provide negative movement tension. Figure 16.1 shows placing concrete around reinforcing bars.

Spring-action Hammer

Spring-action hammers are normally used on thin concrete. This method is most appropriate for total demolition. Cutting concrete along a boundary for a cutout is recommended when spring-action hammers are used. Micro-cracking along with exposed reinforcing steel is likely to occur.

Handled Impact Breakers

Handled impact breakers work well for limited concrete removal. These are basically jack hammer. The breaker may be powered by compressed air, hydraulic pressure;

Fig. 16.1: Placing concrete around reinforcing bars

self-contained gasoline engines, or self-contained electric motors, small, shallow sections of concrete can be demolished with great mobility when a hand-held breaker is used.

Both mechanical and welded splices are allowed by code.

16.3.5 Hydro Milling

Hydro milling is sometimes called hydro-demoliton or water-jet blasting. This type of concrete removal is effective on concrete with a depth of up to 15 cm. Sound concrete and reinforcements are not normally damaged when hydro milling is used. However, this method is expensive and can be slow when sound concrete is encountered. Another problem with this method is the likelihood of blowouts. A blowout is a hole through a concrete member. A large amount of portable water must be available. Some units use up to 1,000 gallons of water per hour. Flying rock is also a common danger.

Rotary Head Milling

Rotary head milling is used to remove deteriorated concrete from mass structures. If the compressive strength of concrete is 8,000 psi or greater, rotary head milling is not practical, remaining concrete can suffer from micro-cracking.

When feasible, splices should be located away from any points of maximum tensile stress.

16.3.6 Presplitting

Presplitting depends on wedging forces in a designed pattern of boreholes to produce a controlled cracking of concrete to facilitate removal of concrete by other means. The extent of presplitting planes is affected by the pattern, spacing and depth of boreholes. Chemical expansive agents and hydraulic splitters are used in presplitting. Reinforcing steel significantly decreases the presplitting plane. A loss of control can occur if boreholes are too far apart or are drilled in severely deteriorated concrete.

Chemical Agents

Chemical agents can be used for presplitting concrete. For this to be used the depth of boreholes should be ten times the diameter of boreholes, or more. This method is expensive. However, it works very well when presplitting vertical planes of significant depth. In the early hours of use, chemical agents can blow out boreholes and cause personal injury. It is not uncommon for rotary head milling or mechanical impacting methods to be used in conjunction with chemical presplitting methods to obtain a completed job.

When dealing with spirally reinforced compression members, lap length in a splice must not exceed 30 cm.

Piston-jack Splitters

If you are presplitting a concrete face that is 25 cm or thicker, a piston-jack splitter is a good way to go. Boreholes for this type of work must have a minimum diameter of 7.5–12.5 cm. The cost of using piston-jack splitters can be prohibitive.

Prep Work

The prep work done prior to a repair is one of the most important steps in making successful concrete repairs. Preparation work varies. It depends on the type of repair being made. In general the concrete being prepared should be sound, clean rough-textured and dry. There are exceptions, but this is the normal rule of thumb.

Fig. 16.2: Working with prepared site

Working with a Prepared Site

Normally, all deteriorated concrete should be removed prior to a repair. This is often done with impact tools, either hand-held or boom-mounted (Fig. 16.2). After secondary removal, sand blasting wet or dry, or water-jet blasting can be used to clean the sound concrete. The options for clearing concrete for a repair are:

- Chemical cleaning
- Mechanical cleaning
- Shot blasting
- Blast cleaning
- Acid etching
- Bonding agents

Chemical Cleaning

Chemical cleaning is needed when concrete is contaminated with oil, grease, or dirt. Detergents and other concrete cleaners can be used to rid the concrete of contaminants. However, the cleaners themselves must also be removed before a repair is made. Avoid the use of solvents because they may push containments deeper into the concrete. While muriatic acid is used for etching concrete, it is not very effective for removing grease or oil.

A middle strip is a strip that is bounded by two column strips.

Mechanical Cleaners

Mechanical cleaners include scabless, scarifies, and impact tools. Different heads on the tools allow for different types of abrasive material. In any event, secondary cleaning will be necessary after using mechanical cleaners. This is done with sand blasting wet or dry, or water-jetting.

Shot Blasting

Thin overlay repairs carry out for shot blasting. Thus, is the use of steel shot being blasted against the concrete a uniform surface? Once the blasting is done, the steel shot is gathered by a vaccum and saved for later use. The result is a dry surface that is ready to accept a repair.

Blast Cleaning

Blast cleaning is done with water-jetting and both wet and dry sand blasting. Sand blasting requires the use of an effective oil trap to prevent contamination of concrete surfaces during the cleaning operation. Water jetting equipment with operating pressures of 6,000 to 10,000 psi is commercially available for cleaning concrete.

Acid Etching

Acid etching is used to remove laitance and normal amounts of dirt. Cement paste is removed by the acid. This provides a rough surface. General opinion is that acid etching should be used only when other choices are not suitable.

Every middle strip must be designed and installed to proportion the resistance of the sum of movements assigned to its two half-middle strips.

Bonding Agents

Bonding agents should be used on repairs that are less than 5 cm thick. Thicker repairs can be made without bonding agents when the receiving surface is properly prepared. Many types of bonding agents are available, always refer to manufacturer's recommendation for use of their product.

Reinforcing Steel

Corrosion is the enemy of reinforcing steel. When the steel reinforcing material has to be replaced, concrete must be removed from around the existing reinforcement. A jack hammer is the tool of choice in most such cases. All weak, damaged, and easily removable concrete should be removed. Circumstances may allow existing reinforcing steel to be cleaned and left in place. To do this, at least one-half of the existing steel must not be corroded.

If an air compressor is used to clean corrosion, make sure the compressor is either an oil-free compressor or a compressor that has an oil trap. Otherwise, the concrete may be stained with blowing oil. Dry sand blasting is the best method for cleaning reinforcing steel. Wet sand blasting and water-jetting can also be effective.

When walls or columns are built integrally with a slab system, they must be constructed to resist factored loads on the slab system.

Anchors

Anchors require the drilling of holes. This is usually done with a rotary carbide-tipped drill or a diamond-studded drill bit. Using a jack hammer to create anchor holes is not recommended because of potential damage to in place concrete. Anchor holes should be cleaned out and protected from debris entering the holes. Bonded anchors can be headed or headless bolts, threaded rods, or deformed reinforcing bars. These anchors may be either grouted anchors or chemical anchors. Grouted anchors are embedded in predrilled holes with Portland cement, Portland cement and sand, or other commercially available premixed grout. An expansive grout additive and acceleration are commonly used with cementitious grouts. Chemical anchors are embedded in predrilled holes with two-component polyesters, vinyl esters, or epoxies. These anchors come in glass capsules, plastic cartridges, tubes, and bulk. Once inserted into a predrilled hole, the chemical anchor casing is broken or opened. The two components mix. Epoxy products are used in bulk systems. The epoxy is mixed in a pot or pumped through a mixer and injected into a hole.

Expansion anchors are placed in predrilled holes and expanded by tightening a nut, hammering the anchor, or expanding into an undercut in the concrete. Expansion anchors that rely on side-point contact to create frictional resistance should not be used where anchors are subjected to vibratory loads. Certain wedge-type anchors do not perform well when subjected to impact loads. Undercut anchors are suitable for dynamic and impact loads. After any anchors are placed and ready for use, the anchor's strength should be tested. This is a step that should not be overlooked. It is desirable to specify a maximum displacement in addition to the minimum load capacity.

17

Strategies and Techniques for Repair/Rehabilitation/ Retrofitting of Structures

The main purpose of repair is to bring back the architectural shape of the building, so that all services work and function properly. Repair does not pretend to improve the structural strength of the building and can be very deceptive for meeting the strength requirements of the future earthquake. The action will include the following:

- Patching up of the defects, such as cracks, fall of plaster, etc.
- Repairing doors, windows, replacement of glass panes
- Checking and repairing electric wiring
- Checking and repairing gas pipes, water pipes, and plumbing services
- Rebuilding nonstructural walls, smoke chimneys, boundary walls, etc.
- Replastering of walls, as required
- Rearranging disturbed roofing tiles
- Relaying cracked flooring
- Redecoration: White/colour washing, distemper and painting, etc.

17.1.1 Restoration/Rehabilitation

- It is the restitution of the strength the building had before the damage occurred. This type of action should be undertaken, when there is evidence that the structural damage can be attributed to exceptional phenomenon that are not likely to happen again and that the original strength provides an adequate level of safety.
- The main purpose of restoration is to carry out structural repairs to load bearing elements. It may involve cutting portions of the elements and rebuilding them or simply adding more structural material, so that the original strength is more or less restored. The process may involve temporary supports, underpinning, etc. Some of the approaches are brought out below:
  - i. Removal of cracked masonry walls, piers and rebuilding them in richer mortar. Use of nonshrinking mortar, like polymer modified mortar will be preferable.
  - ii. Addition of reinforcing mesh on both faces of the cracked wall, holding it to the wall through spikes or bolts and then covering it suitably. Several alternatives have been in use.
  - iii. Injecting epoxy like material, which is strong in tension, into the cracks in walls, columns, beams, etc.

17.1.2 Strengthening/Retrofitting of Existing Buildings (Objectives)

i. Increasing lateral strength in one or more directions, by reinforcement or increasing wall areas or the number of walls and columns.

ii. Giving unity to the structure by providing proper connection between its resisting elements, in such a way that inertia forces generated by the vibration of building can be transmitted to the members that have the ability to resist them. Typical important aspects are the connections between roofs or floors and walls, between intersecting walls and between walls and foundations.

iii. Eliminating features that are sources of weakness or that produce concentrations of stresses in some members, abrupt changes of stiffness from one floor to other, concentration of large masses, large opening in walls without a proper peripheral reinforcement are examples of defect of this kind.

iv. Avoiding the possibility of brittle modes of failure by proper reinforcement and connection of resisting members. Since its cost may be as high as 50 to 60% of the cost of rebuilding, the justification of such strengthening should be fully considered.

The extent of modification should be found using the principles of strengthening and in accordance with local factors applicable to each building.

17.2 TECHNIQUES TO RESTORE ORIGINAL STRENGTH

While considering restoration work, it is important to realize that even fine cracks in load bearing members which are un-reinforced, like masonry and plain concrete reduce their resistance very largely. Therefore, all cracks must be located and marked carefully and critical ones fully repaired either by injecting strong cement or chemical grout or by providing external bondage. The techniques are described below along with other restoration measures:

i. **Small cracks:** If the cracks are reasonably small (opening width = 0.075 cm), the technique to restore the original tensile strength of the cracked element is by pressure injection of epoxy. The procedure is as follows:

The external surfaces are cleaned of non-structural materials and plastic injections ports are placed along the surface of the cracks on both sides of member and are secured in place with an epoxy sealant. The centre to centre spacing of these ports may be applied equal to thickness of the element. After the sealant has cured, a low viscosity epoxy resin is injected into one port at a time, beginning at lowest part of the crack in case it is horizontal. The resin is injected till it is seen flowing from the opposite sides of the member at the corresponding port or from the next higher port on the same side of member. The injection port should be closed at this stage and injection equipment moved to the next port and so on. The smaller the crack, higher is the pressure or more spaced should be the ports so as to obtain complete penetration of the epoxy material throughout the depth and width of member. Larger cracks will permit larger port spacing, depending upon width of member. The technique is appropriate for all types of structural elements, viz. beams, columns, walls and floors. The resin is injected till it is seen flowing from the opposite sides of the member at the corresponding port or from the next higher port on the same side of member. The injection port should be closed at this stage and injection equipment moved to the next port and so on.

ii. **Larger cracks and crushed concrete:** For cracks wider than 6 mm or regions in which the concrete or masonry has crushed, a treatment other than injection is adopted. The following procedures may be adopted:

- The loose material is removed and replaced with materials like expansive cement mortar, polymer modified mortar, quick setting cement or gypsum cement mortar.

- Where found necessary, additional shear or flexural reinforcement is provided in the region of repairs. This reinforcement could be covered by mortar to give further strength as well as protection to the reinforcement.
- In areas of very severe damage, replacement of the member can be carried out.
- In case of damage to walls and floor, steel mesh be provided on the outside of the surface and nailed or bolted to the wall. Then it can be covered with plaster or microconcrete.

iii. Fractured, excessively damaged member: In case of severely damaged RC member, it is possible that the reinforcement would have buckled, or elongated or excessive yielding may have occurred. This element can be repaired by replacing the old portion of steel with new steel using butt welding or lap welding. Additional stirrup ties are to be added in the damaged portion, before concreting so as to confine the concrete and enclose the longitudinal bars to prevent their buckling in future. In some cases it may be necessary to anchor additional steel into existing concrete.

17.3 STRENGTHENING RC MEMBERS

Suggestions for strengthening are:

- RC columns can be best strengthened by jacketing, and by providing additional cage of longitudinal and lateral tie reinforcement around the columns and casting a concrete ring, the desired strength and ductility can thus be built up.
- Jacketing a reinforced concrete beam can also be done in the above manner. For holding the stirrup in this case, holes will have to be drilled through the slab.
- Similar technique could be used for strengthening RC shear walls.

The diagrammatic representation of gunite process and example of grouting process are shown in Figs 17.1 and 17.2 respectively.

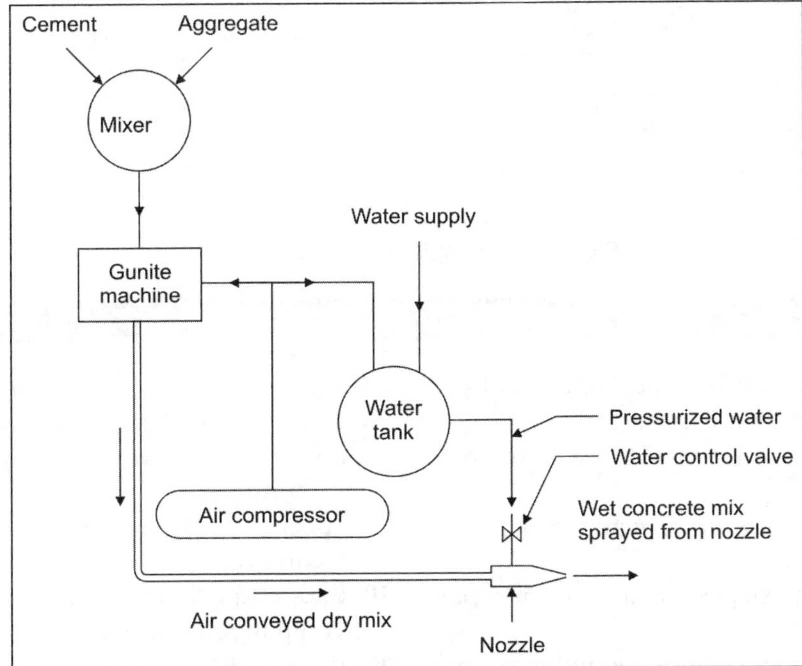

Fig. 17.1: Diagrammatic representation of gunite/grouting process

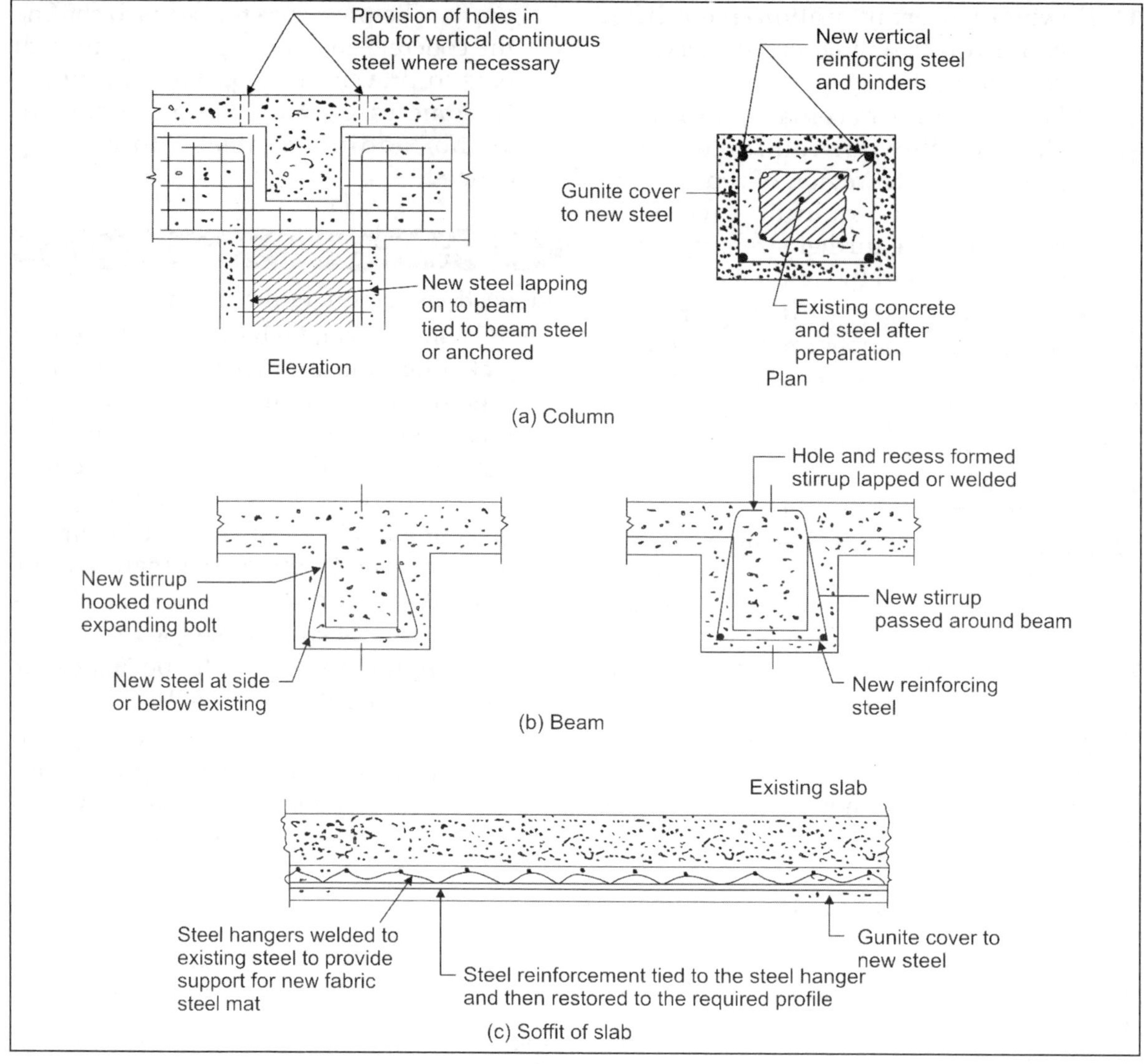

Fig. 17.2: Examples of gunite process

SHORT ANSWER QUESTIONS

1. How do you get economical rehabilitation?
2. How do you select repair methods?
3. What are the objectives followed in repair procedure?
4. What are factors on which repair procedure depend?
5. What are the stages carried out in repair procedures?
6. What are the steps involved in the surface preparation for repair?
7. Write about autogenous healing.
8. What are the design criteria suggested by researchers who work in plate bending techniques?
9. Specify the limitations of plate bending technique.
10. Explain in short about acid etching.
11. Write in short about chalking.
12. What is the use of vacuum impregnation?
13. Define sprayed concrete.

14. How do you repair reinforcement corrosion damaged concrete by sprayed concrete?
15. Outline the format of surface seals.
16. Write about leak sealing in concrete pipelines.
17. What is meant by preplaced aggregate concrete?
18. What is ground floor slabs?
19. What is a suspended floor?
20. Define toppings and screeds.
21. Write about gravity-fed crack sealing.
22. Define flexible sealing.
23. What is the function of overlays?
24. How do you repair dormant cracks?
25. How do you repair leaking cracks?
26. What is the use of preplaced aggregate concrete?
27. List the limitations of preplaced aggregate concreting method.
28. Define slab jacking technique.
29. What are the advantages of slab jacking technique?
30. Write down the disadvantages and limitations of slab jacking technique.
31. Define the term shotcrete.
32. What is the purpose of using shotcrete?
33. What are the advantages of shotcrete?
34. Write down the disadvantages of shotcrete.
35. What is tremie concrete?
36. Define sealers.
37. Write about chloride extraction and alkalization of concrete.
38. How do you repair cracks in concrete?
39. Classify the types of repair.
40. Explain the corrective measures for cracks.
41. Explain the externally bonded mild steel plates as strengthening technique.
42. Explain the procedure of strengthening with external reinforcement.
43. What is the function of overlaps?
44. What is the purpose of surface coatings?
45. What is meant by anticarbonation coatings?
46. How does coatings protect cracked concrete?
47. How is leak sealing applied?
48. Explain leak sealing conventional method.
49. Explain leak sealing injection techniques.
50. What are the factors that decide the rehabilitation techniques?
51. Explain shotcrete/gunite.
52. What is meant by dry pack?
53. Write about preplaced aggregate concrete.
54. What are the stages carried out in repair process?
55. What are the aspects comprised in pretreatment?
56. How do you remove loose material on the surface of concrete?
57. Explain the grinding process.
58. What is the purpose of sand blasting?
59. What is the use of mortar replacement system?
60. What is meant by grouting?
61. What is the use of prepacked concrete?
62. How do you use polymer impregnation in repair of fractured elements?
63. Explain vacuum impregnation in repair of fractured elements.
64. Explain vacuum impregnation process in concrete repair.
65. Explain resin injection technique.
66. What is the function of externally bonded mild steel plates?
67. What is meant by durable repair system?
68. Define the procedure of repair to re-repair.
69. List out the safety measures in dams.
70. Name the rehabilitation techniques for earthquake damage.

THEORY QUESTIONS

1. What do you mean by repair, rehabilitation and retrofitting? What is the difference between them? Explain in details.
2. Explain with neat sketches about the various techniques used in the repair of cracks.
3. Explain in detail about the plate bonding strengthening technique.
4. Explain with sketches about the fiber sheet bonding strengthening technique.
5. What is ferrocement? Explain about ferrocement jacketing.
6. Explain the procedure of epoxy coating of rebar.
7. With the help of a simple sketch, explain guniting/shotcreting.
8. How do you strengthen the various structural elements? Explain in detail with sketches.
9. Explain in detail about acid etching and resurfacing.
10. Explain with sketches the retrofitting of damaged RC members by polymer modified concrete and ferrocement.
11. Explain with neat sketches about the jacketing techniques for the repair of seismically affected beam–column joint in a building.
12. What are the different types of plates/laminates used for strengthening of structural elements? Discuss the different strengthening techniques and their relative merits.
13. Give a complete procedure of epoxy injection to structural crack repair in RC structural elements.
14. Explain the classification of cracks and how the cracks can be measured and reported.
15. Explain coating systems adopted in concreted repairs.
16. Explain the replacement concrete and preplaced aggregate concrete techniques.
17. Explain grouting and dry packing concrete techniques.
18. Explain the considerations in structural repairs in reinforced concrete structures.
19. Explain strengthening techniques with external reinforcement.
20. Explain section enlargement and short span strengthening techniques.
21. Explain external post-tension strengthening techniques.
22. Explain the durable repair system of concrete structures.

Part 5
Retrofitting of Structures

18. Techniques for Strengthening/Retrofitting
19. Repair/Retrofit of Nonengineered Structures
20. Retrofit of Historical and Heritage Structures
21. Retrofit of Reinforced Concrete Structures
22. Retrofit of Foundations
23. Retrofit Using FRP Composites and Base Isolation Technique

Techniques for Strengthening/Retrofitting

18.1 INTRODUCTION

A significant number of facilities were constructed during the first half of the 20th century using reinforced or prestressed concrete materials. Now, at the beginning of the next century, many of these buildings have reached the end of their planned service life, and deterioration in the form of steel corrosion, concrete cracking, and spalling is observed frequently. In addition, many of these structures were built to carry loads that are significantly smaller than the current needs. Because of these factors, many structural engineers are faced with the challenge of evaluating and implementing effective and economical repair and strengthening programs.

Unfortunately, there is no single solution that offers a simple, straightforward method for all repair and strengthening projects. Further, the processes of repair and retrofit of existing structures are complicated because most of these structures are occupied, and much of the mainstream construction community's expertise is centered on new construction. However, success can be achieved, if the repair and strengthening systems are tailored to serve a structure's intended use without interfering with its occupants or function. The key to success is a combination of the different design skills and application techniques—structural strengthening and structural repair—necessary for such projects. As such, the engineer must relay his or her expertise in using mechanical and structural behavior principles to develop a comprehensive retrofit solution.

Strengthening measures are required in structures when they are required to accommodate increased loads. Also, when there are changes in the use of structures, individual supports and walls may need to be removed. This leads to a redistribution of forces and the need for local reinforcement. In addition, structural strengthening may become necessary owing to wear and deterioration arising from normal usage or environmental factors.

18.1.1 Need for Strengthening

Concrete structures need to be strengthened for any of the following reasons:

- Load increases due to higher live loads, increased wheel loads, installations of heavy machinery, or vibrations.
- Damage to structural parts due to ageing of construction materials or fire damage, corrosion of the steel reinforcement, and/or impact of vehicles.

- Improvements in suitability for use to limitation of deflections, reduction of stress in steel reinforcement and/or reduction of crack width.
- Modification of structural system due to the elimination of walls/columns and/or openings cut through slabs.
- Errors in planning or construction due to insufficient design dimensions and/or insufficient reinforcing steel.

18.1.2 Terms of Repair

Concrete experts commonly use the terms structural repair and strengthening to describe building renovation activities. Although the two terms sound similar, they refer to slightly different concepts. Structural repair describes the process of reconstruction and renewal of a facility or its structural elements. This involves determining the origin of distress, removing damaged materials and causes of distress, as well as selecting and applying appropriate repair materials that extend a structure's life.

Structural strengthening, on the other hand, describes the process of upgrading the structural system of an existing building to improve performance under existing loads or to increase the strength of structural components to carry additional loads. For upgrade projects, design engineers must deal with structures in which every element carries a share of the existing load. The effects of strengthening or removing part or all of a structural element, such as penetrations or deteriorated materials must be analyzed carefully to determine their influence on the global behavior of the structure. Failure to do so may overstress the structural elements surrounding the affected area, which can lead to a bigger problem and even localized failure.

18.1.3 Structural Concrete Repair

Although durable, buildings constructed using reinforced and prestressed concrete have a finite service life. When exposed to harsh environments, deicing salts, and other chemicals, these structures may experience significant deterioration, which typically occurs in the form of steel corrosion, concrete spalls, delamination, and cracks.

Interestingly, one of the most severe and widespread problem in concrete is the internal damage caused by the corrosive action of external chlorides on reinforcing or prestressing steel embedded in concrete. Corrosion problems are caused by a corrosion process by-product (rust) that expands up to eight times its original volume. This expansion creates internal pressure, which causes the concrete to crack and spall, resulting in a reduction of the effective area of steel reinforcement and reduced structural capacity of the affected member. If not addressed at early stages, corrosion will continue to grow rapidly, ultimately creating a safety issue of falling concrete and loss of strength.

Figure 18.1 shows an unsuccessful concrete repair in structural member. The repair materials started to delaminate just a few months after the first repair was completed. Adequate surface preparation could have prevented this repair failure. The assessment, design and implementation of a durable repair to an existing structure are indeed more complex than for new construction. In addition to the unknown state of existing structural materials, the degree to which repair materials and the existing material will act as a composite and share loads must be addressed. Prior to establishing a repair strategy, the concrete-repair expert must diagnose the problem's root cause, which enables prescribing repairs

Fig. 18.1: Unsuccessful concrete repair

that are long lasting and durable. Failing to follow this process may result in a frustrating, but common, cyclical outcome known as "repairing the repair".

18.1.4 Structural Repair Techniques for Reinforced Concrete

Structural repair describes the process of reconstruction and renewal of a facility or its structural elements. This involves determining the origin of the distress, removing damaged materials and cause of distress, as well as selecting and applying appropriate repair materials that extend the structure's useful life. In addition to proper repair techniques, regularly scheduled periodical maintenance is essential for long term measures.

Essential Preparation

A crucial step in achieving a durable repair is surface preparation. Multiple close-up observations of spalled concrete that had been repaired previously indicate that in many cases repairs have failed not because of material or technology, but because of the poor quality of surface preparation. The care with which deteriorated concrete is removed and the concrete and steel reinforcement surfaces are prepared often will determine whether a repair project will turnout to be successful. Equipment selection is important, because the concrete removal method should not weaken or crack the surrounding sound concrete. Also care should be exercised to avoid further damage to the reinforcing and prestressing steel. This is achieved by using only industry-specified, lightweight demolition equipment. The effect of concrete removal on the structural integrity should be investigated carefully. A temporary shoring system may be required to relieve the loads on repaired elements in cases where removal of concrete or corroded reinforcing steel is significant enough to affect the structures load-carrying capacity.

Once concrete has been removed, the reinforcing steel should be carefully inspected to determine whether the steel should be simply cleaned, repaired, or replaced. The final step in concrete and steel surface preparation is cleaning using abrasive or water-blasting techniques to remove loose materials and achieve an open-pore structure for the exposed concrete substrate and to remove the rust from the steel. This results in an appropriate surface to bond the new repair material and stop the corrosion process.

Cost of Concrete Repair Versus Time

The cost of repairs becomes exponentially more expensive when preventative maintenance is not scheduled regularly or budgeted for a concrete facility. There are three observed phases describing the natural evolution of the concrete deterioration process and the influence of maintenance on this process, viz. preventative maintenance, repair, and replacement. During the preventative maintenance phase, the owner may spend a fixed, annual maintenance cost to install systems, such as protective coatings, to slow down the deterioration process. Money spent in this phase will delay the ingress of aggressive materials, thus delaying the start of active deterioration (repair phase). In the repair phase, the concrete deterioration has begun, and the repair cost curve increases exponentially overtime. The reason for the rapid increase in cost is that once aggressive materials that cause deterioration have permeated sufficiently into the concrete (a process that may take 20 to 30 years), the deterioration occurs throughout the structure. However, total replacement of the structure may not be an option because of interruption to the function of the structure. Incurring additional costs in the early years of a structure's life to ensure well protected concrete and addressing deterioration problems as soon as they are observed would delay excessive deterioration and may increase significantly the service life of a structure.

18.2 STRUCTURAL CONCRETE STRENGTHENING

Many buildings that originally were constructed for a specific use now are being renovated or upgraded for a different application that may require higher load-carrying capacity. As a result of these higher load demands, existing structures need to be reassessed and may require strengthening to meet heavier load requirements. In general, structural strengthening may become necessary because of codal changes, seismic upgrade, deficiencies that develop because of environmental effects (such as corrosion), changes in use that increase service loads, or deficiencies within the structure caused by errors in design or construction. The structural upgrade of concrete structures can be achieved using one of many different upgrading methods, such as span shortening, external composites, externally bonded steel, external or internal post-tensioning systems, section enlargement, or a combination of these techniques. Similar to concrete repair, strengthening systems must perform in a composite manner with an existing structure to be effective and to share the applied loads. The strengthening is generally done by adopting: (i) jacketing technique and (ii) externally bonding technique.

Strengthening or stiffening of members, such as columns and girders is usually achieved either by replacing poor quality or defective material by better quality material or by adding additional material to the member. In either case, the new material will usually be reinforcing steel, high quality concrete, thin steel plates and straps or various combinations of these materials. The main difficulty in this type of operation is to achieve continuity of structural action between the original material and the new material. The various techniques of bolting, gluing, dowelling and keying have been developed to provide positive force transfer and composite action.

18.2.1 Column Strengthening

One of the simplest and most effective methods for strengthening a column in an existing building is to partially load the column by jacketing between floors and then insert two or more props to carry portion of the axial load. The props are usually rolled steel sections which may subsequently be encased in concrete to improve fire protection and appearances. The disadvantages of this method are that considerable floor space is lost, and that the props may not be effective in transferring moment unless positive connection details are introduced at both ends, e.g. in the form of end plates bolted through holes drilled in the floors.

18.2.2 Strengthening and Stiffening of Beams and Girders

Commonly used methods for strengthening flexural members include the following:

- Provision of additional concrete on the compressive face
- Addition of tensile reinforcement with a cast-in-place or gunited cover
- Bolting or providing steel plates or straps to the surface of the member.

The main problem of strengthening, stiffening and repair operation is to ensure good composite action between old and new material. Bolting of new concrete or steel to the existing member is effective but time consuming and expensive. For this reason, gluing procedures have been developed which provide strong and durable connections between concrete and surface straps and plates.

18.2.3 Strengthening and Stiffening of Slabs

The simplest and cheapest procedure of strengthening and stiffening of slabs is by providing additional props either in span or in the vicinity of existing columns and walls. However, this solution will often be unacceptable for architectural reason, even though props placed quite close to existing columns may be sufficient to bring the structure to an acceptable standard of serviceability.

18.2.4 Jacketing Technique

Jacketing is the process of fastening a durable material over concrete and filling the gap with a grout that provides needed performance characteristics. The materials used for jacket are metal, rubber, plastics, ferrocement and concrete. A typical jacketing process is shown in Fig. 18.2. This restores structural values, protects the reinforcements from exposures to the harmful elements and improves the appearance of the original concrete. Jacketing materials may also be secured to concrete, or by gravity. The method of securing employed will depend upon the pressure, the material used and the positioning of the jacketing material. Fiber glass reinforced plastics, ferrocement and other materials, such as polypropylene can be used for jacketing.

The column jacket can also be used for increasing the punching shear strength of column slab connections by using it as a column capital. When the jacket is provided around the periphery of the column, it is termed a collar. In most of the applications, the main function of the collar is to transfer vertical load to the column. Circular reinforcement can be used for load transfer. The practice of transferring load through dowel bars embedded into columns or shear keys has a disadvantage in that they require drilling of holes for dowels or cutting shear keys which are costly and time consuming and can damage the existing column. Reinforcement encircling the column can be used to transfer the load through shear friction. The expansion of collar as it slides along the roughened surface causes the tensioning of circular reinforcement resulting in radial compression, which provides normal force needed for load transfer. The shear transfer strength is provided by both frictional resistance to sliding and dowel action of reinforcement crossing the crack.

The collar can also be used as mid-column bearing surface, acting as circumferential beam to distribute the concentrated load around the column. The collar is subjected to shear and bending along the collar circumference as well as direct bearing stress under concentrated load. Thus, in addition to shear transfer reinforcement, the collar should be provided with reinforcement for shear and moment within the collar. The repair can be used as an alternate load path from the column to the collar and then to the connecting structural component. Column collars can be provided below the slab to act as column capital to improve punching shear strength of the slab column connection.

18.2.5 Externally Bonding Technique

In this, suitable laminates are pasted externally in the flexural zone to improve upon the strength and deformation characteristics of distressed structural members. The laminates may be fabricated from anyone of the following material: Steel, stainless steel, FRP, ferrocement, and SIFCON and SIMCON. The laminates are bonded to the distressed structural elements using epoxy resin. Design criteria suggested by researchers that worked in plate bending techniques are:

- The plate/laminates width to thickness ratio should not be greater than 50.
- The central axis depth should not be greater than 0.4 times the effective depth of member.

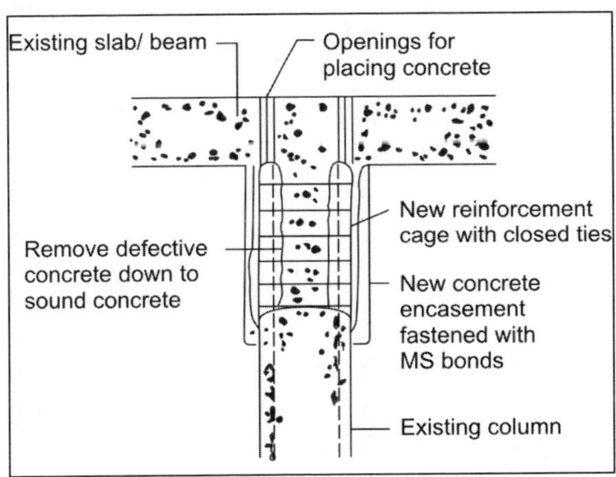

Fig. 18.2: Process of jacketing technique

Strengthening with external bonding of plates has the following limitations:

- Difficulty in manipulating the plates in field conditions
- Deterioration of the bend steel concrete interface
- Lack of proper formation of joints.

A typical process of plate bonding technique using steel plates is shown in Fig. 18.3.

Externally Bonded Mild Steel Plates

The technique of gluing mild steel plates to the soffit of reinforced concrete beams and slabs can be applied to improve the structural performance of the existing structures. It increases the strength and rigidity and also reduces the flexural crack width in the concrete. The technique can be used to enhance the load carrying capacity either due to increase in load requirements or design or construction deficiencies. This technique has many advantages when applied to bridges. The plates can be easily handled and placed, thus allowing the procedure to be carried out relatively quick with minimal disruption to traffic. However, this technique is not recommended for application to bridges with unformed concrete or those affected by corrosion unless the source of the problem is rectified. The solution of an appropriate adhesive and surface preparation of the bonding surface are the most critical parameters for the successful implementation of this technique. The practical considerations to facilitate the placement of plates include the temperature requirements for mixing, applying and curing adhesive should be consistent with the existing conditions in the field and the adhesive should develop its strength rapidly.

Bonded Steel Elements

Strengthening concrete members by using bonded steel plates was developed in the 1960s. With this method, steel elements are glued to the concrete surface by a two-component epoxy adhesive to create a composite system

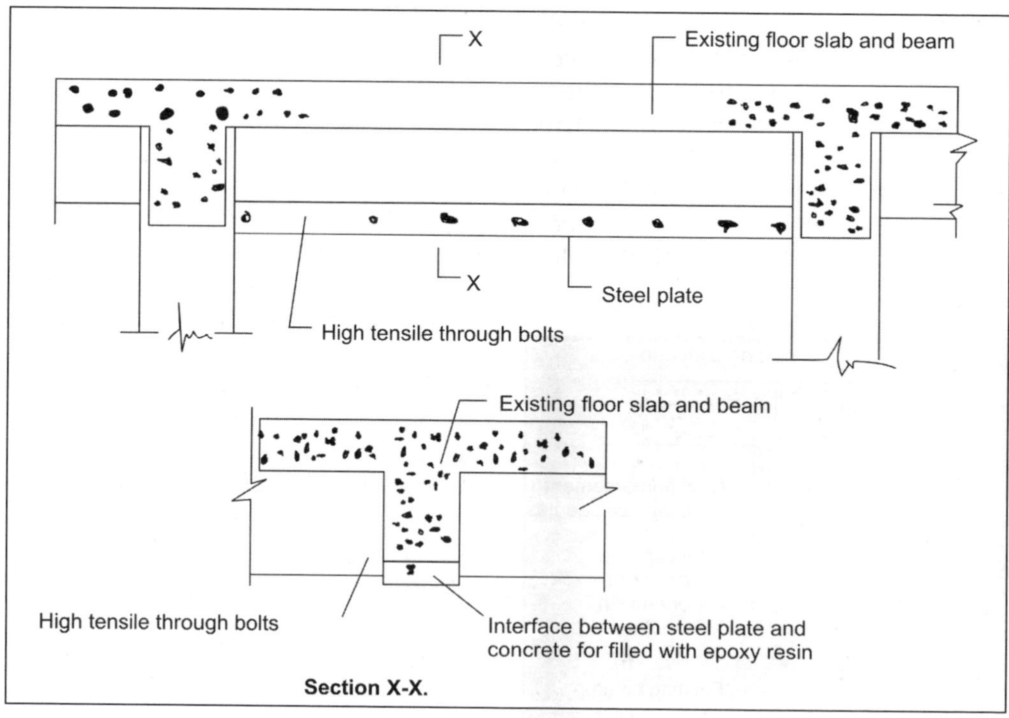

Fig. 18.3: Process of plate bonding technique by steel plates

and improve shear or flexural strength. The steel elements can be steel plates, channels, angles, or built-up members. In addition to epoxy adhesive, mechanical anchors typically are used to ensure that the steel element will share external loads in case of adhesive; mechanical anchors typically are used to ensure that the steel element will share external loads in case of adhesive failure. The exposed steel elements must be protected with a suitable system immediately following installation, and regardless of the specified corrosion protection system, its long-term durability properties and maintenance requirements must be considered fully. A roof strengthened with FRP and steel plates externally is shown in Fig. 18.4.

The roof consisted of prestressed concrete and hollow planks. Installation of the skylights required cutting openings in the planks that would reduce their load-carrying capacity. This issue was resolved by designing a hybrid strengthening system composed of FRP fabric and steel elements. The externally bonded FRP strengthened the planks adjacent to the one to be cut, while the steel elements tied the plank to the adjacent ones, thus creating a new unit consisting of three planks with adequate capacity. In addition to the fast application of this system, this was a cost effective solution that was aesthetically pleasing also.

Fig. 18.4: Externally bonded steel plates with FRP

18.3 STRENGTHENING WITH EXTERNAL REINFORCEMENT

Fiber composites are becoming increasingly important in the construction industry with a number of applications in many areas. The Fibers, such as glass, carbon and aramid can be introduced in a certain position, volume and direction in the matrix to obtain the maximum efficiency. Other advantages offered by the Fiber reinforced are greater efficiency in construction when compared with the more conventional materials. Some important applications of FRP are as follows:

- Use of FRP castings to confine new concrete beams and columns or to strengthen beams in seismic area
- Development of prestressing tendons made of parallel alignment of continuous Fibers
- Strengthening of concrete beams with unidirectional Fiber reinforced concrete sheets bonded on their tension faces by the use of epoxy adhesives.

18.3.1 Composites

Fiber reinforced polymer (FRP) systems are high-strength, lightweight reinforcement in the form of paper thin fabric sheets, thin laminates, or bars that are bonded to concrete members with epoxy adhesive to increase their load carrying capacity. These systems have been used extensively in the aerospace, automotive, and sport equipment industries, and now are becoming a mainstream technology for the structural upgrade of concrete structures. Important characteristics of FRPs for structural repair and strengthening are lower cost, and aesthetic appeal.

As with any other externally bonded system, the bond between the FRP system and the existing concrete is critical, and surface preparation is very important. Typically, installation is achieved by applying an epoxy adhesive to the prepared surface, installing the FRP reinforcement into the epoxy and, when required, applying a second layer of the epoxy adhesive. After curing, the FRP composite will add capacity to the element because it has a tensile strength up to 10 times that of steel.

Figure 18.5 shows a schematic for the structural strengthening of a utility tunnel. The utility tunnel roof originally functioned as a pedestrian walkway. A new dormitory structure required the walkway to be the primary access for emergency vehicles. Analysis of the tunnel's top slab revealed that it did not have adequate strength to carry loads from fire trucks and other emergency vehicles. A structurally efficient, easy to install, and cost effective strengthening option was achieved by using externally bonded FRP sheets. The strengthening solution consisted of carbon FRP sheets bonded to the bottom of the slab, serving as additional bottom tension reinforcement. In addition, the overhanging portions of the slab were strengthened using carbon FRP bars epoxy-bonded in grooves made on the slab's top side. The latter technique is more appropriate than FRP sheets since the bars were bonded below the surface, thereby avoiding traffic damage to the externally bonded reinforcement.

In addition to FRP, steel reinforced polymer composites (SRP) may be used as externally bonded reinforcement. This steel-based, innovative strengthening system (known as hardwire) was developed and first introduced in the market in 2002. Hardwire is a low-cost, reinforcement system consisting of ultra-high-strength steel wires that are twisted together to form reinforcing steel cores approximately 9 mm in diameter. The steel wires have a tensile strength equal to about 10 times more than the strength of conventional structural steel, and the same in elastic modulus. This strengthening system can be applied using epoxy or cementation materials and can be used to increase the shear and flexural capacity of structural elements.

18.4 SHORT SPANNING

Span shortening is accomplished by installing additional supports underneath existing members. Appropriate materials for span shortening include structural steel members and cast-in place reinforced concrete members, which are simple to install. Connections can be designed easily using bolts and adhesive anchors. The structural steel system shown was installed on a parking deck to shorten the span and carry part of the load, transferring it to the existing supporting system. On the downside, such applications may result in loss of space and reduced headroom.

18.5 EXTERNAL POST-TENSIONING

The external post-tensioning technique has been used effectively to increase the flexural and shear capacity of both reinforced and prestressed concreted members since the 1950s. With this type of upgrading, active external forces are applied to the structural member using post-tensioned (stressed) cables to resist new loads. Because of the minimal, additional weight of the repair system, this technique is effective and economical, and has been employed with great success to correct excessive deflections and cracking in beams and slabs, parking structures, and cantilevered members.

The post-tensioning forces are delivered by means of standard prestressing tendons or

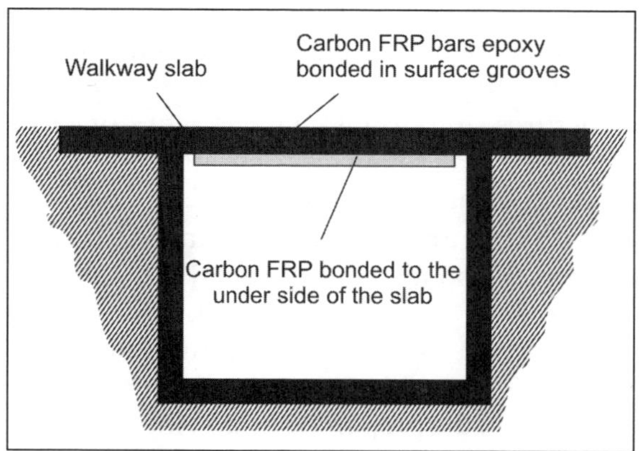

Fig. 18.5: Strengthening with carbon FRP sheets

high strength steel rods, usually located outside the original section. The tendons are connected to the structure at anchor points, typically located at the ends of the member. End-anchors can be made of steel fixtures bolted to the structural member, or reinforced concrete blocks that are cast into place. The desired uplift force is provided by deviation blocks, fastened at the high or low points of the structural element. Prior to external prestressing, all existing cracks are epoxy-injected and spalls are patched to ensure that prestressing forces are distributed uniformly across the section of the member.

Figure 18.6 illustrates an external post-tensioning system used to strengthen prestressed double tees damaged by vehicular impact. Four double tee stems on an overpass located were damaged when the driver of an over-height truck failed to observe the posted height restriction. The four stems suffered excessive concrete cracking and spalling, and damage occurred to some of the internal prestressing steel.

Proposed solutions included replacing the damaged double tees with new ones and installing a steel frame underneath for support. Both options would render the overpass out of service for a longer than desired period. The option of an external post-tensioning system was more economical, required less time to complete, and allowed for a strengthening system that provided active forces. Therefore, it was more compatible with the existing construction. After all cracks were injected, the sides of the stems were formed and new concrete was cast to restore the integrity of the

installed and after the concrete cured the external strands were stressed according to the engineer specified forces.

18.6 SECTION ENLARGEMENT

This method of strengthening involves placing additional bonded reinforced concrete to an existing structural member in the form of an overlay or a jacket. With section enlargement, columns, beams, slabs, and walls can be enlarged to increase their load-carrying capacity or stiffness. A typical enlargement is approximately 5–7.5 cm for slabs and 7.5–12.5 cm for beams and columns.

Figure 18.7 depicts details of a section enlargement used to increase the capacity of a main girder in a university parking garage. The girder was re-evaluated because of a change in the required loading and found to be deficient in flexure and shear. To correct the deficiency, additional flexural and shear steel were added. The entire beam then was formed and a 10 cm jacket of concrete was cast to enlarge the section.

18.7 STRENGTHENING BY SIMCON

The experimental results demonstrate that SIMCON exhibits improved properties in tension, compression, flexure and shear even when comparatively low fiber volume fraction fiber mats are used. Furthermore, since fiber-mats are prepacked in the plant, distribution and orientation of fibers can be more

Fig. 18.6: External post-tensioning technique

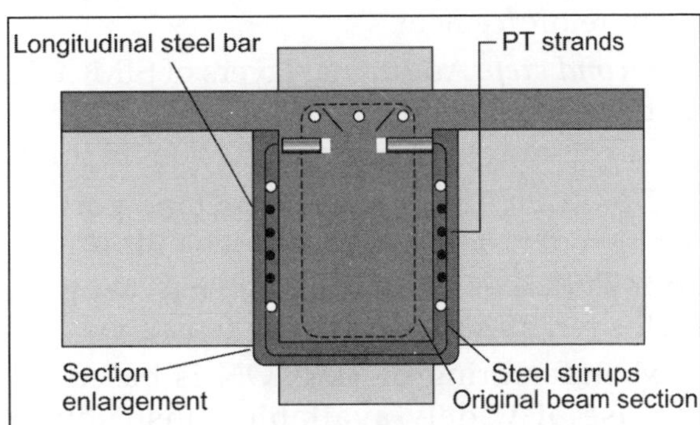

Fig. 18.7: Section enlargement technique

accurately controlled than in the case with short discontinuous fiber HPFRCs. These characteristics allow for the manufacturing of a unique cement-based fiber composite that can have different yet easily controllable properties in the longitudinal and transversal directions. These material characteristics are desirable in repair/retrofit of structural elements, such as columns, which require a high increase in strength and toughness in the transverse direction while increasing only ductility but not strength in the longitudinal direction, i.e. "moment-carrying" direction.

The investigation also demonstrates that SIMCON has considerable potential for both seismic repair/retrofit as well as the development of novel, high-performance composite structural systems. In a retrofit situation, continuous SIMCON fiber-mats, delivered in large rolls, can be easily installed by wrapping around members to be rehabilitated.

In new construction of high performance composite frames, SIMCON is well suited for manufacturing high strength, high ductility, and thin stay-in-place formwork elements that eliminate the need for secondary and most of the primary reinforcement. The presence of a SIMCON layer led to optimization of member dimensions, amount of reinforcement and member weight.

First step: Rebars wrapped in SIMCON are put along the column to provide moment continuity through the joint region, as well as replacement of concrete with SIMCON in the anchorage region of the discontinuous bottom beam reinforcement.

Second step: Additional layers of SIMCON mat are added to increase moment capacity at the column and beam zones facing the joint.

Third step: The entire column (and portion of adjacent beams) is jacketed with SIMCON. Formwork is put next and mat is injected with high-strength slurry (say 14,000 psi).

Manufacturing of SIMCON is based on the use of widely available construction equipment and building expertise, and can

thus be relatively easily introduced into the field without major retraining and changes in existing construction practices. Hence, this novel type of HPFRCC provides some unique new ways of developing durable and cost-effective high performance infrastructural systems, essential for the economic well-being of the nation in the next century. Slurry infiltrated mat concrete (SIMCON) can be used in: (i) seismic retrofit, (ii) the development of a novel, partially-cast-in-place high performance composite frame, and (iii) the development of a "self-stressing" SIMCON stay-in-place formwork that can provide active confinement after the core of the member has been cast-in-place.

18.8 DAM SAFETY: CONCRETE REPAIR/ RETROFITTING TECHNIQUES

Concrete is an inexpensive, durable, strong and basic building material often used in dams for core walls, spillways, stilling basins, control towers, and slope protection. However, poor workmanship, construction procedures, and construction materials may cause imperfections that later require repair. Any long-term deterioration or damage to concrete structures caused by flowing water, ice, or other natural forces must be corrected. Neglecting to perform periodic maintenance and repairs to concrete structures as they occur could result in failure of the structure from either a structural or hydraulic standpoint. This, in turn, may threaten the continued safe operation and use of the dam.

18.8.1 Consideration

Floor or wall movement, extensive cracking, improper alignments, settlement, joint displacement, and extensive undermining are signs of major structural problems. In situations where concrete replacement solutions are required to repair deteriorated concrete, it is recommended that a professional engineer be trained to perform an inspection to assess the concrete's overall condition, and determine the extent of any structural damage and necessary remedial measures.

Typically, it is found that drainage systems are needed to relieve excessive water pressures under floors and behind walls. In addition, reinforcing steel must also be properly designed to handle tension zones and shear and bending forces in structural concrete produced by any external loading (including the weight of the structure). Therefore, the finished product in any concrete repair procedure should consist of a structure that is durable and able to withstand the effects of service conditions, such as weathering, chemical action, and wear. Because of their complex nature, major structural repairs that require professional advice are not addressed here.

18.8.2 Repair Methods

Before any type of concrete repair is attempted, it is essential that all factors governing the deterioration or failure of the concrete structure are identified. This is required so that the appropriate remedial measures can be undertaken in the repair design to help correct the problem and prevent it from occurring in the future. The following techniques require expert and experienced assistance for the best results. The particular method of repair will depend on the size of the job and the type of repair required.

- **The dry-pack method:** The dry-pack method can be used on small holes in new concrete which have a depth equal to or greater that the surface diameter. Preparation of a dry-pack mix typically consists of about 1 part Portland cement and 2½ parts sand to be mixed with water. Then enough water is to be added to produce a mortar that will stick together. Once the desired consistency is reached, the mortar is ready to be packed into the hole using thin layers.
- **Concrete replacement:** Concrete replacement is required when one-half to one square foot areas or larger extend entirely

through the concrete sections or where the depth of damaged concrete exceeds 15 cm. When this occurs, normal concrete placement methods should be used. Repair will be more effective if tied in with existing reinforcing steel (rebar). This type of repair will require the assistance of a professional engineer experienced in concrete construction.

- **Replacement of unformed concrete:** The replacement of damaged or deteriorated area in horizontal slabs involves no special procedures other than those used in good construction practices for placement of new slabs. Repair work can be bonded to old concrete by use of a bond coat made of equal amounts of sand and cement. It should have the consistency of whipped cream and should be applied immediately ahead of concrete placement so that it will not set or dry out. Latex emulsions with Portland cement and epoxy resins are also used as bonding coats.
- **Shotcrete (gunite):** A popular concrete replacement technique for repairing large areas of severely deteriorated concrete and spalled vertical and overhead faces is the use of pneumatically placed concrete or shotcrete. Shotcrete consists of a mixture of moistened cement and fine aggregate (sand) that is sprayed onto the repair area under pressure. Professional assistance and sophisticated equipment are needed to apply shotcrete.
- **Prepacked aggregate with intruded mortar:** This special commercial technique has been used for massive repairs, particularly for underwater repairs of piers and abutments. The process consists of the following steps:
  - Removing the deteriorated concrete
  - Forming the sections to be repaired
  - Prepacking the repairing area with coarse aggregate
  - Pressure grouting the voids between the aggregate particles with a cement or sand–cement mortar.

- **Synthetic patches:** One of the most recent developments in concrete repair has been the use of synthetic materials for bonding and patching. Epoxy-resin compounds are used extensively because of their high bonding properties and great strength. In applying epoxy-resin patching mortars, a bonding coat of the epoxy resin is thoroughly brushed on to the base of the old concrete. The mortar is then immediately applied and troweled to the elevation of the surrounding material.

Conclusion: Before attempting to repair a deteriorated concrete surface, all unsound concrete should be removed by sawing or chipping and the patch area thoroughly cleaned. A sawed edge is superior to a chipped edge, and sawing is generally less costly than mechanical chipping. Before concrete is ordered for placing, adequate inspection should be performed to ensure the following:

- Foundations are properly prepared and ready to receive the concrete.
- Construction joints are clean and free from defective concrete
- Forms are grout-tight, amply strong, and set to their true alignment and grade
- All reinforcement steel and embedded parts are clean, in their correct position, and securely held in place
- Adequate concrete delivery equipment and facilities are on the job, ready to go
- Capable of completing the placement without addition of unplanned construction.

18.9 GENERAL GUIDELINES FOR SEISMIC REHABILITATION OF EXISTING BUILDINGS

Advances in earthquake related technology during the past few decades have led to a realization that seismic risk to life and property can be reduced significantly by improving seismic performance of existing seismically deficient buildings. Detailed post-earthquake investigations of building failures have provided engineers with considerable information concerning the details of building design and construction that enhance earthquake resistance.

18.9.1 Seismic Vulnerability

The vulnerability of a building subjected to an earthquake is dependent on seismic deficiency of that building relative to a required performance objective. The seismic deficiency is defined as a condition that will prevent a building from meeting the required performance objective. Thus, a building evaluated to provide full occupancy immediately after an event may have significantly more deficiencies than the same building evaluated to prevent collapse. Depending on the vulnerability assessment, a building can be condemned and demolished, rehabilitated to increase its capacity, or modified so that the seismic demand on the building can be reduced. Thus, structural rehabilitation of a building can be accomplished in a variety of ways, each with specific merits and limitations related to improving seismic deficiencies.

18.9.2 Common Seismic Deficiencies

Regardless of the evaluation method used, failure to meet the stipulated performance objective implies certain seismic deficiencies. These deficiencies are described below.

Global Strength

Global strength typically refers to the lateral strength of the vertically oriented lateral force resisting system. For degrading structural systems characterized by a negative post yield slope on the pushover curve, a minimum strength requirement may apply. In certain cases, the strength will also affect the total expected inelastic displacement and added strength may reduce nonlinear demands into acceptable ranges. A deficiency in global strength is common in older buildings either due to a complete lack of seismic design or a design to an early building code with inadequate strength requirements. If prescriptive equivalent lateral force methods or linear static procedures have been used for evaluation, inadequate strength will directly relate to unacceptable demand-to-capacity ratios within elements of the lateral force resisting system.

Global Stiffness

Global stiffness refers to the stiffness of the entire lateral force-resisting system although the lack of stiffness may not be critical at all levels. For example, in buildings with narrow walls, critical drift level occurs in the upper floors. Conversely, critical drifts most often occur in the lowest levels in frame buildings. Stiffness must be added in such a way that drifts are efficiently reduced in the critical levels. Although strength and stiffness are often controlled by the same existing elements or the same retrofit techniques, the two deficiencies are typically considered separately. Failure to meet evaluation standards is often the result of a building placing excess drift demands on existing poorly detailed components.

Configuration

This deficiency category covers configuration irregularities that adversely affect performance. In codes for new buildings, these configuration features are often divided into plan and vertical irregularities. Plan irregularities are features that may place extraordinary demands on elements due to torsional response or the shape of the diaphragm. Vertical irregularities are created by uneven vertical distribution of mass or stiffness between floors that may result in concentration of force or displacement at certain levels. In older existing buildings, such irregularities are seldom taken into consideration in the original design and, therefore, normally require rehabilitation measures to mitigate.

Load Path

A discontinuity in the load path, or inadequate strength in the load path, may be considered overarching because this deficiency will prevent the positive attributes of the seismic system from being effective. The load path is typically considered to extend from each mass in the building to the supporting soil. For example, for a panel of cladding, this path would include its connection to the supporting floor or floors, the diaphragm and collectors that deliver the load to components of the primary lateral force resisting system (walls, braces, frames, etc.), continuity of these components to the foundation, and finally the transfer of loads between foundation and soil. Many load path deficiencies may be considered to be part of another element. For example, an inadequate construction joint in a shear wall could be considered a load path deficiency or a shear wall deficiency in the category of global strength.

Inadequate Component Detailing

Detailing, in this context, refers to design decisions that affect a component's or system's behavior beyond the strength determined by nominal demand, often in the nonlinear range. An example of a detailing deficiency is poor confinement in concrete gravity columns. Often in older concrete buildings, the expected drift from the design event will exceed the deformation capacity of such columns, potentially leading to degradation and collapse. Although the primary gravity load design is adequate, the post elastic behavior is not, most often due to inadequate configuration and spacing of ties. Identification of detailing deficiencies is significant in selection of mitigation strategies because acceptable performance often may be achieved by local adjustment of detailing rather than by adding new lateral force resisting elements.

In case of gravity concrete columns, acceptable performance can be achieved by enhancing deformation capacity by adding confinement rather than by reducing global deformation demand by adding lateral force-resisting elements.

Diaphragm Deficiencies

The primary purpose of diaphragms in the overall seismic system is to act as a horizontal beam spanning between lateral force resisting elements. Diaphragm deficiencies include such factors as inadequate shear or bending strength, stiffness, or reinforcing around openings or re-entrant corners. Inadequate local shear transfer to lateral force resisting elements or missing or inadequate collectors are categorized as load path deficiencies.

Foundation Deficiencies

Foundation deficiencies can occur within the foundation element itself, or due to inadequate transfer mechanisms between foundation and soil. Element deficiencies include inadequate bending or shear strength of spread foundations and grade beams, inadequate axial capacity or detailing of piles and piers, and weak and degrading connections between piles, piers and caps. Transfer deficiencies include excessive settlement or bearing failure, excessive rotation, inadequate bending or shear strength of spread foundations and grade beams, inadequate axial capacity or detailing of piles and piers, weak and degrading connections between piles, piers and caps, inadequate tension capacity of deep foundations, or loss of bearing capacity due to liquefaction.

Other Deficiencies

Deficiencies that do not fit into one of the categories described above may include:

- Geological hazards
- Adjacent buildings
- Deteriorated structural materials.

18.9.3 Strategies for Rehabilitation Schemes

Technical Considerations

Selected techniques must eliminate deficiencies, preferably more than one deficiency. First one should consider enhancing existing elements, such as shear walls, moment frames, and bracing frames. The deformation compatibility between new elements and existing elements must be considered. In some cases, the application of base isolators or damping devices is the most efficient way to eliminate deficiencies.

Nontechnical Considerations

The solution chosen for rehabilitation is almost always dictated by building-user oriented issues rather than by merely satisfying technical demands. There are five basic issues that are of concern to building owners or users:

- Construction cost
- Seismic performance
- Short-term disruption of occupants
- Effects on long-term functionality of building
- Aesthetics, including consideration of historic preservation.

All of these characteristics are always considered, but an importance will eventually be put on each of them, either consciously or subconsciously, and a combination of weighing factors will determine the scheme chosen.

Cost: Construction cost is always important and is balanced against one or more other considerations deemed significant. However, sometimes other economic considerations, such as cost of disruption to building users, or the value of contents to be seismically protected can be of orders of magnitude larger than construction costs. Thus, cost may be the only criterion applied when choosing among equivalent rehabilitation options.

Seismic performance: Prior to the emphasis on performance based design, perceived qualitative differences between the probable performance of different schemes would be used to assist in choosing a scheme. Specific performance objectives are often set prior to the development of schemes. Objectives that require a limited amount of damage or continued occupancy will severely limit the retrofit methods than can be used and may control the other issues.

Short-term disruption of occupants: When seismic rehabilitation is done at the time of major building remodeling, disruption issue is minimized. However, in cases where the building is partially or completely occupied, this parameter commonly becomes dominant and controls the design.

Effects on long-term functionality of buildings: This characteristic is often judged less important than others. The planning flexibility is only subtlety changed. However, it can be significant in building occupancies that need open spaces, such as retail spaces and parking garages.

Aesthetics: In historic buildings, considerations of preservation of historic fabric usually control the design. Performance objectives are controlled by limitations imposed by preservation. In nonhistoric buildings, aesthetics is commonly stated as a criterion, but is often sacrificed, particularly in favor of minimizing cost and disruption to tenants.

18.9.4 Rehabilitation Techniques

Different building types require different mitigation techniques for a specific seismic deficiency. Depending on building types and associated seismic deficiencies, alternative recommendations are made to satisfy the performance objective of rehabilitation. Rehabilitation techniques are being developed for common building types. They are as follows:

Wood light-frames: One and two detached dwelling of one or more storeys in height.

Multistorey, multi-unit residential wood frames: Large residential buildings with commercial space at the ground floor.

Steel moment frames: Buildings consist of steel beams and columns, and lateral forces are resisted by moment frames.

Steel braced frames: Buildings consist of frame assemblies of steel beams and columns. Lateral forces are resisted by diagonal steel members placed in selected bays.

Steel frames with infill masonry shear walls: Buildings are normally older buildings that consist of gravity frames with unreinforced masonry, tightly infilling the space between columns.

Concrete moment frames: A complete system of concrete beams and columns. Lateral loads are resisted by cast-in-place moment frames.

Concrete shear wall building (bearing wall system): Usually all concrete with flat slab or precast plank floors and concrete bearing walls. Little, if any of the gravity loads are resisted by beams and columns, lateral loads are resisted by shear walls.

Concrete shear wall buildings (gravity frame systems): Buildings have columns and beams or columns and slabs that essentially carry all gravity loads. Lateral loads are resisted by concrete shear walls surrounding shafts, at the building perimeter, or isolated walls placed specifically for lateral resistance.

Concrete frames with infill masonry shear walls: Buildings are normally older buildings that consist of essentially complete gravity frame assemblies of concrete shear walls surrounding shafts, at the building perimeter, or isolated walls placed specifically for lateral resistance.

Concrete frames with infill masonry shear walls: Buildings are normally older buildings that consist of essentially complete gravity frame assemblies of concrete columns and floor systems. The floors can be of a variety of concrete systems including flat plates, two way slabs, and beam and slab. Exterior walls are constructed of unreinforced masonry, tightly infilling the space between columns horizontally and between floor structural elements vertically, such that the infill interacts with the frame to form a lateral force resisting system.

Tilt-up concrete shear wall buildings: Buildings are constructed with perimeter concrete walls cast on the site and tilted up to form the exterior of the building. The majority of these buildings are one storey with wood roof framing; however, a good number of multistorey buildings also exist with composite deck floors and a wood or steel framed roof.

Precast concrete frames with shear walls: Buildings consist of concrete columns, girders, beams and/or slabs that are precast off the site and erected to form a complete gravity load system. This building type has a lateral force resisting system of concrete shear walls, cast-in-place topping slab, reinforced to provide diaphragm action. In other areas, methods of joining floor sections vary and include use of welder insert plates.

Reinforced masonry bearing wall buildings (similar to tilt-up concrete shear wall buildings): Buildings are constructed with reinforced masonry (brick cavity wall or concrete masonry unit) perimeter walls with a wood or metal deck flexible diaphragm.

Reinforced masonry bearing wall buildings (similar to unreinforced masonry bearing wall buildings): Buildings are multistorey, and typically has interior concrete masonry unit walls and shorter diaphragm spans.

Reinforced masonry bearing wall buildings (similar to concrete shear wall buildings with bearing walls): Buildings consist of reinforced masonry walls and concrete slab floors that may be either cast-in-place or precast. This building type is often used for hotels and motels and is similar to the concrete bearing wall type.

Unreinforced masonry bearing wall buildings: Buildings consist of unreinforced masonry bearing walls, usually at the perimeter and usually brick masonry. The floors are typically of wood joints and wood sheathing supported on the walls and on interior post and beam construction.

Rehabilitation methods are recommended for the following five categories of techniques as appropriate.

- Add new elements
- Enhance existing elements
- Improve connections between elements
- Reduce demand
- Removal of deficient elements.

Seismic rehabilitation techniques not necessarily related to a specific building type, such as those related to diaphragms, foundations and nonstructural components and significant global technique that could be applied to any building, such as seismic isolation or addition of damping are included in this category.

Regardless of the experience and experimental knowledge gained in more than 100 years of reinforced concrete construction, structures require repair and/or strengthening because of natural causes, human error, and change in loading conditions. Further, it is important to recognize that concrete repair and strengthening is a scientific art form that involves the use of conventional, cement based materials, as well as new techniques and materials. It is crucial that structural engineers recognize that strengthening assessment and design is infinitely more complex than new construction. Typically, challenges arise because of unknown factors associated with the structural state, such as continuity, load path, and material properties as well as the size and locations of existing reinforcement or prestressing. The degree to which the upgrade system and the existing structural elements share the loads also must be evaluated and addressed properly in the upgrade design, detailing, and implementation procedure. The importance of detailing and its direct effects on the effectiveness and durability of structural upgrades cannot be over emphasized. In fact, inadequate detailing is one factor that can lead to the total failure of a structural repair system.

19

Repair/Retrofit of Nonengineered Structures

19.1 INTRODUCTION

The basic requirement for seismic resistance of buildings is that the different parts of a building should be tied together so that the building acts as one unit under seismic forces. The essential features for resisting forces are as follows:

- The building should be tied to its foundation.
- Horizontal bands of reinforcement are to be provided at plinth level (plinth band); sill level (sill band); lintel level (lintel band) and roof level (roof band). Vertical reinforcing is to be provided at the corners of the building, junctions of walls and around the opening in walls.
- There should be horizontal bracing (bracing in plan) at the roof level for sloping roofs.
- The opening should be small and away from the corners of the walls.
- The sill band is discontinuous at the door openings. Hence, it is to be integrated with the vertical bands. Roof band is required when roof is sloping and made of flexible materials, such as tiles on purlins and rafters. The roof band is not necessary, when the roof is made of flat concrete slabs.

19.2 REPAIR MATERIALS

Grouts: Grout is a mixture of water, cement and optional materials like sand, water reducing admixtures, and expansion agent. The water to cement ratio is around 0.5. Fine sand is used to avoid segregation. The desirable properties of grout are fluidity, minimum segregation, low shrinkage, adequate strength after hardening, good bond with substrate, no detrimental compound and durable.

Epoxy resins: Epoxy resins are used for the following purposes:

- To bond plastic concrete to a hardened concrete surface
- To bond two rigid materials
- For patch work
- For applying a coat over concrete surface to give colour, resistance to penetration of water and chemicals and resistance to abrasion.

Epoxy resins are excellent binding agents. The low viscosity resins can be injected into small cracks. The higher viscosity resins are used for coating and filling larger openings or holes.

Epoxy mortar: The epoxy mortar is made using epoxy resin, sand, cement and water.

The resin is added as an additional binder. It has high compressive strength, high tensile strength and low modulus of elasticity. The polymer particles join and form chain link reinforcement, increasing the tensile strength of the mortar. There is greater plasticity and reduction in shrinkage stress.

Quick setting cement mortars: These are patented mortars generally having two components and are sold in a prepacked state. These may be classified as follows:

- Unmodified cementitious
- Polyester or epoxy resin based
- Polymer modified
- Cement/Pozzolanic modified.

19.2.1 Shotcrete

Shotcrete is a method in which compressed air forces mortar or concrete through a nozzle to be sprayed on a surface of building component, such as a wall, at a high velocity. The materials used in shotcrete are generally same as those used for conventional concrete. The reinforcement provided is welded wire fabric or deformed bars tacked on the existing surface. It is applied either wet or dry process. In wet mix cement and aggregate premixed with water and pump pushes the mixture through the hose and a nozzle. In dry mix process, compressed air propels premixed mortar and damp aggregate, and at the nozzle end, water is added through a separate hose.

Microconcrete: Based on hydraulic shrinkage binders, these readymade formulations are tailored to produce concrete which is flowable and free of shrinkage. They are applied in complicated locations and thin sections, such as concrete jackets.

Fiber reinforced concrete: It has better tensile strength as compared to conventional concrete. They also have improved ductility (energy absorption capacity) and durability. They are being increasingly used for structural strengthening.

Fiber reinforced polymers: FRP composites are made up of a polymer matrix and fibers.

The fiber can be of glass, carbon or aramid. They possess high strength-to-weight ratios, high fatigue strength, high wear resistance, vibration absorption capacity, dimensional stability, high thermal and chemical stability and corrosion resistance. FRP wraps can be used to strengthen structural members.

Ferrocement: It is constructed of cement mortar reinforced with closely spaced layers of small diameter wire mesh. The mesh may be made of steel or other suitable material. The mortar should be compatible with opening size and weight of the mesh. The mortar may contain discontinuous fibers.

19.3 REPAIR TECHNIQUES

The applicability of repair techniques varies depending on the problems to be addressed. The associated techniques normally used are brought out below.

19.3.1 Repointing

First wall should be made wet and all loose debris to be cleared. The joints to be repointed should be racked to a depth which is two times the joint height. Next, fresh mortar should be placed by trowels. The mortar should be non-shrinking types. The repointed portion should be cured properly.

19.3.2 Small Crack Repairs

The cracks that are small in width (< 0.75 mm) can be effectively repaired by pressure injection of epoxy (IS:13935). The surfaces are thoroughly cleaned of loose materials. Injection ports are placed along the length of the cracks on both sides, at intervals approximately equal to the thickness of the member. Low viscosity resin is injected into the ports sequentially, beginning at the port at the lowest level and moving upwards one by one. The resin is pushed through the packer till it is seen flowing from the other end or from a port higher than where it is injected. The port is closed at this juncture and the packer is moved to the next higher port.

Larger cracks require larger packer spacing depending on thickness of the member. Vacuum injection has a typical fill level of 95% and can fill cracks as small as 0.025 mm. A similar technique can be applied to strengthen walls.

19.3.3 Large Cracks with Crushed Material

For cracks with width larger than 6 mm or in regions where brickwork or concrete crushed, the following procedures are suitable:

- Loose material in the crack is removed
- If necessary, the crack is dressed to have a V-groove
- At wide cracks, fillers like flat stone chips can be used
- To prevent widening of cracks, they are normally stitched.

The stitching consists of drilling small holes of diameter 6–10 mm on both sides of the crack, cleaning the holes, filling up these with epoxy mortar and anchoring the legs of stitching dogs (U-shaped steel bars of diameter 3–6 mm with short legs). The stitching dogs can have variable length and orientation. The spacing of the reinforcement should be reduced at the ends of the crack. Stitching will not close the crack, but it prevents further propagation at the ends of the crack. The stitching will stiffen the area near the vicinity of the crack. A stitching technique can be used for stitching cracks parallel to the surface.

19.4 STRENGTHENING OF WALLS

The walls support the roof and upstairs floor, as well as provide resistance to lateral load. The walls can be strengthened by following methods:

- Grouting
- Providing through bond elements
- Reducing size of the opening
- Providing containment reinforcement
- Using ferrocement layers
- Using fiber reinforced polymer (FRP) laminates.

The method of grouting is already discussed under repair techniques. The methods of through bond elements and containment reinforcement are described below. The method of uses of ferrocement layers and FRP laminates are covered under separate chapters.

19.4.1 Through Bond Elements

Through bond elements are those whose length is equal to the wall thickness. The elements can be reinforced concrete blocks spaced at 1 m interval. The steps of placing these elements are as follows (IS:13935):

- The plaster from the surface is removed at the locations of through bonds.
- The stones or masonry units at those locations are removed to make holes.
- 8 mm diameter hooked mild steel bars are placed in the holes.
- The holes are filled up with concrete of mix 1:2:4.
- The areas are cured adequately.

19.4.2 Containment Reinforcement

In this method, first grooves are made in the mortar joints. Next, horizontal, vertical and cross bars are inserted in the grooves, which are subsequently covered by mortar.

19.4.3 Stitching of Perpendicular Walls

Corners and T-junctions in perpendicular walls can be connected by effective 'stitching' of the walls. In this method, first holes are drilled in an inclined pattern. Next, steel bars are placed in the holes. Subsequently, grout is injected to form the bond between the bars and the walls, as well as to provide protection against corrosion of the bars. Generally, 8–10 mm diameter bars of grade Fe 415 are used for stitching. A collar band can be provided at the junction of the two perpendicular walls at the lintel level to enhance the structural integrity. In many buildings, the partition walls are first ones to collapse during any hazards like earthquake. To avoid failure such a wall should be integrated by stitching it with the two longitudinal walls at its ends.

19.5 STRENGTHENING OF PILLARS

In the barrack type of building, pillars made of bricks are placed along the corridor. These pillars support the roof trusses. In absence of any reinforcement, the pillars are normally weak under any lateral forces, leading to failures. Such pillars can be strengthened by concrete jacketing. The additional stirrups can be inserted from the top of the pillars after lifting the roof truss by temporary support. Else each stirrup has to be made of two pieces for insertion. The trusses should be properly anchored to the pillars.

19.6 TECHNIQUES FOR GLOBAL STRENGTHENING

Global strengthening aims to improve the lateral load resistance of the building as a whole. This will relieve overloaded members and ensure better seismic behavior.

19.6.1 Introduction of Walls

Long walls are vulnerable to out-of-plane collapse. For a wall thickness t and cross walls spacing a, the ratio a/t should not exceed 40. To ensure this, cross walls can be added. This type of intervention is possible in long barrack type accommodation used for schools, dormitories or other purposes, where functionality of the interior space can be suitably adjusted. The junction between the new wall and the old wall will be a T-junction.

19.6.2 Introduction of Pilasters or Buttresses

In the case of longitudinal walls of barrack type buildings, pilasters or buttress walls added externally will enhance lateral load resistance. The functionality of the interior space is retained. The buttresses need to be integrated with the existing wall by key stones.

19.6.3 Improvement of the Frame

Most of the nonengineered constructions do not have adequate lateral load resistance. The rural hut made with casuarina and bamboo posts is a typical example. The introduction of braces in the periphery stabilizes the frame. Provision of doors can be accommodated by providing braces on either side. There can be cross braces or knee braces in the vertical planes of the walls or in the horizontal plane at the top of the walls. Instead of using ropes for tying the braces, metal straps and mild steel wares of suitable gauge can make the system better. The vertical posts should be properly anchored with cross sticks tied firmly well below the ground level. This provides good anchorage for the post. The portion of post below ground level should be protected by tar coat.

19.6.4 Splint and Bandage Strengthening Technique

To economize in the retrofit cost while targeting to enhance the integrity of the building, ferrocement with galvanized welded steel wire mesh can be added on the exterior surface of the walls in the form of vertical belts (splints) and horizontal belts (bandages). The vertical belts are placed adjacent to the openings and at the corners. The horizontal belts are placed between the openings and the roof (or upstairs floor). A belt is not required below the roof level, if the roof is made of flat RCC slabs. A horizontal belt is required at the plinth level, when the plinth height is more than 900 mm.

The specification for the mesh size is provided in the IS:13935. The reinforcement for the vertical belts should start at least 300 mm below the plinth level and continued up to the roof belt. The width of mesh for a vertical belt at the corner should be kept a minimum of 250 mm on both the walls, so that it covers a width of a brick. The width of the belt on each side of the corner should be 25 mm extra to the width of the mesh.

Horizontal belts can be also provided in the interior surface of the walls. A vertical reinforcement bar can be provided at each inside corner of a room in place of a vertical belt. The bar should be tied to the wall by dowels spaced at one meter. The bottom of the bar should be anchored 750 mm below the ground level and made continuous through the upstairs floor slab. A minimum cover of 15 mm should be provided with cement mortar. The belts can be anchored to the walls by stainless steel anchors of size 8 mm diameter. The size and spacing of anchors depend on the type selected. The information is normally available from the manufacturer's catalogue. The steps of anchoring are as follows:

- A hole is drilled and cleaned.
- Epoxy or grout is placed in the hole.
- The anchor is inserted with the wire mesh.
- The epoxy is allowed to harden for the anchor to gain the pull out strength.

20

Retrofit of Historical and Heritage Structures

20.1 INTRODUCTION

The principal difference between a heritage structure and a regular structure is that retrofitting techniques cannot be indiscriminately applied with the sole aim of improving structural response to earthquakes. Judicious selection and application of retrofitting techniques, respecting the authenticity and the heritage value of the structure in its entirety is paramount. Use of new materials for repair must be dealt cautiously. The aim of retrofitting is to preserve the historical structure for generations to come.

20.2 BASIC PRINCIPLES OF STRUCTURAL RESTORATION

Structures of architectural heritage present a number of challenges in restoration and retrofit, which limit the application of modern codes and building standards. Recommendations are desirable and necessary to ensure rational methods of analysis and repair methods appropriate to the cultural context. They are intended for those involved in conservation and restoration, but cannot in anyway replace specific knowledge acquired from cultural and scientific texts.

20.3 CONDITION ASSESSMENT

- Nondestructive tests: Thermal method, radar technique, ultrasonic pulse velocity test, vibration test and endoscopy
- Intrusive tests: Core test, *in situ* shear test, bond test and test of masonry prisms
- Numerical modeling techniques.

20.4 STRENGTHENING OF MASONRY WALLS

i. Repair to cracks: The repair procedures for various crack width are different.
  - For crack width less than 1.0 mm "injection with epoxy" is recommended. It may be costly, if damaged area is large.
  - For crack width 0.3–3 mm, injection with cement grout that contains shrinkage reducing admixtures.
  - For crack width more than 10 mm, reconstruction of damaged area with new brick units. The cracks can be sealed with mortar, if the wall thickness is relatively small.
ii. Repointing of bed joints (with steel bars preferably).
iii. Reinforced concrete jacketing (for heavy damage).

iv. Injection grouting
 v. Prestressing
vi. Wall reconstruction
vii. Strengthening using fiber reinforced polymer
viii. Strengthening using shape memory alloys
 ix. Repair of wall corners and intersections
 x. Improving connections of secondary/non-structural elements.

20.5 STRENGTHENING OF ARCHES, VAULTS AND DOMES

Dry stone masonry offers very high strength in compression, but their joints provide limited shear and tensile resistance, as they depend purely on friction. A positive connection between the stone blocks can be achieved using dowels, camps or special tie bars or structural connections inserted through specially prepared holes in the joints, without being visible from outside.

Damages to arches are generally caused by lack of thrust at the springers. The thrust can be achieved by steel ties or by large masonry masses like buttresses at the outside. The ties can be post-tensioned for better effectiveness. A reinforced concrete jacket can be strengthened by pinning technique, where steel pins are introduced by drilling and subsequent grouting.

In domes, external support rings can provide compression to enhance integrity. In corbelled dry blocks domes, an inwards collapse can occur due to loosening of the blocks. Internal support rings or bracing placed in the interior surface can counter the damage.

20.6 STRENGTHENING OF TOWERS AND SPIRES

Towers and minares have the problems similar to that of pillars and walls. Their form, height and slenderness imply a lack of redundancy for redistribution of stresses and lack of energy dissipation capacity. This results in concentration of stresses at base and brittle failure. Repair and strengthening of towers mainly consists of containment of lateral expansion. Reinforcements and prestressed tendons may be required in horizontal and vertical directions.

An effective, short-term reversible intervention to stabilize old masonry towers would be strengthening under dry construction with tie rods. Steel diaphragms are provided at the center of cross-section to prevent inward movement of the walls of a tower. The diaphragms also serve as anchorages for radially arranged tie rods that protrude through putlog holes in the walls. This prevents the walls from bulging outwards. Moderate post-tensioning of the tie rods can be favourable. Vertical prestressing of towers may improve the overall flexural capacity of towers, but it can drastically change (at times for the worse) the dynamic behavior of the tower.

20.7 REDUCTION OF SEISMIC EFFECTS ON STRUCTURES

- **Reduction of forces:** If the structure can be "isolated" either at the foundation level or at the ground floor level from the portion below, then the forces in the members due to ground motion are reduced. The remedial action consists of cutting the structure to create a joint and to allow relative movement of the super- and sub-structures. This technique is called "base isolation".
- **Reduction of mass:** A reduction of induced forces can be achieved by reducing mass, especially at upper levels. Extreme situations may require some amount of demolition to reduce the loading.

20.8 STRENGTHENING OF SOILS AND FOUNDATIONS

Various methods used are as follows:
- Increasing the area of foundation, lowering the foundation level and strengthening the existing foundation.

- Inclusion of structural elements, such as piles, micropiles, and underpinning
- Modifying the effective stress in the soil by drainage or consolidation
- Improving the subsoil by chemical or cement grouting
- Replacing the entire substructure.

20.9 ARCHAEOLOGICAL RECONSTRUCTION

20.9.1 Principles

The archaeological reconstruction is a viable option for a heritage structure destroyed in seismic or other hazards. Archaeological principles give a basis for reconstruction of structures that have been destroyed by natural (earthquakes, volcanoes, floods, tsunami/storms, etc.) or manmade calamities (wars, etc.) from original salvaged materials.

Reconstruction depends on effective use of earlier architectural drawings, verbal documentation, sketches, paintings, and photo-documentation of the structure under consideration. Reuse of salvaged building materials to the maximum extent is imperative to retain authenticity. Therefore, the process relies extensively on documentation of the remains of the structure and their correlation with drawings. In case of missing components, new materials compatible in structural behavior and similar appearance to the parent martial should be used in the reconstruction. Modern tools for documentation, structural analysis, communication (for example, multimedia) and construction practices help the process of archaeological reconstruction.

20.9.2 Procedure of Rubble Clearance

The following is a cause of action for archaeological clearance of rubble:

- Determining and documenting the surroundings of location, where the object is found with clear reference. Classification of the place with clear orientation and designation of the area of the find.
- Evaluation of the individual pieces found; determining whether they are worth recording.

- Clear levelling of the objects found and recording all information pertaining to the find.
- Description, illustration and graphical presentation of the piece found. Decision on storage or filling to be taken.
- Documentation of all stages of work; preparing drawings to map out the place where the object is found in the ground plan and the sectional drawing.
- Identification of the piece found based on all knowledge gained up to the present.
- Archeological evaluation of the knowledge gained towards reconstruction of the structure. Preparation for reutilization of parts/pieces for the new structure.

The aim of reconstruction is to secure as much original materials as possible for the reconstruction work. The objective of clearing the rubble is to obtain information with regard to the following points:

- Original geometry and dimensions, and archeological details
- Details of original construction and materials used
- Interior decoration (stucco, paintings, plastering, lining and finishing)
- Technological problems involved in erecting the building
- Effectiveness and advisability of restoration measures
- Degree of destruction.

20.9.3 Procedure for Archiving Finds

An inventory of all the finds is to be done up to match the pieces wherever possible. A database with electronic pictorial information of all the finds has to be created for future archeological work. The finds are then mapped on to specially prepared drawings of the latest state of the facades and a number of ground plans, sections and photographs. Many of the objects found, can provide valuable information as models and patterns for the reconstruction.

20.9.4 Criteria for Reconstruction

The following are the criteria for reconstruction:

- The structure has to be reconstructed with salvaged, reusable materials from the ruins coupled with new materials that are compatible, using modern techniques of construction.
- The reconstructed structure has to incorporate certain modifications taking into consideration modern day requirements (like electrical services, lifts, plumbing services, etc.).
- Analytical and numerical investigations on the structural behavior of the original structure will help in incorporating seismic strengthening measures in the new structures.
- Issue of weathering of materials must be confronted (for example, waterproofing).
- Durability and precision can be ensured with the use of modern analytical and construction tools and practices.

21

Retrofit of Reinforced Concrete Structures

21.1 INTRODUCTION

Reinforced concrete (RC) buildings include residential, dormitory, institutional, office, commercial and industrial buildings. The multistoreyed buildings have moment resisting frames consisting of a framework of RC beams and columns. The frames are intended to carry the vertical loads. In the seismic regions, these buildings should be augmented with shear walls or braces to carry the lateral loads.

The lateral load resistance of a frame generates from the flexural action of beams and columns and the flexural rigidity of the beam–column joints. There should be adequate number (at least two) of well laid out continuous frames in the two orthogonal directions in order to generate the lateral stiffness and strength of a building. Some existing building configurations that are adequate for resisting gravity loads are not suitable for resisting earthquake forces. In this chapter, first the common deficiencies observed in existing RC buildings for resisting earthquake forces are identified. It is essential to identify the deficiencies in a building before undertaking retrofit. Identification of the deficiencies is also expected to create awareness for future construction.

For convenience, the deficiencies can be broadly classified as global deficiencies and local deficiencies. Global deficiencies refer to the deficiencies which are observed in the building as a whole. These deficiencies are subjective in nature. But for the purpose of evaluation, IS:1893-2002 (Part I) provides quantitative or qualitative definitions for some of the deficiencies.

Local deficiencies refer to the deficiencies in individual members. Each beam and column in a frame should have not only the required strength, but also sufficient deformation capacity. A slab should be properly connected to the frames. The deficiencies can be detected by checking the capacities and studying the detailing of reinforcement in the slabs, beams, columns and joints as shown in the as-built drawings. The capacities can be compared with the force demands calculated from a building analysis. The detailing of reinforcement in the drawings can be compared with the requirements of the code IS:13920-1993.

When a building has deficiencies that have not been accounted for in the structural design, it needs retrofit. The different retrofit strategies can be grouped under global retrofit strategies or local retrofit strategies. A global retrofit strategy targets the performance of the building as a whole under lateral loads.

Addition of new walls, frames or braces, reduction of any irregularity or mass of the building are grouped under global retrofit strategies. These improve the lateral strength and stiffness of the building. If a building is significantly deficient in resisting seismic forces, a global retrofit strategy should first be investigated. A local retrofit strategy targets the strength and ductility of a member, without significantly affecting the lateral strength or stiffness of the building. Repeated use of a local retrofit strategy can improve the ductility in the base shear versus roof displacement behavior of the building. The local retrofit strategies include concrete jacketing or attaching steel plates or wrapping polymer sheets to a column or a beam. Each type of these strategies is described in this chapter, and finally some general remarks on retrofit of RC buildings are provided.

21.2 BUILDING DEFICIENCIES

This section lists some common deficiencies observed in multistoreyed RC buildings in India. The observations were made from the buildings which were damaged or collapsed due to the earthquake at Bhuj, Gujarat, in 2001. A few other observations were made in a project on evaluation and retrofit of RC building (manual for seismic evaluation and retrofit of multistoreyed RC buildings, 2005).

21.2.1 Global Deficiencies

Global deficiencies are the attributes that degrade the lateral load resisting mechanism of a building subjected to an earthquake. Some of the deficiencies are caused by irregularities in the structural configuration (IS:1893-2002). The irregularities are broadly classified as plan irregularities and vertical irregularities. The plan irregularities can be detected by observations and simple calculations based on the plan of a building. Similarly, the vertical irregularities can be detected from the elevation of a building.

The irregularities result in an irregular load path, leading to structural damage and failure. The effects of irregularities may not be detected by the conventional equivalent static analysis. The irregularities are discussed earlier in detail. In the present chapter, the instances of the irregularities in existing buildings are highlighted.

Plan Irregularities

- *Torsional irregularity:* Torsional irregularity is caused by plan asymmetry and/or eccentricity between the center of mass of the floors and the center of rigidity of the frames. It is commonly observed in buildings with overhead water tanks, roof-top swimming pools and heavy auxiliary equipment.

- *Re-entrant corners:* Figure 21.1 shows the plan of an existing residential building with several re-entrant corners.

- *Diaphragm discontinuity:* A rigid floor slab acts as a horizontal diaphragm that mobilizes the frames and shear walls to resist the lateral loads. In many residential buildings there are multiple dwelling units on a given level. In order to provide windows, lift and stair wells, and ensure privacy of each unit, often large cutouts and openings are made in the floor slabs as shown in Fig. 21.1. This leads to diaphragm discontinuity. When the dwelling units on two sides of a building are staggered in elevation, the diaphragm action is reduced.

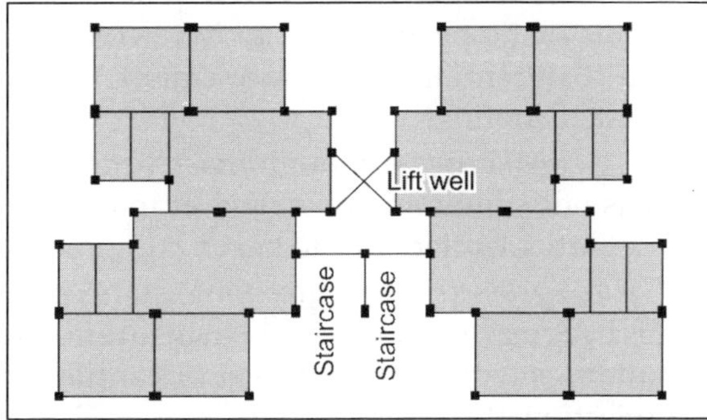

Fig. 21.1: Plan of residential building with diaphragm discontinuity and re-entrant corners

- *Out-of-plane offsets:* An example is when the columns along the perimeter of a building are discontinued at the ground storey as shown in Fig. 21.2. These columns are supported on cantilever overhang beams and are termed floating columns. The offset in load path is from the perimeter frame in the upper storeys to the outer columns in the ground storey. This type of frame may be adequate for gravity loads, but perform poorly when subjected to earthquakes. Instances of such building frames occur where there is a limitation for moving space along the periphery of the building at the ground level.
- *N-parallel systems:* This deficiency may occur in building with nonrectangular grid plans, such as curved buildings.

Vertical Irregularities

- *Stiffness irregularity:* The wall panels within a frame (infill walls) affect the stiffness of a storey. In the recent years, to facilitate parking of vehicles, infill walls are omitted in the ground storey as shown in Fig. 21.2. This type of open ground storey may lead to a soft storey. Irrespective of the number of storeys, one side of the columns is 230 mm (if not both) to flush them with the walls in the upper storeys. Such buildings are commonly referred to be on stilts. If the ground storey is used as shops, infill walls are not placed in the front side to have open front or glazing. Absence of adequate plinth beams or tie beams leads to long columns and differential lateral movement of the isolated footings.
- *Mass irregularity:* Although mass irregularity is not commonly observed, it may exist in a particular floor due to heavy equipment.
- *Vertical geometric irregularity:* Instances of this irregularity are observed in institutional buildings with set-back tower or cantilever projection at the top.
- *In-plane discontinuity:* A floating column is an example of in-plane discontinuity for the

frame in the elevation as shown in Fig. 21.2. A column in the upper storeys that is interrupted at the first floor and supported on a transfer beam, is another example. For a strong transfer beam, the supporting columns may be weak. Such type of beam–column joints is undesirable for seismic resistance. The failure of a column before the formation of hinges in the supported beams is disastrous. Hence, the strong column–weak beam concept is strongly advocated in seismic design.

- *Strength irregularity:* An open ground storey with 230 mm columns may lead to a weak storey, resulting in an undesirable sway mechanism under seismic load. The sway mechanism refers to the movement of the top storeys like a single block with large

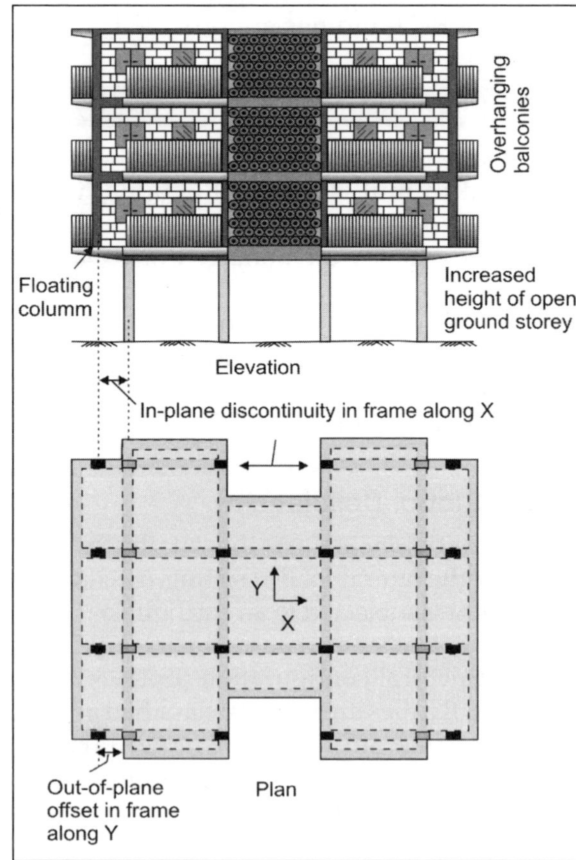

Fig. 21.2: Typical building elevation and plan showing plan and vertical irregularities

deformation of the ground storey columns. The building swings back and forth like an inverted pendulum during an earthquake. In the process, the columns in the ground storey get severely damaged.

21.2.2 Local Deficiencies

Local deficiencies arise due to improper design, faulty detailing, poor construction and poor quality of materials. These lead to the failure of individual members of the building, such as flexural and shear failures of beams, columns and shear walls, crushing or diagonal cracking of masonry walls and failure of beam–column joints or slab–beam or slab–column connections.

Local deficiencies can be minimized by following the ductile detailing requirements specified in IS:13920-1993. The basic principles and guidelines are explained with the aid of Fig. 21.3.

For Columns

- The splicing of longitudinal bars should be at the central half of the column.
- There should be sufficient transverse reinforcement in the form of closed hoops near the joints and at the location of splices

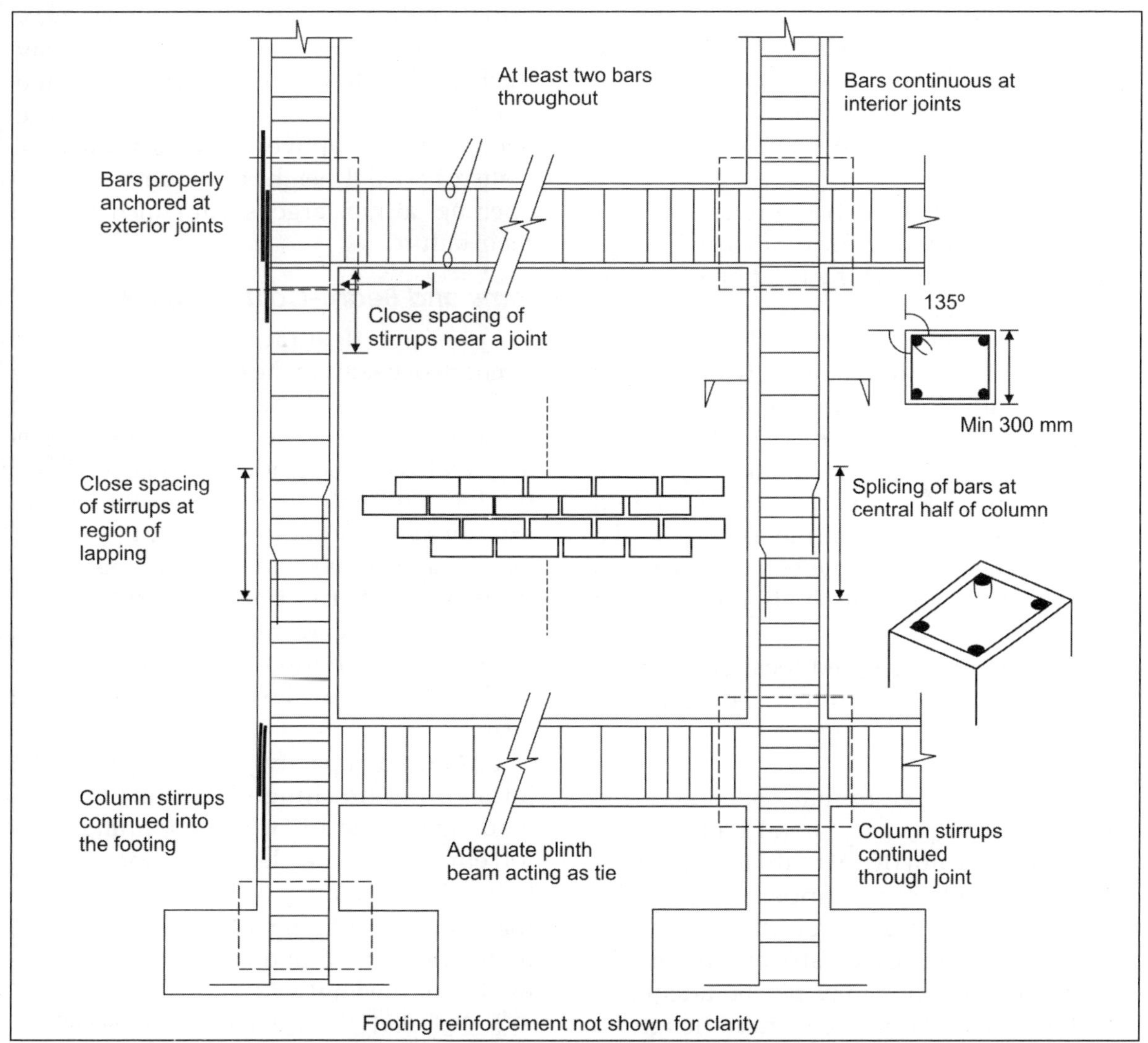

Fig. 21.3: Elevation of frame showing detailing for seismic forces

of longitudinal bars. The amount of hoops near a joint should be such that the supported beams generate their flexural capacities before the column fails in shear.

- The hoops should be continued throughout the joint and into the footing.

Floor Beams

- There should be minimum flexural capacities both for sagging and hogging throughout the length of the beam across a frame. There should be at least two longitudinal bars at the top and two bars at the bottom which are continuous throughout the length and interior joints.
- There should be proper anchorage of the longitudinal bars at the exterior joints. A joint in which there is a beam only on one side in the plane of the frame is termed an exterior joint.
- There should be sufficient transverse reinforcement in the form of closed hoops near the joints and at the location of splices of longitudinal bars. The amount of hoops near a joint should be such that the beam generates its flexural capacity with yielding of the reinforcing bars (rebar) before it fails in shear. The yielding of the rebar under reversed cyclic loading increases the bending (expressed as rotation of section) near a joint, resulting in the formation of a plastic hinge. The ductile behavior leads to the absorption and dissipation of internal energy.

The commonly observed local deficiencies in the members are summarized next.

Columns

- As mentioned before that one side of the columns in the primary lateral load resisting frames is 230 mm (if not both) to flush them with the walls in the upper storeys. The flexural and shear strengths of such a column in the lower storeys of multi-storeyed buildings may not be adequate. Columns with large aspect ratio (length to width ratio) can be inadequate under biaxial moments. Since earthquake ground motion can occur in any direction, the columns which are part of orthogonal frames should be designed for biaxial moments.

- The ties are widely spaced. A tie gets warped (the sides are not in a plane) and the ends are not bent by 135° with adequate length inserted within the core of the section. As a result the longitudinal bars tend to buckle and the confinement of concrete is poor. This leads to failure of the column before the formation of hinges in the supported beams.
- Faulty splicing of rebar is also deterring metal to the formation of hinges. The location of splice in a column just above the floor level with inadequate splice length is inappropriate.
- Short and stiff (captive) columns due to infill walls of partial height or columns next to openings attract larger shear which leads to their failure.

Beams and Beam–Column Joints

- The positive flexural strength and shear strength of beams in the primary lateral load resisting frames at the plinth level or first floor level tend to be inadequate. This is aggravated by the inadequate anchorage of the longitudinal rebar at the joints. In an exterior joint, the bars (especially the bottom bars) may not have adequate hooks. In an interior joint, the bottom bars may be discontinuous. Under moment reversal, the discontinuous bottom bars may pull out leading to loss of flexural strength of the beam.
- The rotation capacity of a beam near the joint (potential hinge location) may be inadequate due to lack of confining reinforcement. This may lead to a sudden shear failure before a hinge is formed.
- Inadequate confinement of reinforcement in a joint leads to undesirable shear deformation of the joint.

The observed deficiencies are schematically shown in Fig. 21.4.

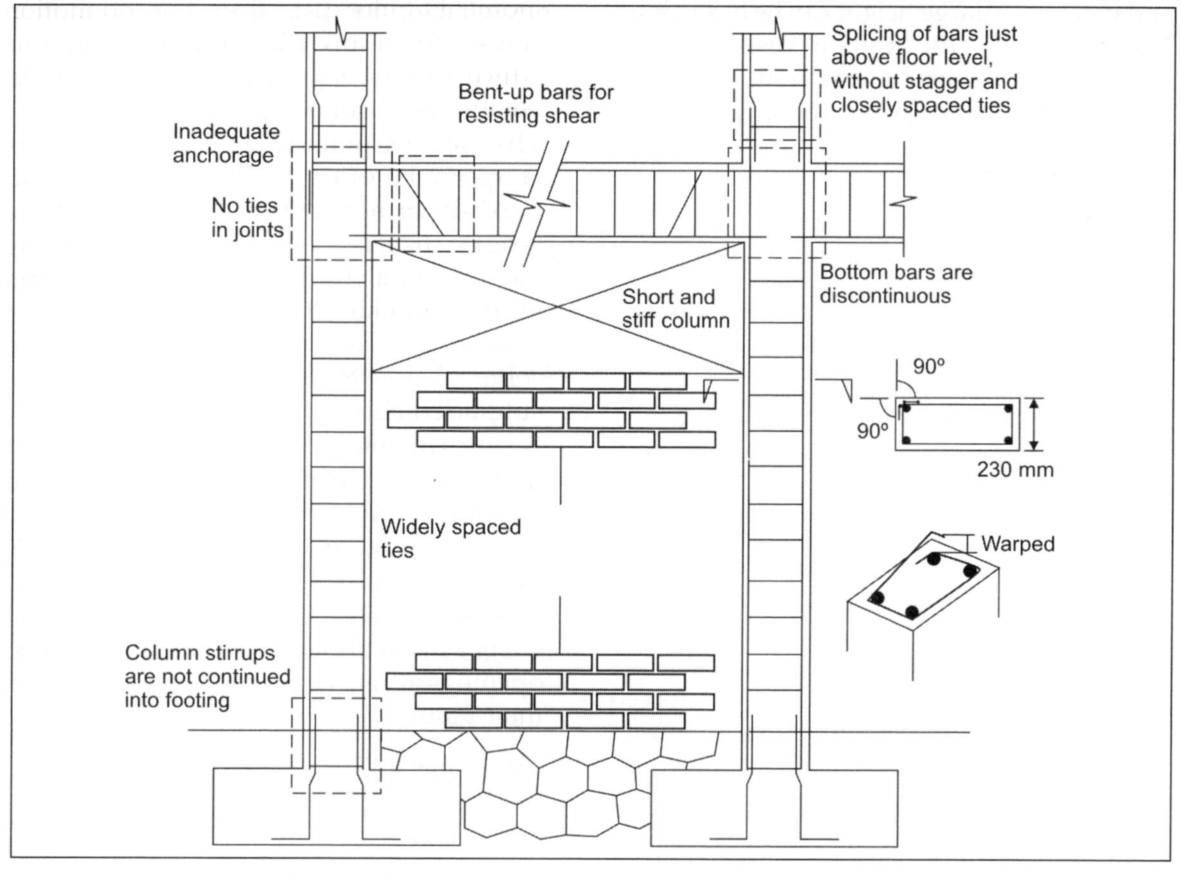

Fig. 21.4: Elevation of a typical frame showing local deficiencies

Slabs and Slab–Column Connections

The following deficiencies affect the diaphragm action of the floor slabs:

- When a floor slab acts as a diaphragm, additional reinforcement is required at the edges. The reinforcement at the edges perpendicular to the direction of lateral force is termed chord reinforcement. If the slab is not supported on edge beams, for example in a flat slab, the requirement of chord reinforcement cannot be neglected.

- The lack of adequate shear reinforcement at the slab–column connections in a flat slab lead to failure of the slab.

Structural Walls

If a wall is not designed adequately, the contribution of the wall in the lateral load resistance of a building is not utilized. The deficiencies that affect the performance of the wall are given below:

i. Lack of adequate boundary members to resist the axial forces due to in-plane bending of a wall.

ii. Inadequate reinforcement at the slab–wall or beam–wall connections. This reduces the integrity of the wall with the building frame.

Foundation

This isolated footings are not properly designed for the seismic forces. The plinth beams or the beams may be absent which makes the footings vulnerable to lateral spread.

Unreinforced Masonry Walls

Due to lack of out-of-plane bending capacity, the falling of masonry blocks from infill walls and parapets at large height can be hazardous. The potential of the walls to resist in-plane lateral loads is not materialized, if there is failure due to out-of-plane bending.

Precast Members

Lack of tie reinforcement in precast members can lead to dislocation of the members and collapse of the building.

21.2.3 Miscellaneous Deficiencies

Deficiencies in Analysis and Sensing

Inadequate tools for proper analysis lead to deficiencies in analysis and design. Some of the common deficiencies are listed below.

- A building is designed only for gravity loads. There is no analysis for seismic loads.
- Neglecting the effect of infill walls: Frames with brittle infills, such as unreinforced masonry, behave similar to braced frames with the infills acting as diagonal compression struts. When the infill walls are neglected in the analysis of a building, a longer time period is estimated by the analysis, thus resulting in lower calculated seismic forces. This concern is addressed in Clause 7.10.3, IS:1893-2002, which recommends a more stringent formulation for calculating the time period of masonry infilled frames. The stiffness of an infill wall influences the location of centre of rigidity of the lateral load resisting system in a storey. If infill walls are present only on one side, neglecting the infill walls in the analysis may lead to overlooking a torsional irregularity.
- Inadequate geotechnical data: The foundations are designed without adequate information of the type of soil, bearing capacity and locations of fill and fault. In the absence of data on faults, amplification effects due to interference of the earthquake waves are neglected.

- Neglecting the P-Δ effect: For a building on soft soil especially, the loss of stiffness during an earthquake leads to an increase in the displacement response. The increased displacement leads to higher eccentricity of the vertical loads, causing additional moment in the columns (P-Δ effect). If the P-Δ effect is not accounted for in the analysis, the member design forces are likely to be underestimated.

Deficient Construction Practices

- Volume batching of concrete, that may lead to increased water content.
- Additional dosage of water to increase slump, resulting in higher water-to-cement ratio.
- Inadequate compaction and curing of concrete, leading to honeycombed and weaker concrete.
- Inadequate cover leading to rebar corrosion.
- Poor quality control in the constituents of concrete.
- Use of re-rolled steel, having inadequate strength and ductility.

The deficiencies in construction practices lead to deviations from design assumptions and calculations.

Lack of Integral Action due to Poor Design

The building performance is degraded due to the absence of tying of the lateral load resisting members. For example, the beams are not framed into the elevator core walls as shown in Fig. 21.5. If the beams are eccentric to column lines with large offsets, then unaccounted torsion is introduced in the columns and beam–column joints.

Deficient Staircase

If the staircase slab is simply supported longitudinally, a collapse of the slab closes the escape route for the occupants.

Pounding of Buildings

Another poor design concept arises from inadequate gap between adjacent building or

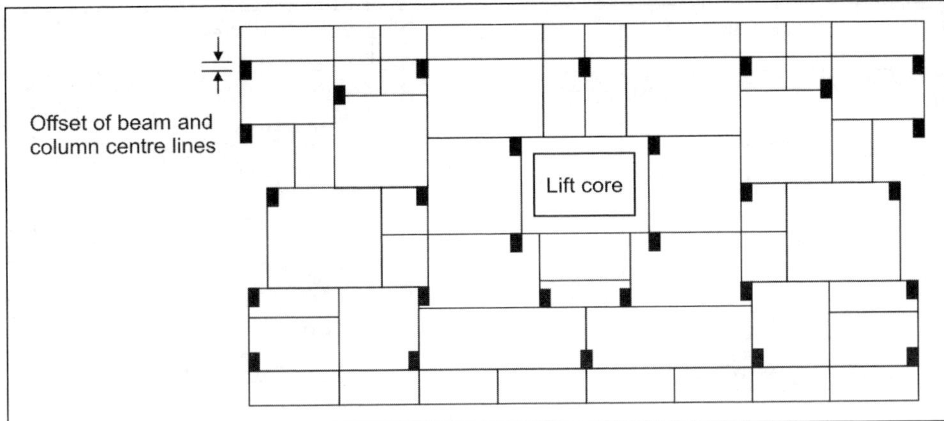

Fig. 21.5: A building plan showing lack of integral action

seismic joints between the segments of a building. When the provided gap is not adequate, the building or segments of a building will collide or "pound" against each other as they respond to the earthquake excitation in out-of-phase motion. This phenomenon transfers forces to the buildings which are unaccounted for in the design. This is more critical when the storey heights in the two adjacent buildings are different. The slabs of one building will pound against the walls or columns of the other.

Lack of Maintenance and Unaccounted Addition or Alteration of the Buildings

Lack of maintenance and unscrupulous addition or alteration without any seismic analysis and design check, can lead to high seismic vulnerability or even collapse, possibly under gravity loads.

21.3 RETROFIT STRATEGIES

A retrofit strategy is a technical option for improving the strength and other attributes of resistance of a building or a member to seismic forces. The retrofit strategies can be classified under global and local strategies. A global retrofit strategy targets the performance of the entire building under lateral loads. A local retrofit strategy targets the seismic resistance of a member, without significantly affecting the overall resistance of the building.

The grouping of the retrofit strategies into local and global need not be mutually exclusive. For example, when a local retrofit strategy is used repeatedly it affects the global seismic resistance of the building. It may be necessary to combine both local and global retrofit strategies under a feasible and economical retrofit scheme.

21.3.1 Global Retrofit Strategies

When a building is found to be severely deficient for the design seismic forces, the first step in seismic retrofit is to strengthen and stiffen the structure by providing additional lateral load resisting elements. Additions of infill walls, shear walls or braces are grouped under global retrofit strategies. A reduction of an irregularity or of the mass of a building can also be considered to be global retrofit strategies. The analysis of a building with a trial retrofit strategy should incorporate the modeling of the additional stiffening members.

Addition of Infill Walls

The lateral stiffness of a storey increases with infill walls. Addition of infill walls in the ground storey is a viable option to retrofit buildings with open ground storeys as shown in Fig. 21.6. Due to the 'strut action' of the infilled walls, the flexural and shear forces and the ductility demand on the ground storey columns are substantially reduced. Of course,

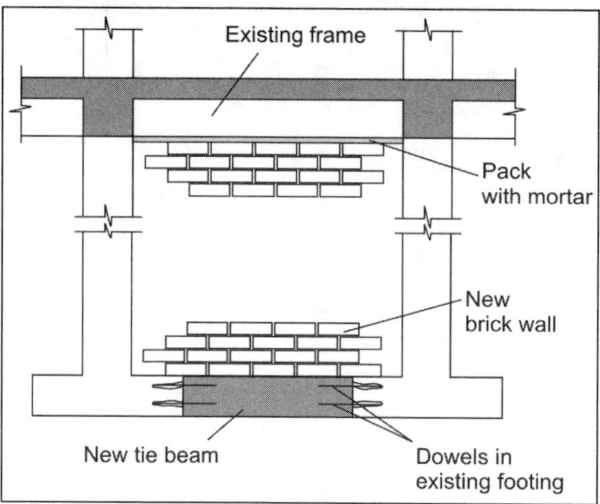

Fig. 21.6: Addition of a masonry infill wall

infill walls do not increase the ductility of the overall response of the building. The new wall should be placed on a foundation or tie beam in absence of plinth beams.

Addition of Shear Walls or Wing Walls or Buttress Walls

Shear walls, wing walls or buttress are added to increase lateral strength and stiffness of a building. The shear walls are effective in building with flat slabs or flat plates. Usually the shear walls are placed within bounding columns as shown in Fig. 21.7a, whereas wing walls are placed adjacent to columns as shown in Fig. 21.7b. The buttress walls are placed on the exterior sides of an existing frame as shown Fig. 21.7c. The critical design issues involved in the addition of such a wall are as follows:

- To integrate the wall to the building for transferring of lateral forces.
- To design the foundations for the new wall.

The disadvantage is that if only one or two walls are introduced, the increase in lateral resistance is concentrated near the new walls. Hence, it is preferred to have distributed and symmetrically placed walls. The shift of the centre of rigidity should not be detrimental. For a buttress wall, the new foundation should be adequate to resist the overturning moment due to the lateral seismic forces without

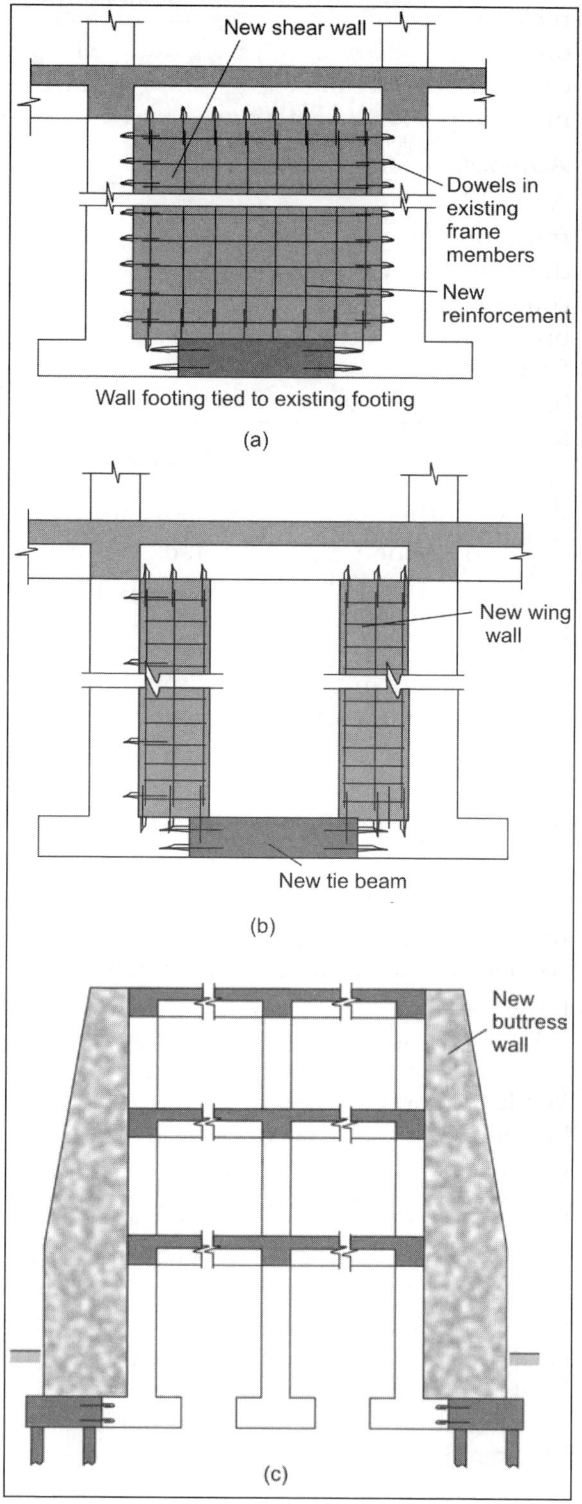

Fig. 21.7: (a) Addition of shear walls (PEMA 172), (b) Addition of wing walls, (c) Addition of buttress walls

rocking or uplift. The stabilizing moment is only due to the self-weight of the wall. This can be as low as compared to the overturning moment.

Addition of Steel Braces

A steel bracing system can be inserted in a frame to provide lateral stiffness, strength, ductility, hysteretic energy dissipation, or any combination of these as shown in Fig. 21.8. The braces are effective for relatively more flexible frames, such as those without infill walls. The braces can be added at the exterior frames with least disruption of the building use. For an open ground storey, the braces can be placed in appropriate bays while maintaining the functional use. Passive energy dissipation devices may be incorporated in the braces to enhance the seismic absorption. The types of bracing, analysis and design of braces are covered in earlier chapters. The connection between the braces and the existing frames is an important consideration of this strategy. One technique of installing braces is to provide a steel frame within the designated RC frame. The steel frame is attached to the RC frame by installing headed anchors in the latter, else the braces can be connected directly to the RC frame. Here, since the braces are connected to the frames at the beam–column joints, the forces resisted by the braces are transferred to the joints in the form of axial forces, both in compression and tension. While the addition of compressive forces may be tolerated, the resulting tensile forces are of concern.

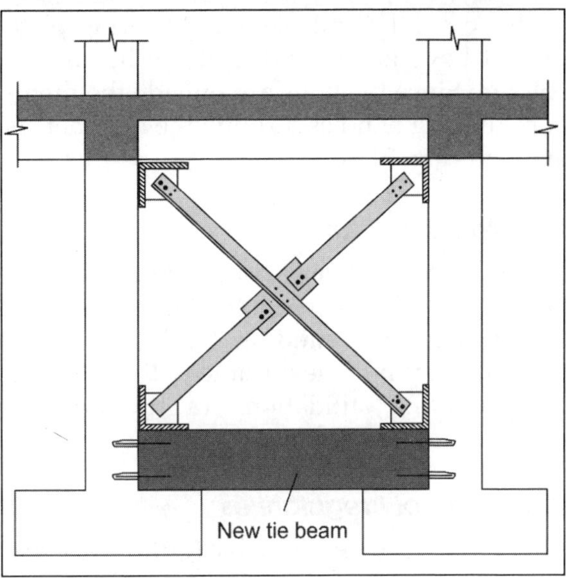

Fig. 21.8: Addition of steel braces

A few types of connections are shown in Fig. 21.9.

Type I: The force in brace is transferred to the frame through the gusset plate, end plates and anchor inserts (Fig. 21.9a).

Type II: This type of connection is similar to type I, except for the method of anchoring the end plates at the joint. An end plate is connected using through bolts which are anchored at the opposite face to a bearing plate. In this type, the widths of end plate and bearing plate are equal to or less than the width of beam or column (Fig. 21.9b).

Type III: This type of connection is similar to type II, except for the location of bolts.

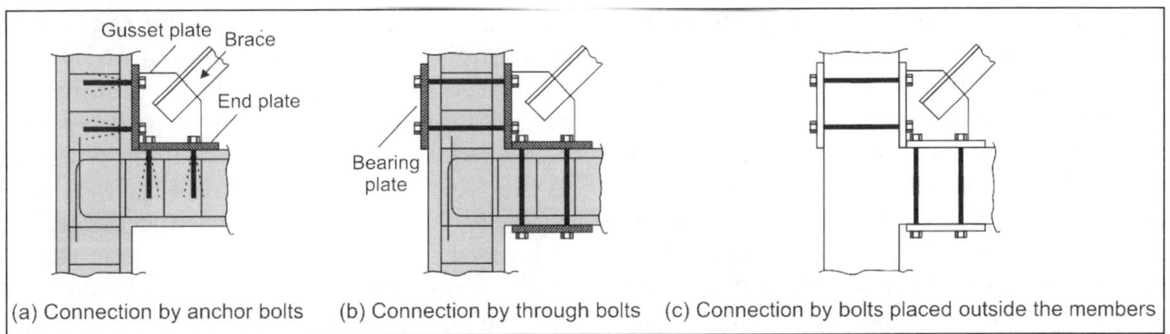

Fig. 21.9: Types of connection of braces to an RC frame

In this type, the end plate and bearing plate project beyond the width of the beam and column. Since the bolts are outside the members, drilling of holes through the members is avoided (Fig. 21.9c).

Addition of Frames

A new frame, either of concrete or steel, can be introduced to increase the lateral strength and stiffness of a building. Similar to a new wall, integrating a new frame to the building and providing foundations are critical design issues.

Reduction of Irregularities

The plan and vertical irregularities are common causes of undesirable performance of a building under an earthquake. Reduction of the irregularities may be sufficient to reduce force and deformation demands in the members to acceptable levels. Addition of infill walls, shear walls or braces can alleviate the deficiency of soft and/or weak storeys. Discontinuous components of the lateral load resisting system, such as floating columns can be extended up to the foundation. In this case, the supporting cantilever beams have to be checked for the sagging moment. The infill walls of partial height can be extended to reduce the vulnerability of short and stiff (captive) columns.

Torsional irregularities can be corrected by the addition of frames or shear walls to balance the distribution of stiffness and mass. An eccentric mass due to an overhead water tank can be relocated. Seismic joints can be introduced to transform a single irregular building into multiple regular structures. A joint can be introduced by providing two rows of columns, one on each side of the joint. The intermediate portion can be covered by a crumple section. Detail of crumple section is provided in IS:4326-1993. Although partial demolition can have impact on the appearance and utility of the building, it can be an effective measure to reduce irregularity.

Reduction of Mass

A reduction in mass of the building results in reduction of the lateral forces. Hence, this option can be considered instead of structural strengthening. The mass can be reduced through demolition of unaccounted additional storeys, replacement of heavy cladding, partitions, terrace or under-floor fillings, or removal of heavy storage and equipment loads, or change in the use of the building.

Energy Dissipation Devices and Base Isolation

Most energy dissipation devices not only supplement damping, but also provide additional lateral stiffness to a building. Base isolation devices reduce the structural force entering a building by elongating the time period of the structure and thus decreasing the base shear force. These devices are described in Chapter 23.

A comparative evaluation of the different global retrofit strategies is provided in Table 21.1

21.3.2 Local Retrofit Strategies

Local retrofit strategies pertain to retrofitting of columns, beams, joints, slabs, walls and foundations. The local retrofit strategies are categorized according to the retrofitted elements. The retrofitting of foundations is separately covered in Chapter 22. The analysis of a building with a trial local retrofit strategy should incorporate the modeling of the retrofitted elements.

The local retrofit strategies fall under three different types: Concrete jacketing, steel jacketing (or use of steel plates) and fiber reinforced polymer (FRP) sheet wrapping. In the present chapter, the first two types are discussed. Chapter 23 is exclusively devoted to the retrofit using FRP. Out of the first two types, each one has its merits and demerits. Table 21.1 provides a comparative evaluation of the strategies with general statements. It may be noted that deviations are expected in individual retrofit projects.

Table 21.1: Comparative evaluation of the global retrofit strategies

Retrofit strategy	Merits	Demerits	Comments
Addition of infill walls	• Increases lateral stiffness of a storey • Can support vertical load if adjacent column fails	• May have premature failure due to crushing of corners or dislodging • Does not increase ductility • Increases weight	• Low cost • Low disruption • Easy to implement
Addition of shear walls, wing walls and buttress walls	• Increase lateral strength and stiffness of the building substantially • May increase ductility • Needs adequate foundation	• May increase design base shear • Increase in lateral resistance is concentrated near the walls • Needs integration of the walls to the building	• High disruption based on location, involves drilling of holes in the existing members
Addition of braces	• Increases lateral strength and stiffness of a storey substantially • Increases ductility	• Connection of braces to an existing frame can be difficult	• Passive energy dissipation devices can be incorporated to increase damping/stiffnes or both
Addition of frames	• Increases lateral strength and stiffness of the building • May increase ductility	• Needs adequate foundation	• Needs integration of the frames to the building

Column Retrofitting

The retrofitting of deficient columns is essential to avoid collapse of a storey. Hence, it is more important than the retrofitting of beams. The columns are retrofitted to increase their flexural and shear strengths, to increase the deformation capacity near the beam–column joints and to strengthen the regions of faulty splicing of longitudinal bars. The columns in an open ground story or next to openings should be prioritized for retrofitting. The retrofitting strategy is based on the "strong column–weak beam" principle of seismic design. During retrofitting, it is preferred to relieve the columns of the existing gravity loads as much as possible, by propping the supported beams. The individual retrofit strategies are described next.

Concrete Jacketing

Concrete jacketing involves addition of a layer of concrete, longitudinal bars and closely spaced ties. The jacket increases both the flexural strength and shear strength of the column. Increase in ductility has been observed; if the thickness of the jacket is small there is no appreciable increase in stiffness. Circular jackets of ferrocement have been found to be effective in enhancing the ductility. The disadvantage of concrete jacketing is the increase in the size of the column. The placement of ties at the beam–column joints is difficult, if not impossible. Drilling holes in the existing beams damages the concrete, especially if the concrete is of poor quality. Although there are disadvantages, the use of concrete jacket is relatively cheap. It is important to note that with the increase in flexural capacity, the shear demand (based on flexural capacity) also increases. The additional ties are provided to meet the shear demand.

There can be several schemes of providing a concrete jacket. A scheme is selected based on the dimensions and required increase in the strength of the existing column, available space of placing the longitudinal bars. To increase the flexural strength, the additional longitudinal bars need to be anchored to the foundation and should be continuous through the floor slab. Usually the required bars are placed at the corners so as to avoid intercepting the beams which are framing into the column. In addition, longitudinal bars may be placed along the sides of the column which are not continuous through the floor. These bars provide lateral restraint to the new ties. A tie cannot be made of a single bar due to the obstruction in placement. It can be constructed of two bars properly anchored to the new longitudinal bars. It is preferred to have 135° hooks with adequate extension at the ends of the bars.

Figure 21.10 shows some possible schemes of concrete jacketing. The sections of existing column and the jacket are shown with lighter and darker shades, respectively. The reinforcement of the existing columns is not shown for clarity. In Fig. 21.10a, each tie is made up of a U-piece and a straight piece. In Fig. 21.10b, the jacket is provided only on two sides of the existing column. In Fig. 21.10c, each tie is made up of two U-pieces. The new longitudinal bars are connected to the existing longitudinal bars by welding Z- or U-shaped bent bar. In Fig. 21.10d, each tie is made up of semioctagonal or L-shaped pieces.

Since the thickness of the jacket is small, casting self-compacting concrete or the use of shotcrete are preferred to conventional concrete. To ensure the composite action of the existing and the new concrete, the options for preparing the surface of the existing concrete are hacking by chisel, roughening by wire buff or using bonding chemicals. Slant shear tests have shown that the preparation of the surface is adequate to develop the bond between the existing and new concrete. Inserting dowels in the existing column at a spacing of 300–500 mm enhances the composite action shown in Fig. 21.10b. The dowels are attached by epoxy pumped into the drilled holes. The length of insertion depends on the type of epoxy and the strength of the existing concrete. The latter can be estimated through nondestructive or intrusive tests as described in Chapter 7. Based on the information of the epoxy from the manufacturer's catalogue and the evaluated concrete strength, the length of insertion is

determined. Of course, drilling holes damage the existing column. Moreover, it has been observed that the presence of dowels increases the disintegration of the concrete jacket during the formation of a plastic hinge under dynamic loads. If the jacket is all around the existing column, then the dowels can be avoided. The shrinkage of the new concrete generates friction with the existing concrete. If the jacket is only partially around the existing column, the existing bars can be exposed at locations. The new bars can be welded to the existing bars using Z- or U-shaped bent bar as shown in Fig. 21.10c. Of course, it has been observed that welding of two different grades of steel increases corrosion.

The minimum specifications for the concrete jacket are as follows:

- The strengths of the new materials must be equal to or greater than those of the existing column. The compressive strength of concrete in the jacket should be at least 5 MPa greater than of the existing concrete.
- For columns where extra longitudinal bars are not required for additional flexural capacity, a minimum of 12 mm diameter bars in the four corners and ties of 8 mm diameter should be provided.
- The minimum thickness of the jacket should be 100 mm.
- The minimum diameter of the ties should be 8 mm and should not be less than of the diameter of the longitudinal bars. The angle of bent of the end of the ties should be 135°.
- The centre-to-centre spacing of the ties should not exceed 200 mm. Preferably, the spacing should not exceed the thickness of the jacket. Close to the beam–column joints, for a height of 1/4th of the clear height of the column, the spacing should not exceed 100 mm.

A simplified analysis for the flexural strength of a retrofitted column can be done by the traditional method of interaction curves (SP:16-1980, "Design Aids for Reinforced Concrete to IS:456-1978, published by the

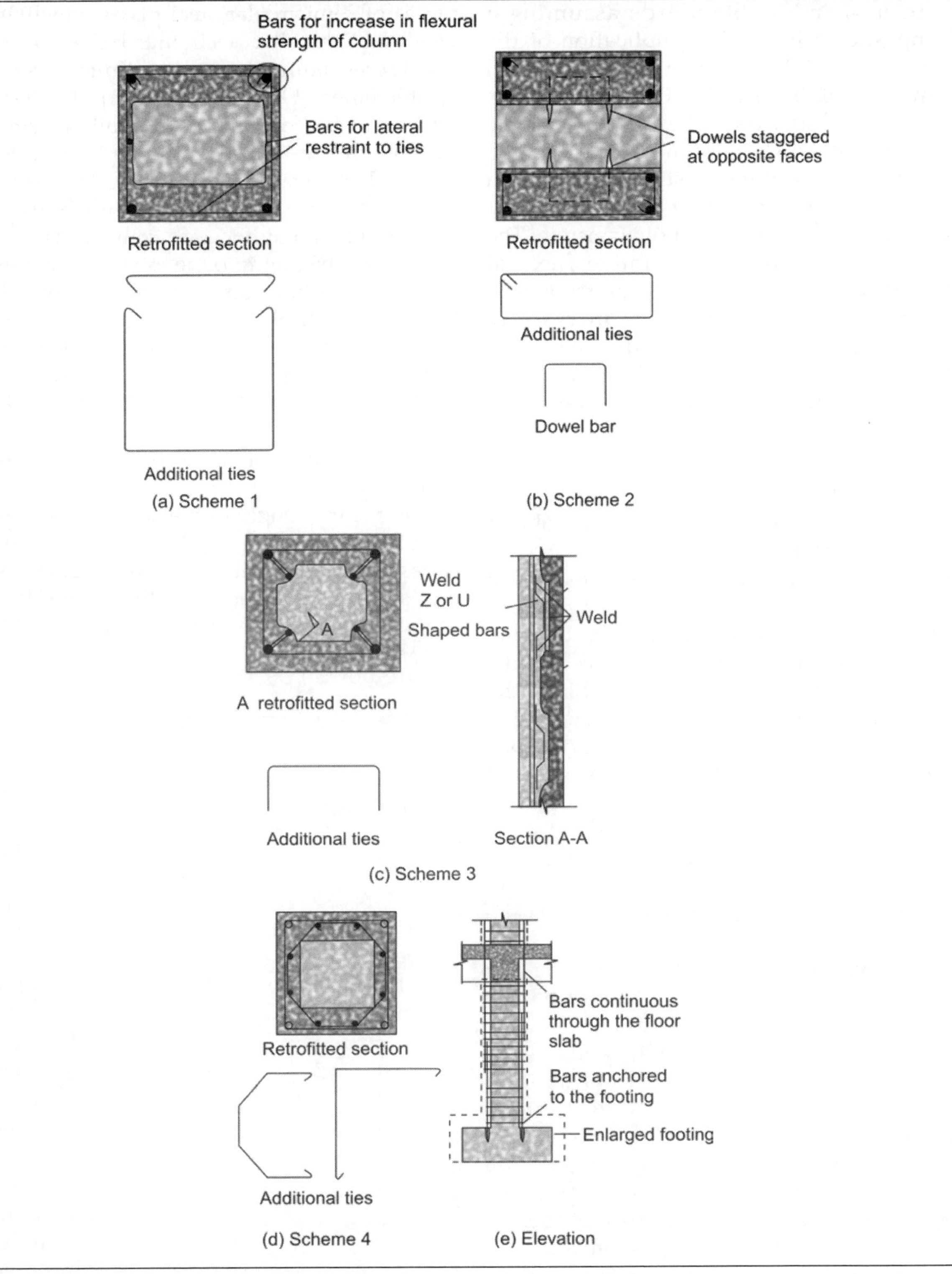

Fig. 21.10: Schemes for concrete jacketing of columns

Bureau of Indian Standards assuming a composite section). The publication of the International Federation for Structural Concrete (FIB Bulletin 24, 2003) recommends such an analysis. Even if the grade of concrete in the jacket is higher, it can be considered to be same as that of the existing section. Such an analysis assumes that there is perfect bond between the new and old concrete. The yield moment and the ultimate flexural capacity can be conservatively limited to 90% of the calculated values. The increase in shear capacity can be calculated based on the amount of additional ties. For the requirement of confinement, only the additional ties are to be considered.

Steel Jacketing

Steel jacketing refers to encasing the column with steel plates and filling the gap with nonshrink grout. The jacket is effective to remedy inadequate shear strength and provide passive confinement to the column. Lateral confining pressure is induced in the concrete as it expands laterally. Since the plates cannot be anchored to the foundation and made continuous through the floor slab, steel jacketing is not used for enhancement of flexural strength. Also, the steel jacket is not designed to carry any axial load. If the shear capacity needs to be enhanced, the jacket is provided throughout the height of the column. A gap of about 25–50 mm is provided at the ends of the jacket so that the jacket does not carry any axial load. For enhancing the confinement of concrete and deformation capacity in the potential plastic hinge regions, the jacket is provided at the top and bottom of the column. Of course, there is no significant increase in the stiffness of a jacketed column. Steel jacketing is also used to strengthen the region of faulty splicing of longitudinal bars. As a temporary measure after an earthquake, a steel jacket can be placed before an engineered scheme is implemented.

Circular jackets are more effective than rectangular jackets. A jacket is made up of two pieces of semicircular steel plates which are welded at the site. A circular jacket can be considered equivalent to continuous hoop reinforcement. Of course, circular jackets may not be suitable for columns in a building since the columns are mostly rectangular in cross section. In a rectangular jacket, steel plates are welded to corner angles. Anchor bolts or through bolts may increase confinement but it involves drilling into the existing concrete. A simpler form of strengthening is to weld batten plates to the corner angles. This form is referred to as steel profile jacketing as opposed to the encasement provided by continuous plates. Figure 21.11 shows the different techniques of providing a steel jacket. The steel plates need protection against corrosion and fire.

The shear strength of the jacket (V_j) can be calculated by considering the jacket to act as a series of independent square ties of thickness and spacing t_{sj}, where t_{sj} is the thickness of the plates. For rectangular columns,

$$V_j = \frac{A_{sj} f_{sj} d_{sj}}{S_{sj}}$$

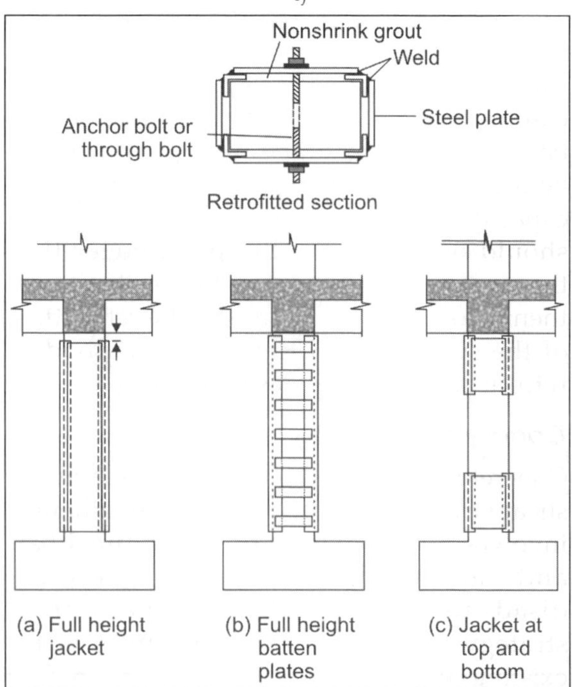

Fig. 21.11: Techniques for steel jacketing of columns

Here, A_{sj} is the total area of the assumed square tie, $A_{sj} = 2t_{sj}^2$, f_{sj} is the allowable stress of the jacket, d_{sj} is the depth of the jacket (can be equated to the transverse depth of the column) and S_{sj} is the spacing between the square ties, $S_{sj} = t_{sj}$. The allowable stress of the jacket can be assumed to be half of its yield stress. The required thickness t_{sj} can be calculated from the required value of V_j.

Fiber Reinforced Polymer Sheet Wrapping

Fiber reinforced polymer (FRP) has desirable physical properties like high tensile strength to weight ratio and corrosion resistance. FRP sheets are thin, light and flexible enough to be inserted behind pipes and other services ducts, thus facilitating installation. In retrofitting a column with FRP sheets, there is increase in ductility due to confinement without noticeable increase in the size. The main drawbacks of FRP are the high cost, brittle behavior and inadequate fire resistance. The details of retrofitting using FRP sheets are covered in Chapter 23.

Beam Retrofitting

The beams are retrofitted to increase their positive flexural strength, shear strength and the deformation capacity near the beam-column joints. The lack of adequate bottom bars and their anchorage at the joints needs to be addressed. Usually the negative flexural capacity is not enhanced since the retrofitting should not make the beams stronger than the supporting columns. Of course, the strengthening of beams may involve the retrofitting of the supporting columns. The individual retrofit strategies are described next.

Concrete Jacketing

Concrete is added to increase the flexural and shear strengths of a beam. The strengthening involves the placement of longitudinal bars and closely spaced stirrups. There are disadvantages in this traditional retrofit strategy. First, the drilling of holes in the existing concrete can weaken the section if the width is small and the concrete is not of good quality. Second, the new concrete requires proper bonding to the existing concrete. In the soffit of a beam, the bleed water from the new concrete creates a weak cement paste at the interface. If the new concrete is not placed all around, restrained shrinkage at the interface induces tensile stress in the new concrete. Third, addition of concrete increases the size and weight of the beam.

Instead of conventional concrete, fiber reinforced concrete can be used for retrofit. In addition to strength, this leads to the increase of energy absorption capacity. It is important to note that with the increase in flexural capacity, the shear demand (based on flexural capacity) also increases. Additional stirrups need to be provided to meet the shear demand.

There are a few schemes for concrete jacketing as shown in Fig. 21.12. The difficulty lies in anchoring the new bars. A scheme is selected based on the deficiency and the available space for placing the bars. The schemes given in IS:13935-1993 involves drilling holes in the existing beam. But drilling holes for the stirrups at close spacing damages the beam, especially if the concrete is of poor quality. The additional longitudinal bars should be continuous through the beam–column joints. The bars are placed at the corners so as to avoid intercepting the transverse beams. Additional longitudinal bars may be placed at the sides for checking temperature and shrinkage cracks. These bars need not be continuous through the joints.

If the beams support a masonry wall, then closed stirrups may not be possible. In such a situation the ends of U-stirrups may be threaded and anchored by nuts at the top surface of the slab as shown in Fig. 21.12a. If there is no wall above the beam or the beam supports a removable partition wall, a pair of U-stirrups can be welded to the existing stirrups. The scheme shown in Fig. 21.12d may be adequate for strengthening of gravity loads. But unless the slab is thick, the stirrups will not be properly anchored to sustain the dynamic seismic forces.

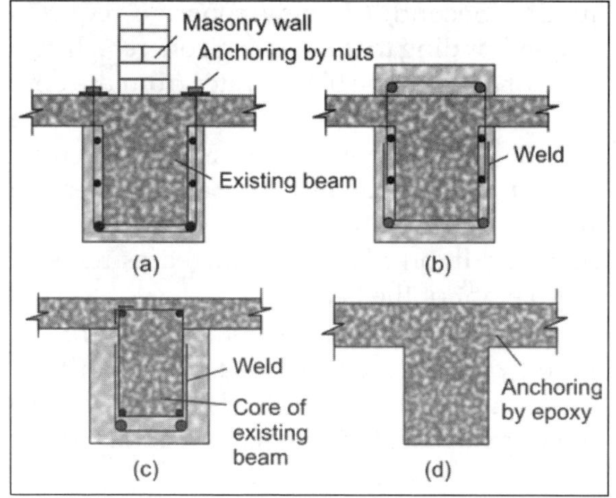

Fig. 21.12: Schemes for concrete jacketing of beams

Similar to concrete jacketing of columns, the use of self-compacting concrete or shotcrete is preferred to conventional concrete. The surface of the existing concrete is to be prepared to ensure the composite action of the existing and the new concrete.

A simplified analysis for the flexural strength of a retrofitted section can be performed by the traditional method of beam analysis. Even if the grade of concrete in the jacket is higher, it can be considered to be same as that of the existing section. Such an analysis assumes a perfect bond between the new and existing concrete. For a rigorous analysis considering different grades of concrete, a layered approach is required.

Bonding Steel Plates

The techniques of bonding mild steel plates to beams are used to improve their flexural and shear strengths. The addition of steel plate is rapid to apply, does not reduce the storey clear height significantly and can be applied while the building is in use. The plates are attached to the tension face of a beam to increase the shear strength.

The plates can be attached by adhesives or bolts. The plates attached by adhesives are prone to premature debonding and hence, the beam tends to have a brittle failure. A beam with plates attached by bolts tends to have a

ductile failure. But the use of bolts involves drilling in the existing concrete, which may weaken the section, if the width is less or if the concrete is not of good quality. Any exposed steel plate is prone to corrosion and fire. Hence, adequate protection is required for beams retrofitted with steel plates.

The analysis for flexural strength of beams with plates bonded to the tension face is based on satisfying the equilibrium and compatibility equations and the constitutive relationships. It models the adhesive failure due to the stress concentration at the location of plate cut-off. The essential features of the model are as follows:

a. The steel plate is assumed to act integrally with the concrete beam and conventional beam theory is used to determine the flexural capacity.

b. The normal and shear stresses at the interface of concrete and the plate at the location of plate cut-off are calculated to check the failure of the adhesive.

The strain and stress diagrams and internal forces at the ultimate limit state for a tension face plated beam of rectangular cross-section are shown in Fig. 21.13.

The depth of neutral axis (x_u) can be found from the equilibrium equation as follows:

$$X_u = f_{st} A_{st} + f_{pt} b_p t_p$$
$$= 0.36 f_{ck} b$$

The ultimate moment capacity (M_{UR}) of the plated section is given by the following expression:

$$M_{UR} = 0.36 f_{ck} b x_u (d - 0.416) + f_{pt} b_p d_p (d_c + t/2)$$

Here,

b, D, d: Width, overall depth and effective depth of the original section

b_p, t_p, d_p: Width, thickness and effective depth of the steel plate

f_{ck}: Characteristic cube strength of concrete (in MPa)

f_{st}, f_{pt}: Stresses in internal rebar and external plate respectively, corresponding to the

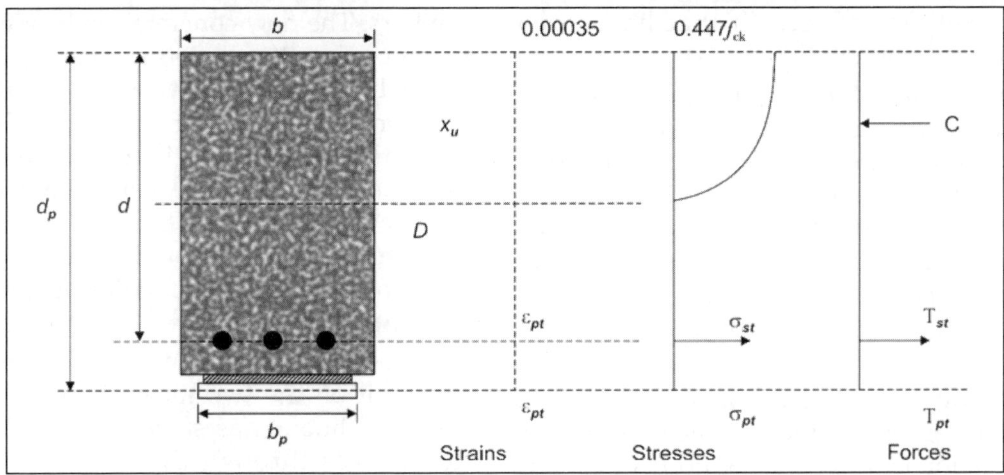

Fig. 21.13: Strain and stress diagrams and internal forces of a tension-face plated beam

respective strains. These can be calculated from the compatibility equations and the constitutive relationships for steel of the rebar and plate. For an under-reinforced section, $f_{st} = f_y/1.15 = 0.87f_y$, where f_y is the yield stress of the rebar. If the plate thickness is such that it is found to yield before the concrete crushes then $f_{st} = f_{yp}/1.1 = 0.9f_{yp}$, where f_{yp} is the yield stress of the plate.

The material safety factor for rolled steel plates is 1.1 as per IS: 800 (draft), "Indian standard code of practice for general construction in steel".

Since retrofitting is required mostly near the beam–column joints, the equations for a rectangular section are applicable. Of course a similar set of equations can be derived for a flanged section when the span region of a beam requires strengthening. The width of the plate (b_p) is limited to the width of the beam (b). The thickness of the plate (t_p) is limited such that the section does not become over-reinforced. The adhesive failure is checked by the following equation:

$$\tau_0 + \sigma_0 \tan 28° \le c_{\text{all}} \qquad \text{...(9.4)}$$

Here, τ_0 and σ_0 are the shear and normal stresses respectively at the interface of concrete and the plate at the location of plate cut-off. The allowable coefficient of cohesion for the adhesive is denoted as c_{all}. The expression of τ_0 and σ_0 are based on the shear force at the location of plate cut-off, elastic modulus of the plate and the elastic and shear moduli of the adhesive.

The beams strengthened by plates bonded to the side face are subjected to the following modes of failure: Fracture of bolts, buckling of plates and splitting of concrete. Barnes *et al* (2001) derived expressions for the enhancement of shear capacity of such beams.

FRP Wrapping

Like steel plates, FRP laminates are attached to beams to increase their flexural and shear strengths. The details of retrofitting using FRP sheets are covered in Chapter 23.

Beam–Column Joint Retrofitting

The retrofitting of a beam–column joint aims to increase its shear capacity and effective confinement. Since access to a joint is not readily available, retrofitting a joint is difficult. The strengthening is carried out along with that for the adjacent columns and beams. The retrofitting should aim at the following improvements:

- To prevent the pull out of any discontinuous bottom bars in the beams
- To reduce the deterioration of the joint under cyclic loading

The methods of retrofit that have been investigated are as follows.

Concrete jacketing: A joint can be strengthened by placing this through drilled holes in the adjacent beams. To avoid drilling holes in the beams, a steel cage can be fabricated around the additional vertical bars in the column to provide confinement. A simpler option is a concrete fillet at the joint to shift the potential hinge region of the beam away from the column face.

Steel jacketing: If space is available, steel jacketing can be used to enhance the performance of joints. A simpler option is to attach plates in the form of brackets at the soffits of the beams.

FRP wrapping: In the use of FRP sheets to strengthen the joints, the considerations are number of layers, orientation and anchorage of the FRP sheets and the preparation of surface of the existing concrete.

Wall Retrofitting

A concrete shear wall can be retrofitted by adding new concrete with adequate boundary members. The new concrete can be added by shotcrete. For the composite action, dowels need to be provided between the existing and new concrete as shown in Fig. 21.14. The foundation of the wall is to be sufficiently strengthened to resist the overturning moment without rocking or uplift. The reinforcement of the foundation is not shown for clarity. Details of retrofitting foundation are provided in Chapter 22.

A cross-tie connecting the layers of reinforcement on the two faces of the new wall should be hooked as shown in Fig. 21.14. It should pass through diagonally opposite quadrants formed by the vertical and horizontal bars.

For the masonry infill walls, whose failure can cause injury, additional concrete with wire mesh or FRP sheets can be used to strengthen for out-of-plane bending. Steel braces fitted to the RC frame have been to check the out-of-plane bending of walls. Steel strips bolted to the walls have been investigated.

A comparative evaluation of the different *local* retrofit strategies is provided in Table 21.2.

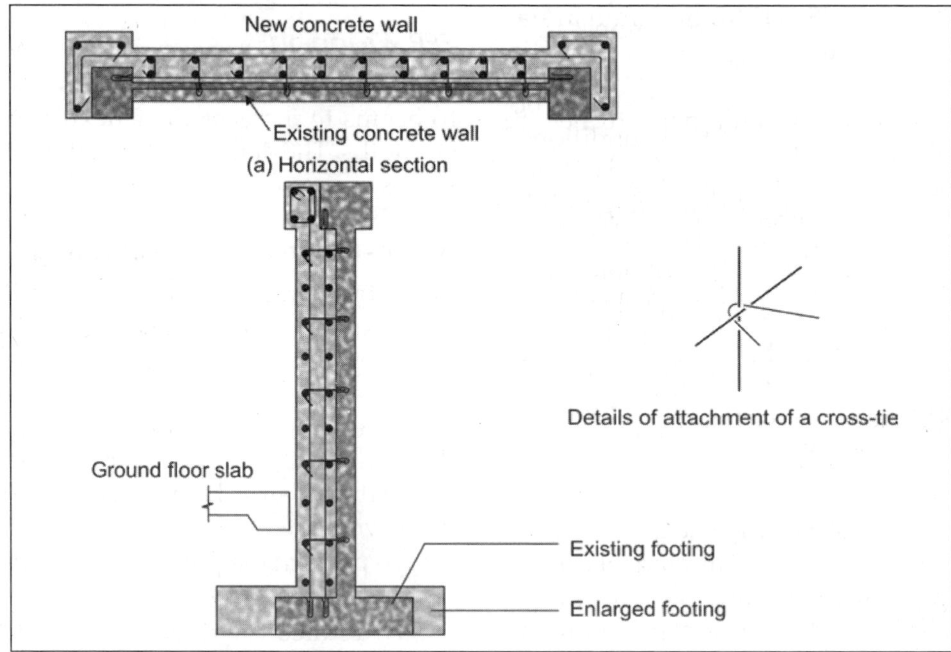

Fig. 21.14: Strengthening a wall using concrete

Table 21.2: Comparative evaluation of the local retrofit strategies

Retrofit strategy	Merits	Demerits	Comments
Concrete jacketing	• Increases flexural and shear strengths and ductility of the member • Easy to analyze • Compatible with original substrate	• Size of member increases • Anchoring of bars for flexural strength; involves drilling of holes in the existing concrete • Needs preparation of the surface of existing member	• Low cost • High disruption • Experience of traditional RC construction is adequate
Steel jacketing columns	• Increases shear strength and ductility • Minimal increase in size	• Cannot be used for increasing the flexural strength • Needs protection against corrosion and fire	• Can be used as a temporary measure after an earthquake • Cost can be high • Low disruption • Needs skilled labour
Bonding steel plates to beams	• Increases either flexural or shear strength • Minimal increase in size	• Use of bolts involves drilling in the existing concrete • Needs protection against corrosion and fire	• More suitable for strengthening against gravity loads • Cost can be high • Low disruption·
Fibre reinforced polymer wrapping	• Increases ductility • May increase flexural or shear strength • Minimal increase in size • Rapid installation	• Needs protection against fire	• Cost can be high • Low disruption • Needs skilled labour

21.4 GENERAL REMARKS

Importance of Seismic Evaluation

Considering the cost of retrofit, it is imperative to have seismic evaluations of a building both for the existing and retrofitted conditions to justify the selected retrofit strategies. When a new member is added to an existing building under a global retrofit strategy, the load transfer and the compatibility of deformation between the new and the existing elements are crucial. The local transfer should be judged from the analysis of the building. The compatibility of deformation should be ensured by proper detailing of the connections of the existing and new members.

Selection of a Retrofit Strategy

When a building is severely deficient for the design seismic forces, it is preferred to a select a global retrofit strategy to strengthen and stiffen the structure. Next, if deficiencies still exist in the members, local retrofit strategies are to be selected. Beyond this recommendation, it is not prudent to prescribe a retrofit strategy as a generic application. Each retrofit strategy has merits and demerits depending upon the project. A retrofit strategy is to be selected after careful considerations of the cost and constructability. Proper design of a retrofit strategy is essential. The failure mode in a member after retrofitting should not become brittle. A global retrofit strategy that involves a shift in either of the centre of mass or centre of rigidity, should be checked for torsional irregularity. Any alteration of the load path has to be carefully detected and any overstressed member has to be identified. Additional demand on the foundation has to be accounted for.

Quality of Construction

Retrofit aims at overcoming the deficiencies of an existing building. The quality of construction for a successful retrofit scheme cannot be overemphasized. Any sort of patch work will be a wasted effort.

21.5 SUMMARY

The deficiencies commonly observed in reinforced concrete multistoreyed buildings are illustrated. The seismic detailing of the reinforcing bars is explained briefly. The faults in the detailing are highlighted. The retrofit strategies are described under global and local strategies. The global strategies cover addition of walls and braces to mitigate the effect of open ground storey. The local strategies include concrete and steel jacketing. Comparative evaluations of the strategies are placed in tabular form. Finally, some remarks are incorporated for general awareness.

22

Retrofit of Foundations

The loads from a building get transmitted to the soil through the foundation. A seismic retrofit of a building includes strengthening of inadequate foundations or supplementing with new foundations. This chapter covers the important aspects of deficiencies of foundation, analysis and assessment of foundation, the types of intervention to strengthen the foundation and methods of execution. The types of intervention include strengthening rubble masonry foundation, enlarging the area of reinforced concrete footing, underpinning the foundation, drilling micropiles, strengthening of piles and base plates. The descriptions are supported with schematic sketches for clarity. The methods of execution briefly describe the types of shoring, temporary supports and underpinning.

All buildings have to be supported on adequate foundations. The foundations transmit the loads to the soil underneath. Hence, the safety of the buildings depends critically on the foundations. A builder has options for the material and construction of the building, such as masonry, timber, concrete or steel. However, in many cases, there is little choice with regard to the site of the building. The site may have unsatisfactory soil condition. In a few cases, the soil condition may be improved by ground improvement techniques. But in most cases, for economy, the prevailing soil condition is adopted with a suitable foundation.

Soil has a wide range of characteristics depending on the nature of formation. The soil that forms at the location of the parent rock (residual soil) is different from that forming at a different location (transported soil). Even at a given site, the soil taken from two locations from the same stratum can show widely varying properties. The properties are also affected considerably by environmental changes and vibrations during an earthquake. Therefore, it is important to investigate the site before choosing a suitable foundation for a new building or retrofitting the foundation for an existing building.

Some types of clay which are very hard when dry, lose their bearing capacity when wet. In India, vast areas have black cotton soil which contains expansive clay mineral. The latter is responsible for excessive swelling and shrinkage. In the coastal areas, sandy soil with high water table is susceptible to liquefaction.

22.2.1 Types of Foundation

The foundations of structures can broadly be grouped as either shallow or deep. The essential features of these two types of foundation are discussed below.

Shallow Foundations

- Column or wall footings and rafts are classified as shallow foundations. A shallow foundation transmits the loads from the structure to a soil stratum at a relatively small depth from the ground surface.
- The location and the depth at which a shallow foundation is placed should have adequate bearing capacity. The foundation should not be affected by volume changes of the soil caused by weathering, expulsion of soil from beneath or due to the foundations of adjoining structures.
- The settlement, especially any differential settlement, tends to cause damages to the structure. Hence, both the settlement and differential settlement should be within limits.

The above requirements of location, adequate bearing capacity and permissible settlement, should be satisfied individually for each footing.

Deep Foundations

- Piles and well foundations are classified as deep foundations. In piles, the load from the structure is supported by frictional resistance around the piles and/or by end bearing when the piles rest on rock or a hard stratum.
- Unlike a shallow foundation constructed after excavating the soil up to the foundation bed, *in situ* construction of a deep foundation makes it impossible to visually inspect the quality.
- For buildings on expansive clay, such as black cotton soil, under-reamed piles are most suitable. A pile with a bulb at the bottom is taken to a suitable depth where seasonal moisture variation is less.

It is essential to diagnose the deficiencies in the foundation before undertaking retrofit. Also, the highlighting of deficiencies is expected to create awareness for future construction. The causes of deficiencies are listed below:

- The foundation may not have adequate strength as per the requirements of the current seismic code IS:1893-2002. Based on the increased seismic forces, the forces on the foundation will increase. Very often, old buildings are found to have been designed for gravity loads only.
- The foundation can deteriorate due to the ground water and/or soil containing aggressive chemicals. Industrial pollutants, such as sulphates and chlorides are responsible for deterioration of concrete. In such situations, chemical analyses of the soil and water must be undertaken.
- On many occasions, settlement of the soil causes damage to the building.

22.4 CONDITION ASSESSMENT OF FOUNDATIONS

In a retrofit project, the condition of the foundation can be a determining factor as to whether a building can be retrofitted. It is necessary to check the foundation for the following:

- Adequacy of soil strata against resistance to liquefaction
- Stability of the area, particularly for buildings retained on earth or hill slopes
- Maximum bearing pressure on the foundation including the earthquake effect.
- Structural adequacy of the foundation elements, such as footing slabs, pile caps, piles, etc.

The condition assessment of buildings is covered in earlier chapters. Here, some information is provided pertinent to foundations. The condition assessment includes some of the following depending upon the nature of building.

- Investigation of construction records and archives for soil conditions.
- Fresh soil investigation including sampling and testing. Measurements of ground-water level and pore water pressure.
- Survey of the foundation and foundation walls. If required, a few typical footings may be exposed by test pits. The surveys examine any deterioration of the material.
- Report of settlement, formation of cracks and tilting of the walls, heave in the adjacent areas and vertical/horizontal alignment of the foundation.

22.5 METHODS OF ANALYSIS

22.5.1 Analysis of the Building

In the analysis of a building, traditionally the soil-structure interaction is neglected. The bottom of the ground storey walls or columns are considered to be fixed or pinned depending upon the type of foundation. Recommendations on the modeling of the base conditions are given in earlier chapters. Once the forces at the base are obtained from the analysis, the foundation can be analysed separately. For a refined analysis of important building, the soil-structure interaction can be incorporated. The interaction refers to the effect of the soil deformation on the forces in the building.

The following approaches are usually adopted for modeling the soil foundation system.

a. **Shallow foundations:** The soil beneath the shallow foundations can be modeled using spring supports (referred to as Winkler's spring supports) with stiffness calculated based on the modulus of subgrade reaction (simply, subgrade modulus). Typical ranges of subgrade modulus are provided by bowles (2001).

b. **Pile foundations:** A pile group can be modeled as a column. The following concepts are used in the model.

- *Depth of fixity:* The piles are assumed to be fixed at certain depth below the pile cap. The depth of fixity depends on the stiffness of the soil (represented by the subgrade modulus) and the fixity condition of the piles at the pile cap. IS:2911 provides recommendation for the depth of fixity.
- *Soil springs:* Soil springs are assumed at the interface of each pile and the soil. The stiffness of a spring is derived from the subgrade modulus.

The forces on the foundation are subsequently obtained from the analysis of the building model.

22.5.2 Analysis of the Foundation

a. **Shallow foundations:** The foundation may be considered to be either rigid or flexible depending on the relative stiffness of the foundation with respect to the soil. For a rigid foundation, the soil pressure is linear and the computations of the bending moment and shear force in the footing slab or raft are simpler. For a flexible foundation, the analysis is done using Winkler's spring supports. Although the analysis needs more effort, the bending moment is usually less compared to that obtained from the analysis based on rigid foundation.

b. **Pile foundations:** In the case of a pile group, the forces on individual piles can be evaluated using the finite element method. In this method, a pile is modeled using beam elements and the pile cap is modeled as a thick plate. The forces from the column are applied on top of the pile cap. From the analysis, a pile is checked for the maximum compression force, uplift force, bending moment and shear force.

22.6 TYPES OF INTERVENTIONS

If the existing foundation does not meet the demand, retrofitting needs to be carried out. It is, of course, difficult to carry out the operation while the building is in-service.

The following difficulties are to be envisaged during retrofitting of foundations:

- Evacuation of the entire area or part of the area in phases
- Breaking the ground floor slab inside the building and paved areas outside the building
- Restricted movement during intervention due to temporary supports
- Restricted headroom for mobilizing large construction equipment, such as piling rig
- Noise/vibrations in the building.

In spite of the above difficulties, retrofitting has to be carried out in view of the safety of the building for future use. To improve the inadequate bearing capacity, reduce excessive settlement or potential for liquefaction, the zone below the existing foundation can be improved by injection of grout or chemical into the ground. This is described in earlier chapters. The following interventions are related with the foundation alone.

22.6.1 Strengthening Rubble Masonry Foundation

The method of underpinning can be used only if the masonry has sufficient strength for intervention. For underpinning, both sides of the wall are excavated. The material underneath the foundation is removed in segments (say, for about a length of 1.5 m). Reinforcing bars are introduced below the wall and concrete is cast. A layer of expansive mortar can be provided between the masonry and the new concrete, as shown in Fig. 22.1.

A masonry foundation can be strengthened by additional reinforcement and concrete as shown in Fig. 22.2. First, both sides of the wall are exposed by excavating the soil. Holes are drilled at intervals of 0.5–1.0 m. Reinforcing bars are placed in the holes. After cleaning all loose material, the surface of the wall is covered with concrete. Instead of conventional concrete, shotcrete can be applied. The bars must be protected from corrosion by a

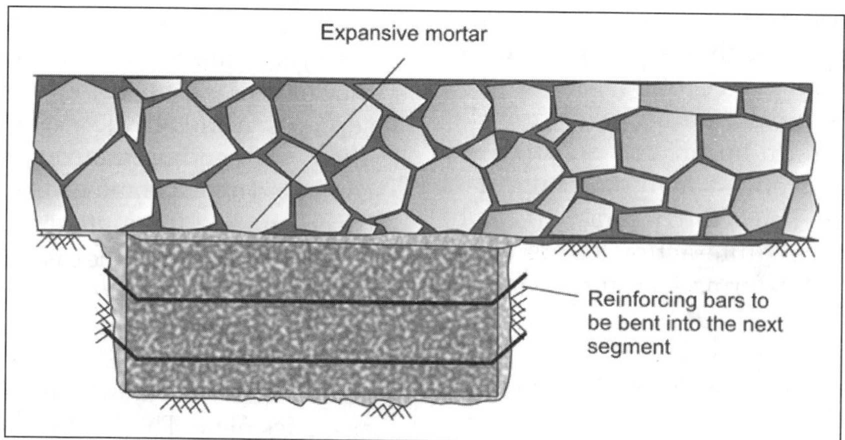

Fig. 22.1: Underpinning of rubble masonry foundation (elevation)

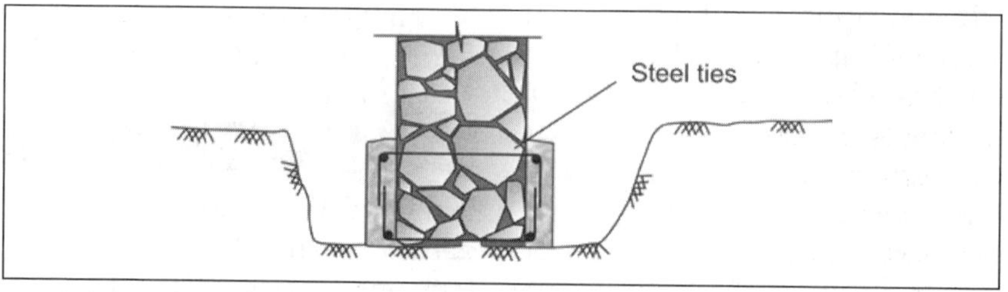

Fig. 22.2: Strengthening of rubble masonry foundation by reinforcement

concrete cover of 40 mm. The wall can be strengthened by grout injection.

Another method of strengthening is to provide reinforced concrete (RC) beams on either side of the wall. The beams can be tied at regular intervals as shown in Fig. 22.3. Typical dimensions of the beams and reinforcement are provided in IS:13935. If necessary, the beams can also be supported on piles.

22.6.2 Enlarging the Area of Footing

In some cases, the foundation can be strengthened by enlarging the size of the RC footing as shown in Fig. 22.4. The footing can be a wall footing or a column footing. The following procedure can be adopted:

- Excavate around the foundation up to the bottom of the plain concrete (PC) leveling course over the required width.
- Expose the cover of the footing along the vertical face and top surface.
- Drill horizontally along the perimeter to attach dowel bars.
- Cast additional PC leveling course for the extended area of the footing.

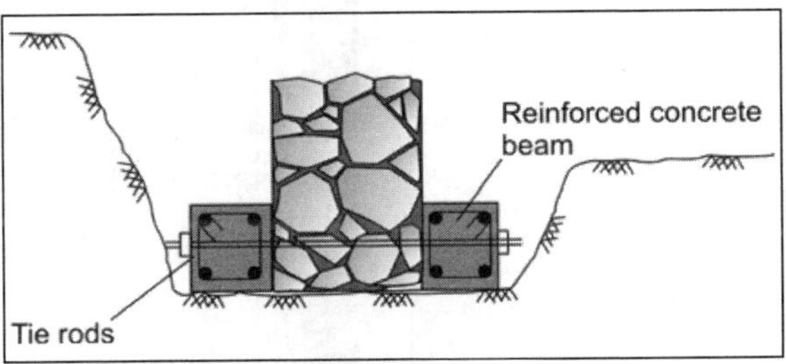

Fig. 22.3: Strengthening of rubble masonry foundation by beams

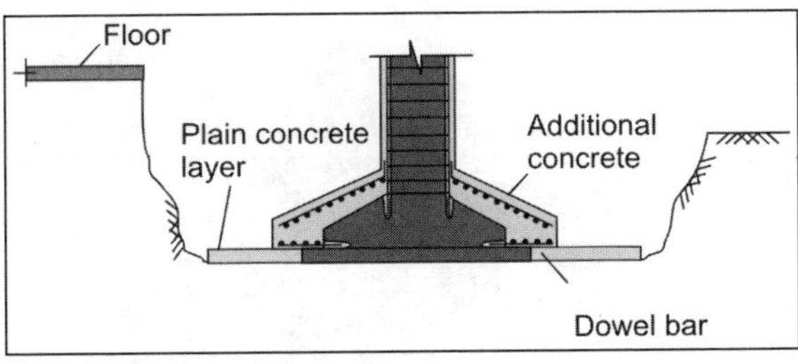

Fig. 22.4: Increasing the area of footing

- Attach dowel bars with epoxy grout into the holes drilled.
- Apply a suitable polymer bonding agent over the exposed surface of the footing.
- Arrange the rebar for the additional area as per the design.
- Cast the additional concrete against formwork with proper compaction.
- Remove the formwork after initial set.
- Cure the foundation by covering the concrete with wet gunny hags. Do not flood the area with water, as this may weaken the founding strata.
- Strengthen the column, if required, by external jacketing.
- Backfill the foundation with selected cohesionless soil in layers, well compacted up to the ground level.
- Complete the plinth filling and flooring or paving.

Along with the footing, the column can be strengthened by concrete jacket. This has been covered in Chapter 21.

22.6.3 Strengthening by Micropiles (Root Piles)

Micropiles are small diameter (100–200 mm) piles that can be drilled through the existing foundations, either vertical or inclined. A schematic sketch of these piles is shown in Fig. 22.5. The piles extend much beyond the foundation level and transfer the loads to the deeper level over a wider width. After a hole is drilled, it is grouted with cement slurry and reinforcing bars are inserted before the initial setting of the grout.

22.6.4 Underpinning with Piles

When the strata below the foundation are found to be very weak and increasing the area of a footing is not sufficient, it becomes necessary to underpin the foundation. This is achieved by transferring the load of the existing structure to a deeper level after installing additional piles.

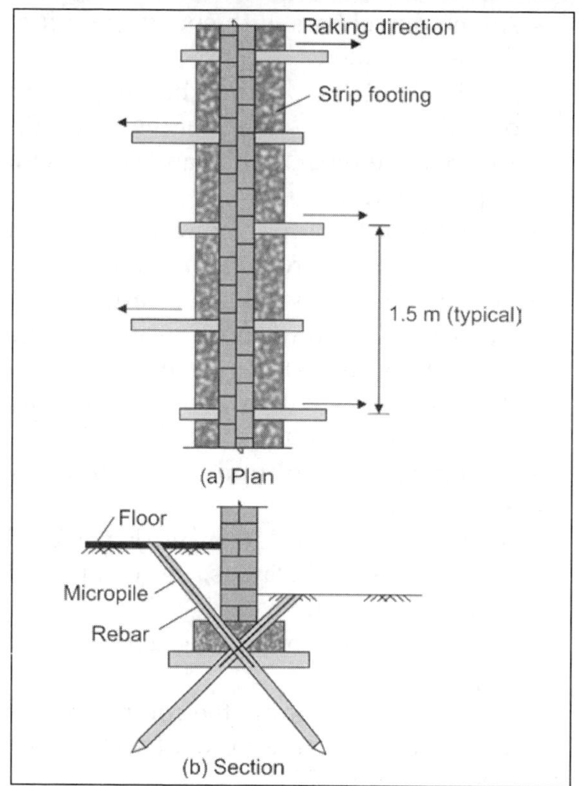

Fig. 22.5: Strengthening by micropiles

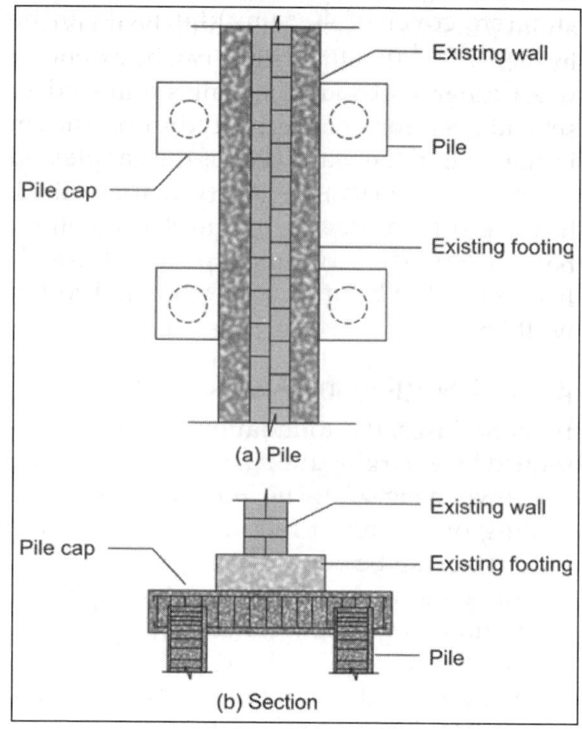

Fig. 22.6: Arrangement of piles for strip foundation

The piles are connected to the existing footing. The procedure of foundation is separately explained for masonry walls and RC columns.

a. Continuous strip foundation under masonry walls: Figure 22.6 shows that the piles can be provided in pairs at intervals. If the wall is close to the property line, the pile cap is provided from one side as shown in Fig. 22.7. In this case, the external pile may be subjected to tension and hence, should be designed accordingly.

In absence of existing RC footing, it is necessary to introduce a longitudinal plinth beam to transfer the concentrated reactions from the pile caps uniformly over the length of the wall. Introduction of the plinth beam is complicated and needs careful planning. After temporarily shoring the roof, intermediate floor and wall, 50 percent of the width of the wall over the height of the designed plinth beam is removed in alternate stretches of

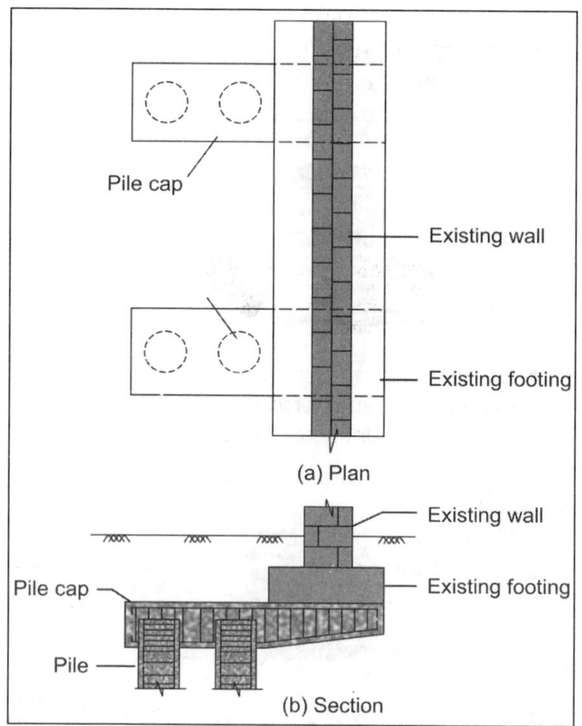

Fig. 22.7: Alternative arrangement of piles for strip foundation

about 3 m length each. The plinth beam is cast in such a way that the stirrup can be extended when the other half of the wall is removed. A schematic sketch of the procedure is shown in the case study at the end of the chapter.

Depending on the rigidity of the plinth beam, needle beams are cast under the plinth beam with centre-to-centre spacing of 2–4 m. Each needle beam has a length equal to the width of the strip footing plus 1 m. The needle beams are supported on pile foundation. Based on the load, the piles can be small under-reamed piles. The following construction sequence can be adopted:

- Prop the roof, intermediate floor and walls.
- Excavate the floor inside the building and the ground outside the building up to the bottom of the proposed plinth beam.
- Break the wall over the required height of the plinth beam and for half the thickness of the wall.
- Introduce the reinforcement cage.
- Cast half of the plinth beam, preferably with self-compacting concrete. Use steel formwork (preferably) along the plinth wall and compact the concrete adequately.
- Within 3–7 days, remove the other half of the wall. Extend the reinforcement from the previous cage, provide additional longitudinal reinforcement as required, and cast the remaining half of the plinth beam.
- At the selected locations for needle beams, excavate both inside and outside of the building to the required dimensions. The depth of the trench should be up to bottom of the needle beam plus 50 mm for the PC leveling course.
- Construct the pile foundation for each end of the needle beam. The cut-off level of these piles should be 50 mm above the PC level.
- Break the wall directly below the plinth beam for the required cross-sectional area of the needle beam.

- After completion of the piles, provide PC over the required area for receiving the needle beams.
- Arrange the reinforcement of the needle beams integrating with the reinforcement of the piles.
- Provide formwork for the vertical faces of the needle beams.
- Cast the needle beams to a level such that top surfaces are in contact with the bottom of the plinth beam.
- Cure the needle beams and plinth beams.
- Backfill the excavation on both sides of the wall with well compacted cohesionless soil.
- Complete the flooring and paving.

b. Foundation for reinforced concrete builing: In the case of a reinforced concrete framed structure supported on individual footings, the number and size of additional piles required under a column is calculated based on the analysis of the structure. The additional piles are symmetrically placed around the edge of the existing foundation to ensure effective transfer of load. After construction of the piles, the size of the footing is increased both in plan and thickness with additional reinforcement. Dowel bars are provided between the existing and new concrete. A schematic sketch of the arrangement is shown in Fig. 22.8.

The following steps may be adopted for construction:

- Prop the portion of the building around the column.
- Expose the existing footing or pile cap up to the PC level. The excavated area should be adequate to accommodate the additional piles and the pile cap.
- Chip-off the cover of the footing/pile cap along the vertical face and top surface.
- Drill the holes through the vertical face to provide necessary dowel bars.
- Construct piles at required locations. The cut-off level of the piles should be 50 mm above the PC level.

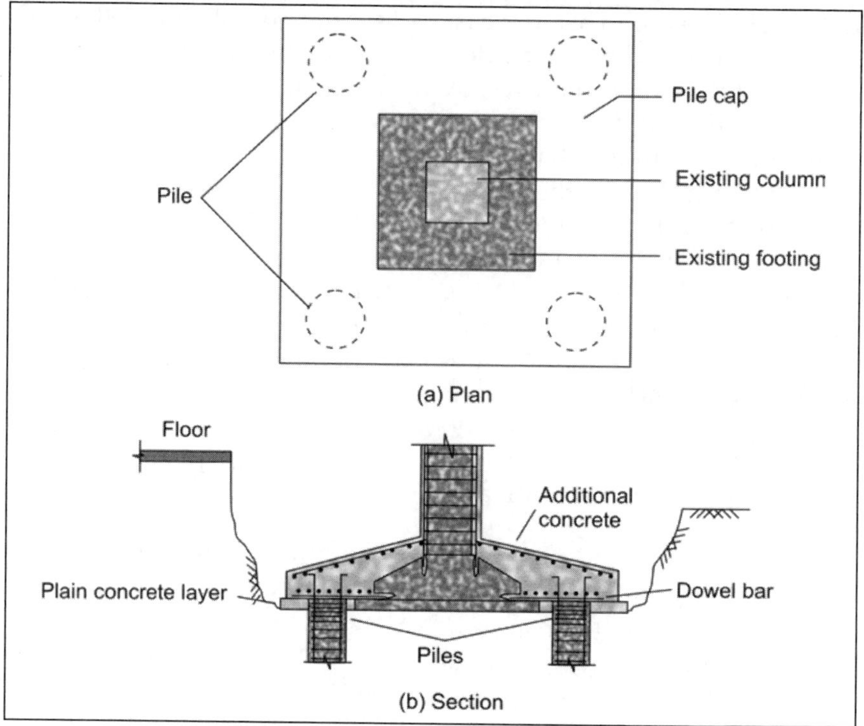

Fig. 22.8: Underpinning of reinforced footing with piles

- Extend the PC course over the required area of the additional pile cap.
- Insert dowel bars in the holes drilled and bond them with epoxy.
- Provide additional reinforcement for the footing as required.
- Erect formwork.
- Cast the concrete and compact the same.
- Remove the formwork after 24 hours and cure the concrete surface.
- Backfill the foundation with well compacted cohesionless soils.
- Complete the flooring.

22.6.5 Strengthening of Piles

For building on piles, the piles may be inadequate to resist the design lateral loads calculated as per the current seismic code. Also, the piles may be deficient just below the pile cap, where deterioration and damage to piles are common. The strengthening of piles can be as follows:

- Dig down in segments to expose the piles to a level where the defects are visible as shown in Fig. 22.9a.
- Expose the cover to an extent such that the damages and corrosion are removed. Provide a base cap at the bottom level to ensure continuity of the additional reinforcement for the piles as shown in Fig. 22.9b.
- Provide additional reinforcement and concrete jacket around the pile based on the forces from the analysis of the building as shown in Fig. 22.9c.
- Backfill the foundation with well compacted cohesionless soils as shown in Fig. 22.9d.
- Complete the flooring.

22.6.6 Strengthening of Base Plates

Base plates and anchors bolts designed for gravity loads can fail under tension due to lateral loads as shown in Figs 22.10 and 22.11. The strengthening involves increase in the size of the base plate and anchoring it to the new

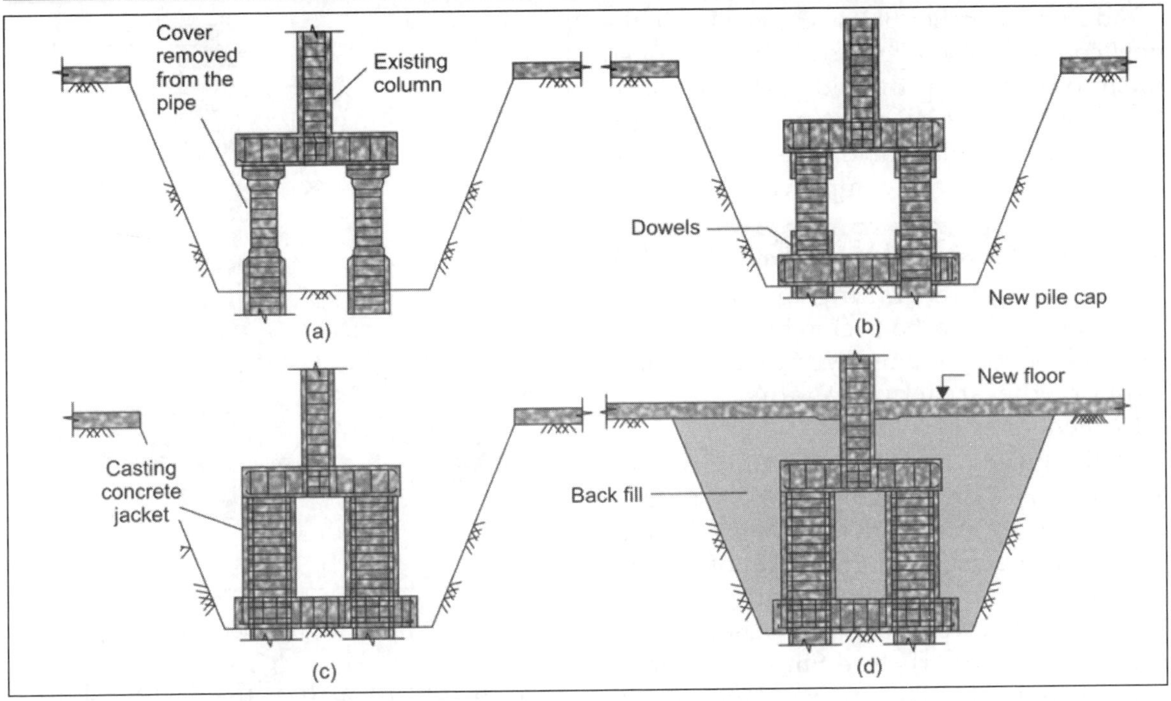

Fig. 22.9: Strengthening of piles

Fig. 22.10: Damage in a base plate

Fig. 22.11: Damage in an anchor bolt

foundation as shown in Fig. 22.12. The new foundation can consist of piles and a pile cap. If there is an existing pile foundation, it can be strengthened as described in the earlier section.

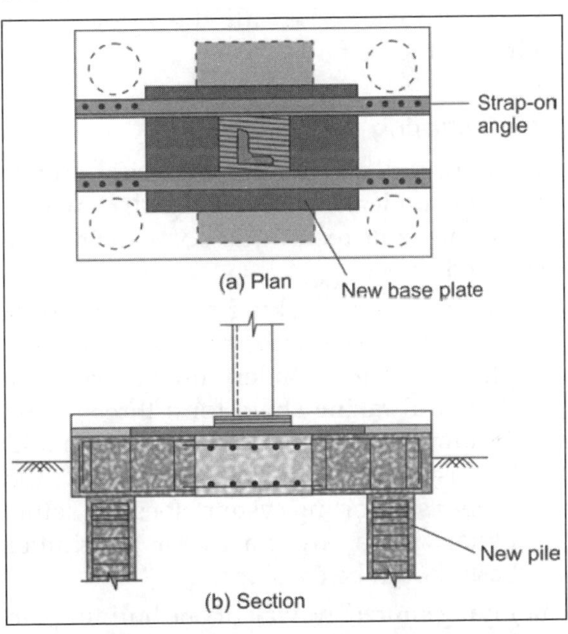

Fig. 22.12: Anchoring base on to new pile cap

Additional anchor bolts are installed in the new concrete. These anchor bolts are bolted to strap-on to keep the old and new plates together. Since the lever arm between the new bolts is more, the lateral load resistance of the base plate significantly improves.

22.7 METHODS OF EXECUTION

Before any underpinning operation, the foundation should be preferably relieved of the load. Shoring is employed to temporarily support the structure against possible settlement or even collapse.

Both shoring and underpinning require specialized expertise. Hence, these should be undertaken by firms with knowledgeable and experienced persons. A great deal of shoring work can be minimized by carefully studying the inherent strength and redistribution capabilities of the structure. Since a reinforced concrete framed building can redistribute the vertical loads, they do not need extensive shoring. On the other hand, buildings with load bearing walls need extensive shoring.

When the load bearing walls are in good condition, they can be secured by temporary horizontal ties and vertical braces. The ties are generally installed at all the floor levels including the roof.

22.7.1 Shoring

The shoring depends on the specific project and type of underpinning envisaged. However, a few common techniques of shoring are discussed below:

a. **Raking shore:** Raking shore is used to support external walls from going out-of-plumb during the underpinning operation. A typical raking shore for a three-storey building is shown in Fig. 22.13. It should be ensured that the soil underneath the timber supporting system does not settle. Otherwise, a larger area for the timber base should be considered.

b. **Flying shore:** The wall of one building can be supported by the wall of the adjacent

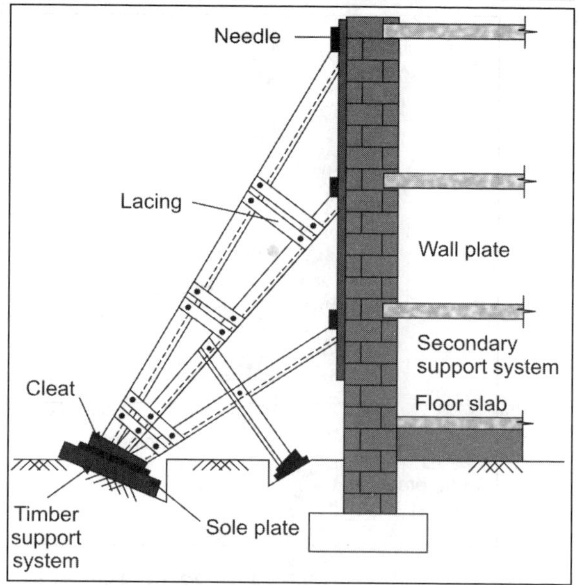

Fig. 22.13: Typical arrangement for raking shores

building by providing flying shores, as shown in Fig. 22.14. The distance between the two buildings should be small for effective shoring. Flying shores cannot support the weight of the wall. However, they prevent bulging and out-of-plane movement of the wall.

c. **Dead shores:** Dead shores are vertical struts that relieve the load on the foundation by providing alternate load path to the ground. Dead shores enable to clear the area for underpinning operations. The needle beams are passed through holes cut in the wall. They transfer the superimposed load to temporary foundations as shown in Fig. 22.15.

22.7.2 Use of Needle Beams

A column in a framed building can be supported by needle beams as shown in Fig. 12.16. The loads can be transferred to the needle beams through a steel collar by wedge action or channel embedded in the column. The needle beams are supported away from the area to be excavated.

Instead of the above schemes, the first floor beams can be supported on all the four sides

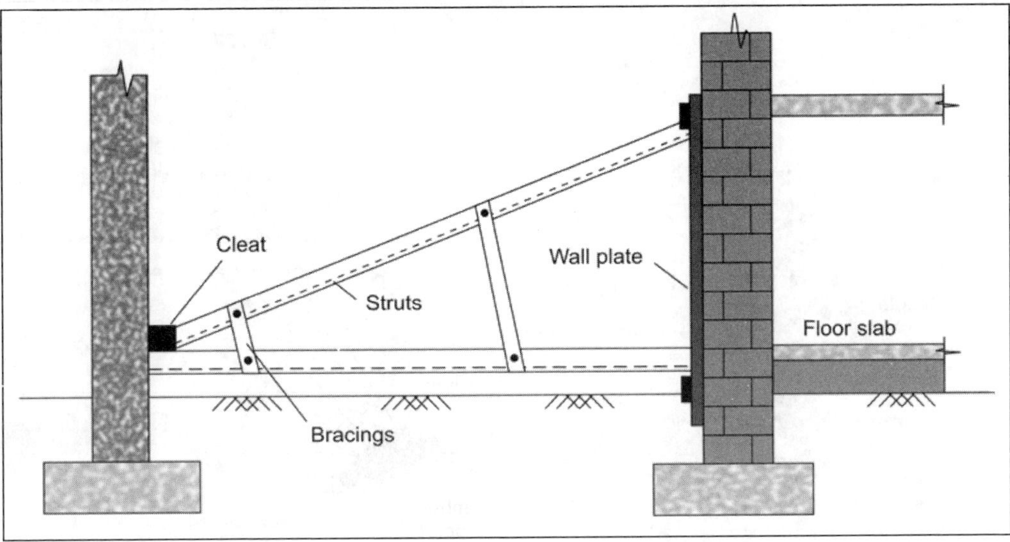

Fig. 22.14: Typical arrangement for flying shores

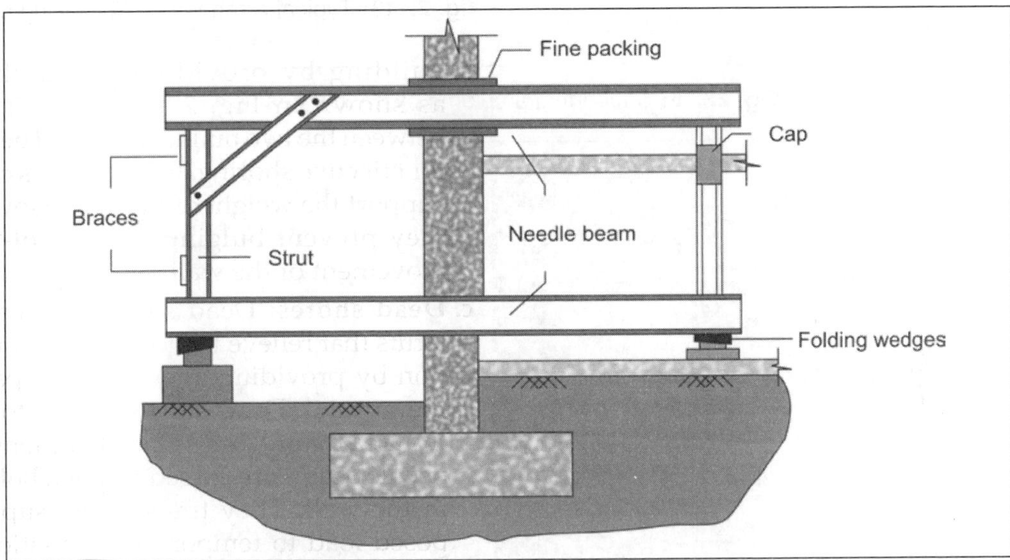

Fig. 22.15: Typical arrangement for dead shores

and the column can be left hanging as shown in Fig. 22.17. Once the column load is removed, the excavation for underpinning can commence.

22.7.3 Methods of Underpinning

Once underpinning is selected as the retrofit strategy, the method of underpinning and temporary support should be worked out. The methods of underpinning are described below.

a. Underpinning continuous strip foundation: Generally, a masonry wall can be unsupported over a length of 1–1.5 m, as shown in Fig. 22.18. Figure 22.19 shows the sequence of the pits for underpinning. The segments are numbered 1–6 based on the sequence. The work can simultaneously proceed on segments having the same number. After the excavation, concrete is placed in the pit up to a depth of

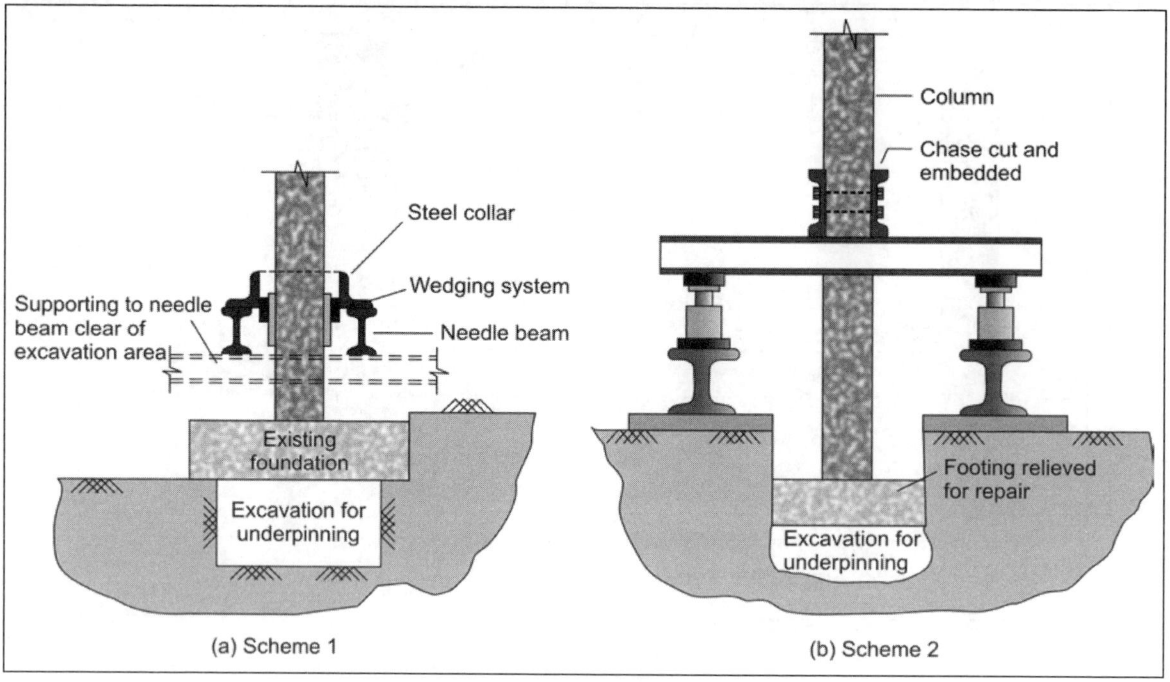

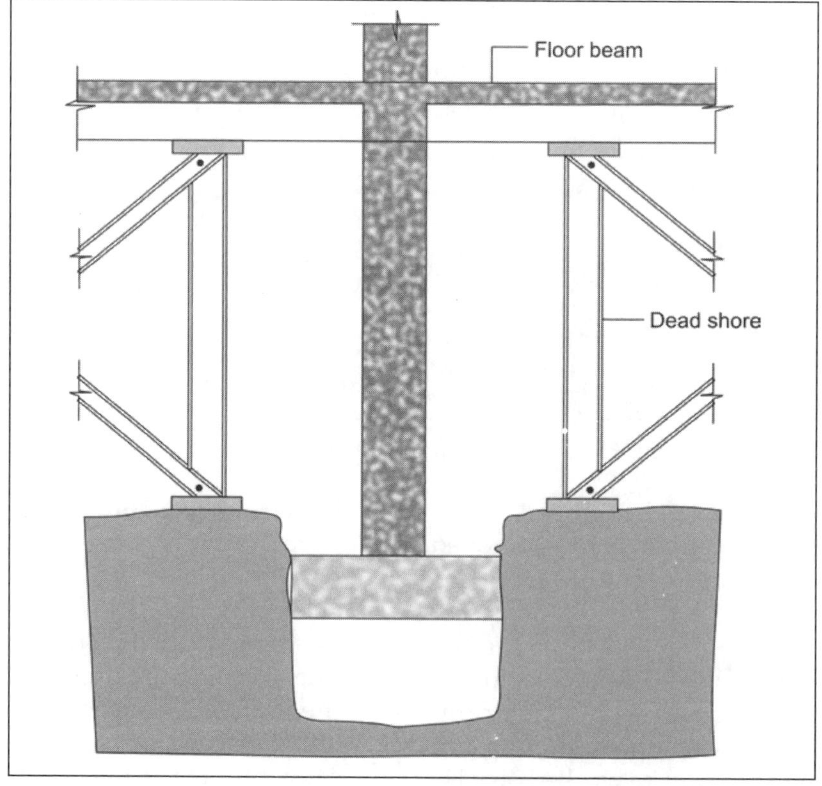

Fig. 22.16: Supporting a column by needle beams

Fig. 22.17: Supporting floor beams by dead shores

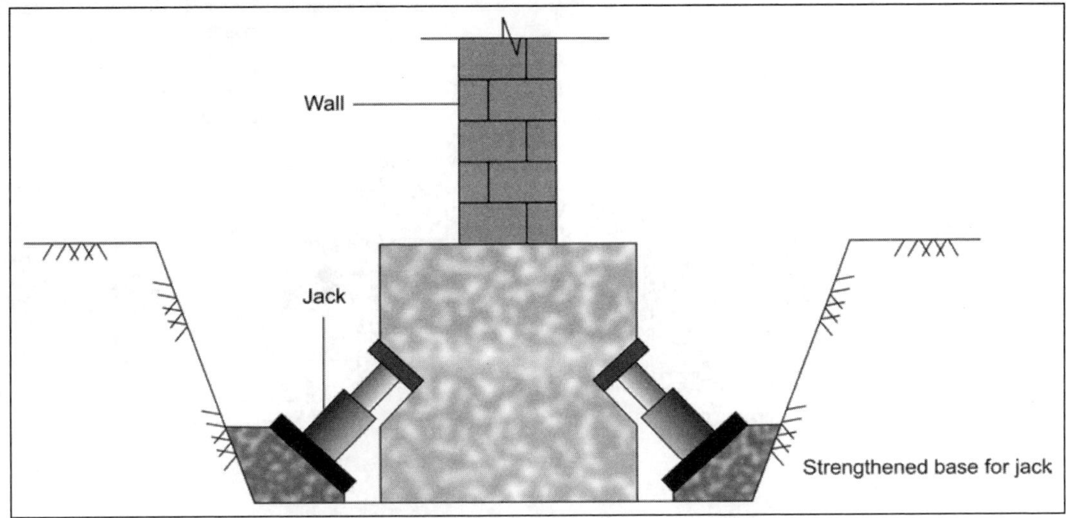

Fig. 22.18: Method of supporting the wall

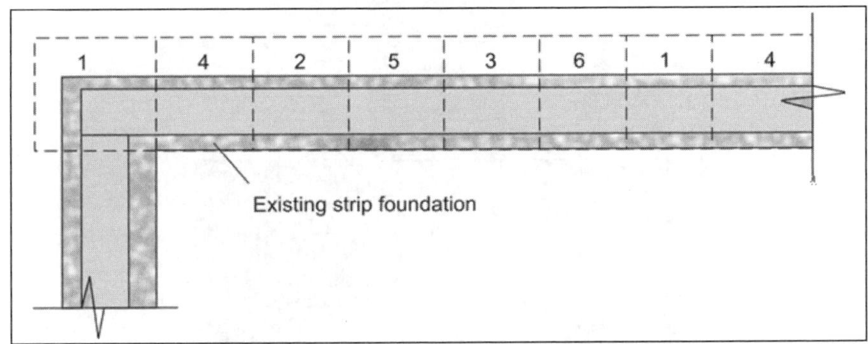

Fig. 22.19: Sequence of pits for underpinning strip foundation

50–100 mm from the underside of the existing foundation. Any gap between the new concrete and the existing foundation can be filled with expanding mortar. Horizontal reinforcing bars of short lengths can be provided longitudinally between the segments. Polymeric agents, provided between adjacent segments, improve the bond. After the last pit is completed, it is worthwhile to undertake pressure grouting to fill up any void in between the segments.

b. **Underpinning with piers or pile groups:** Pier or pile group foundation (including well foundation) may be a desirable method of underpinning at deep excavation sites. First, the load on the existing foundation is relieved by inserting needle beams as explained in the previous section. Next, additional beams are provided to transfer the load from the existing foundation to the piers or pile groups. The beams can be inserted in one of the following ways:

• Support the wall for the full length between the piers, as shown in Fig. 22.20a.

• Insert the beams into openings made in the foundation and supporting them on piers, as shown in Fig. 22.20b.

• Insert precast concrete blocks into the openings made in the foundation and make it continuous between the piers, as shown in Fig. 22.20c.

After connecting the piers to the inserted beams, the needle beams can be carefully removed to transfer the load to the piers.

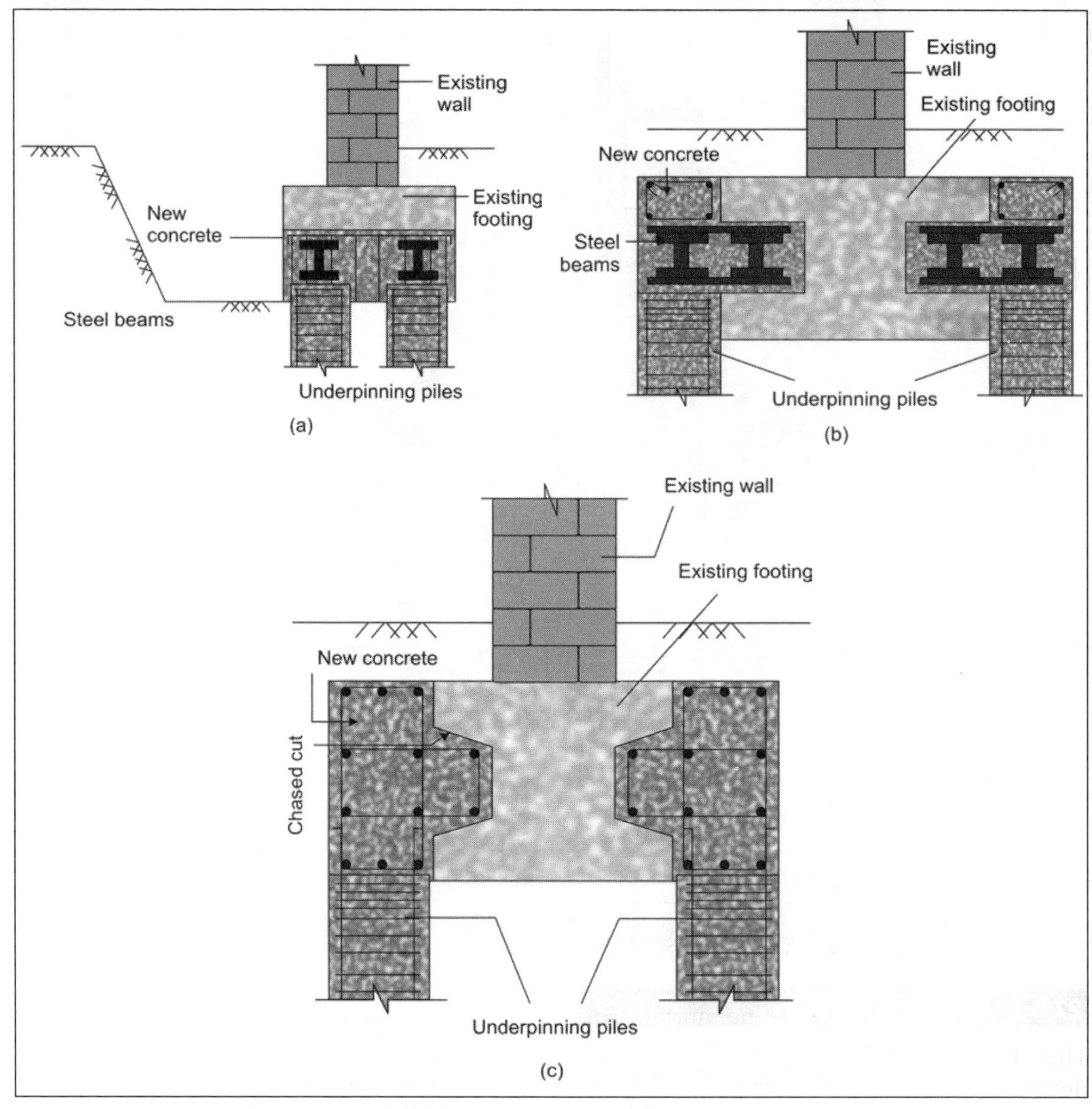

Fig. 22.20: Underpinning with pier/piles and beams

22.8 IMPLEMENTED CASE

A laboratory building of a polytechnic college had undergone excessive settlement, with tilt and wide cracks in the walls due to inadequacy of the foundation. The structure was considered to be unsafe and alternative arrangement was made to run the laboratory classes. The site had 4–6 m thick filled-up soil, which was of very low shear strength and high compressibility. The foundation was strengthened by introducing a new plinth beam and under-reamed piles. After strengthening, the building has been in operation without any sign of further settlement or cracks. Figure 22.21 shows elevations of the existing and strengthened foundations. Figure 22.22 shows the sequence of underpinning.

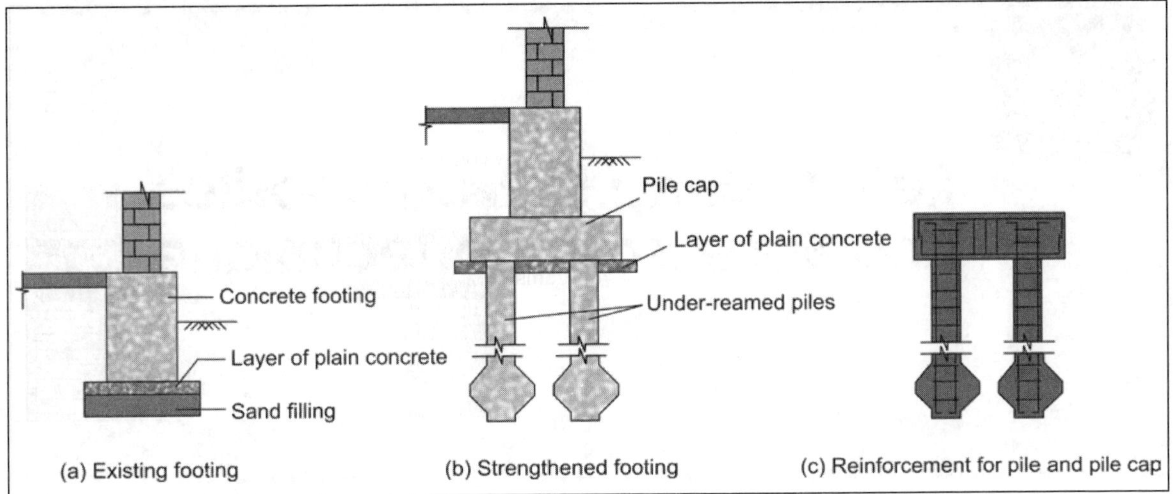

Fig. 22.21: Foundation before and after underpinning

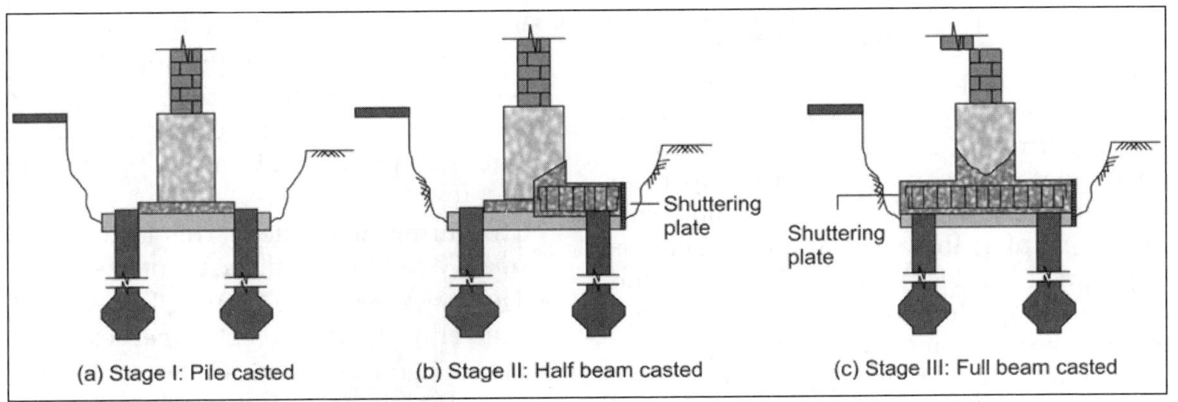

Fig. 22.22: Sequence of underpinning operation

22.9 SUMMARY

This chapter covers the important aspects of deficiencies of foundation, analysis and assessment of foundation, the types of intervention to strengthen the foundation and the methods of execution. The types of intervention include strengthening rubble masonry foundation, enlarging the area of reinforced concrete footing, underpinning the foundation, drilling micropiles, strengthening of piles and base plates. The methods of execution briefly describe the types of shoring, temporary supports and underpinning.

23
Retrofit Using FRP Composites and Base Isolation Technique

23.1 OVERVIEW

Fiber reinforced polymer (FRP) is a composite material consisting of polymeric resin reinforced with high strength fibers. The composite material is available in the form of sheets (or fabrics), preformed shapes and bars. In this chapter, the application of sheets is highlighted. The FRP sheets are thin, light and flexible enough to be inserted behind pipes, electrical and mechanical ducts, thus facilitating installation. The chapter highlights the advantages of using FRP, the constituent materials, the types of FRP composites and the technique of bonding the laminates on a substrate.

For strengthening of masonry walls, sketches of the configurations of the FRP laminates are provided. The procedures for analysis of a retrofitted wall for out-of-plane bending and in-plane shear are elucidated. The use of FRP in retrofitting reinforced concrete beams is explained under strengthening for flexure and shear. The use of FRP in increasing the shear strength and confinement of concrete in a column are briefly covered. Equations relevant in the analysis of a retrofitted section are provided.

23.2 INTRODUCTION

The fiber reinforced polymer (FRP) composites are useful for repair, rehabilitation and retrofit of structures for the following reasons:

- The FRP sheets are light and flexible, which facilitates installation. It does not need drilling of concrete or masonry. There is less disruption during strengthening.
- The curing time is less. This leads to reduced downtime to the users of a building.
- The sheets have high strength-to-weight ratio and superior creep properties.
- The material is chemically inert and has resistance against electrochemical corrosion.
- There is good fatigue strength, which is suitable for fluctuating loads.

The fiber reinforced polymer (FRP) composites are used in the repair of buildings, water tanks, bridges, marine structures, etc. This chapter covers the use of FRP in the repair and retrofit of masonry and reinforced concrete buildings. Even though the materials used in FRP are relatively expensive as compared to the traditional strengthening materials, such as concrete and steel, the labor, equipment and construction costs are often lower. Strengthening by FRP can be used in areas with limited access, where traditional techniques, such as concrete and steel jacketing would be impractical.

23.2.1 Constituent Materials

FRP consists of primarily two components, polymeric resin and high strength nonmetallic fibers. The resin acts as the matrix (the bulk component) and the fibers as the reinforcement. There can also be additives and fillers. Resins, such as epoxy and polyesters have been used in different techniques of repair and retrofit. The availability of high strength resins has increased the use of FRP. The resins must have the following characteristics:

- Compatibility and adhesion with the substrate, such as concrete or masonry
- Compatibility and viscous enough for adhesion with the fibers
- Not too viscous to retain the filling ability so that voids are not present
- Development of appropriate mechanical properties for the composite
- Resistance to environmental effects, like moisture, salt water, temperature and other chemicals normally associated with corrosion.

Typical values of the properties of epoxy resin are provided in Table 23.1. However, a particular resin has to be tested before application. The fibers can be of glass, carbon or aramid. Among them, glass and carbon fibers are common. The glass and carbon fibers are made into fabric form like unidirectional cloth as shown in Fig. 23.1 and woven roving mat as shown in Fig. 23.2. Glass fibers have lower stiffness and cost as compared to carbon fibers. They are suitable in low cost seismic retrofit applications. Typical values of the properties of glass fiber are given in Table 23.2

(a) Carbon fibers

(b) Glass fibers

Fig. 23.1: Unidirectional cloth

(a) Carbon fibers

(b) Glass fibers

Fig. 23.2: Woven roving mat

Table 23.1: Typical values of the properties of epoxy resin

Tensile strength	55–130 N/mm^2
Tensile modulus of elasticity	2800–4200 N/mm^2
Flexural strength	125 N/mm^2
Flexural modulus	2000–4200 N/mm^2
Impact strength	0.1–1.0 J/m of notch
Poisson's ratio	0.2–0.33

Table 23.2: Typical values of the properties of glass fibre

Density	0.9 kg/m^3
Tensile strength	1700 N/mm^2
Tensile modulus of elasticity	75,000 N/mm^2

The fibers in the sheet or fabric can be oriented as follows:

- Unidirectional fiber sheets, where the fibers run predominantly in one direction
- Multidirectional fiber sheets or fabrics, where the fibers are oriented in at least two directions.

The final composites are elastic up to failure and do not exhibit plasticity. The ultimate strain is high which is desirable in seismic retrofit applications. The axial stress versus strain curves of concrete cylinders wrapped with several layers of FRP laminates show increased strength and ductility due to confinement.

23.2.2 Types of FRP Composites Based on Installation

The FRP composites can be categorized based on how they are delivered to the site and installed.

Dry Lay-up FRP Composites

In this method, first the substrate is coated with resin. Next, the dry FRP sheets are bonded on the substrate and a fresh coating of resin may be applied. The composite is subsequently cured.

Wet Lay-up FRP Composites

First, the FRP sheets are saturated with resin and then bonded on the substrate. It is then cured as per the specifications.

Preimpregnated FRP Composites

These consist of fiber sheets that are pre-impregnated (prepreg) with a saturating resin. The prepreg composites are bonded to the substrate with or without additional resin application, depending upon the specific system requirements. The fiber two varieties are wound or mechanically applied on the surface. The composites are subsequently cured. The prepreg composites are widely used in aerospace, automobile and ship building applications.

Precured FRP Composites

These consist of a wide variety of shapes manufacture offsite in the supplier's facility and shipped to the site. They are bonded to the substrate with resin. Pultruded and precured FRP sheets are frequently used for repair and retrofit. The pultruded sheets are manufactured by the pultrusion process. In this process, the fibers are pulled through a bath of resin and passed through preforming fixtures.

23.2.3 Selection of FRP Composites

For a particular project, the suitability of using FRP composites for seismic retrofit should be assessed prior to selecting the product. To assess the suitability of FRP, the designer should evaluate the existing building for lateral strength, identify the deficiencies and their causes, and determine the condition of the substrate. A thorough field investigation of the dimensions of the members, compressive strength and soundness of the concrete or masonry, location, size and cause of cracks, spalling of plaster or concrete cover, location and extent of corrosion of reinforcing bars, etc. should be undertaken. The topics are covered under preliminary evaluation, condition assessment and seismic analysis covered in earlier chapters.

23.2.4 Installation of FRP Composites

The concrete or masonry surfaces to which the FRP laminates are to be applied should be freshly exposed and loose unsound materials have to be removed. Surface preparation can be done using sand or water blasting techniques. All dust, dirt, oil, curing compound, existing coatings and any other matter that could interfere with the bond of the FRP composites to the substrate should be removed. Bugholes and other small surface voids should be completely filled with putty materials. The surface of application should be smooth and convex.

The following points are important:

- Concavity in the surface should be eliminated. A stretched FRP sheet will lose contact with the surface around depressions.
- Sharp edges and corners should be avoided. The corners should be rounded to a minimum radius of 25 mm. The sheets are susceptible to tearing at these locations due to stress concentration.
- Since the installation procedure depends on the product, the instructions attached with the product should be followed carefully.

The typical lay up of an FRP composite on the substrate is shown in Fig. 23.3. The FRP systems are commonly installed using dry fiber fabrics/sheets and a saturating resin. The saturating resin is applied uniformly to the whole prepared surface. The fibers should be gently pressed into the uncured saturating resin in a manner recommended by the composite manufacturer. Entrapped air between layers should be released or rolled out before the resin sets. Sufficient resin should be applied to achieve full saturation of the fibers. Successive layers of resin should be applied to achieve full saturation of the fibers. Successive layers of resin should be placed before the complete curing of the previous layer. The sequence of steps for installation of FRP sheets on a concrete substrate are shown in Fig. 23.4. After the resin is completely cured, a top coating is provided.

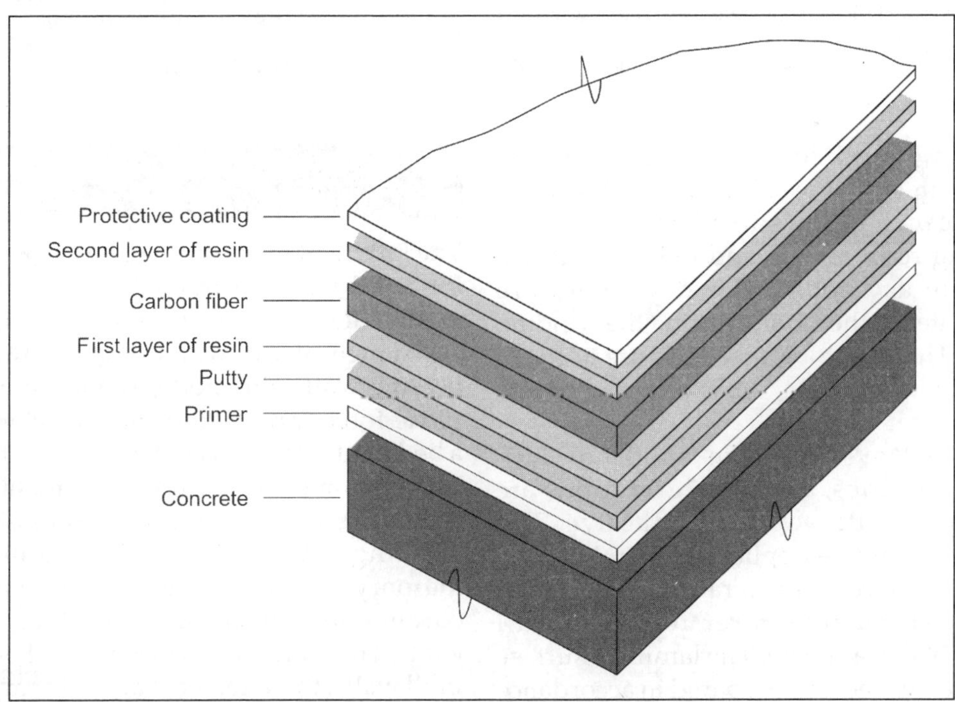

Fig. 23.3: Typical lay up of FRP composite (extruded view)

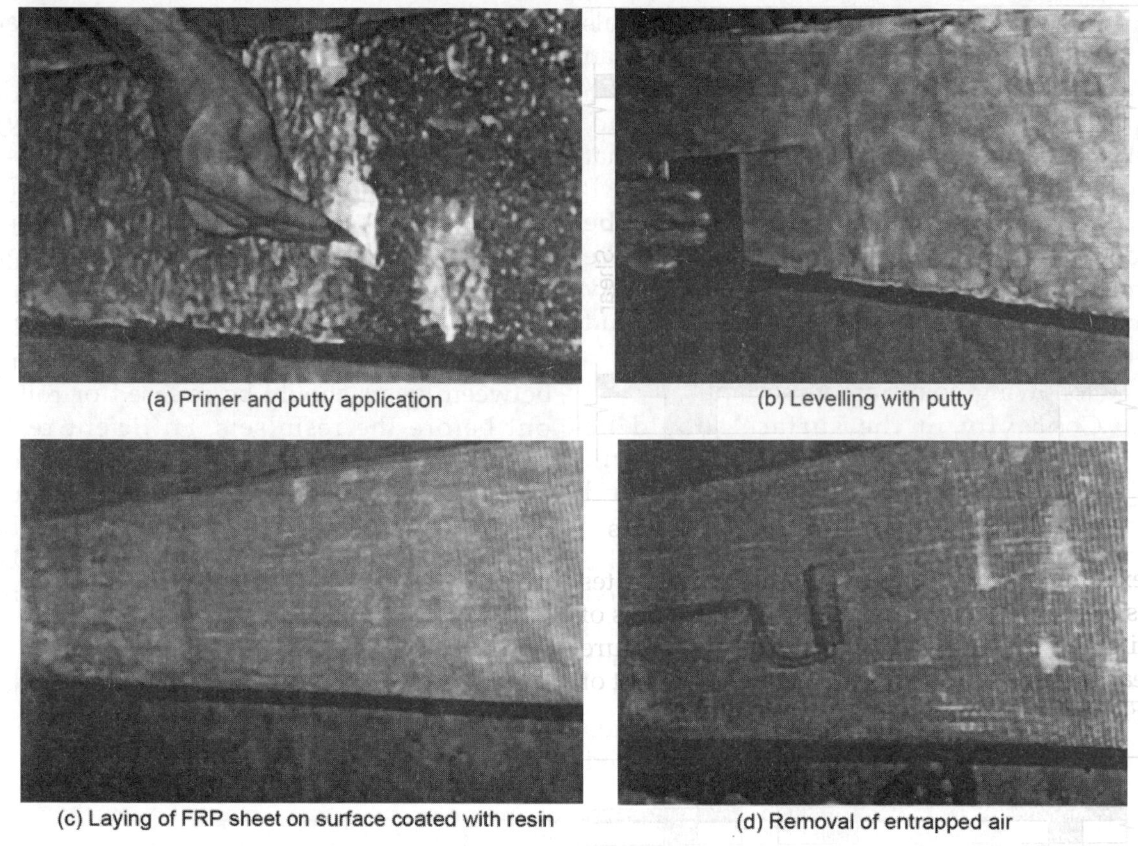

Fig. 23.4: Installation of FRP sheet on concrete substrate

Machine applied systems utilize pre-impregnated (prepreg) tows or dry fiber tows. The prepreg tows are impregnated with saturating resin off site and delivered to the work site as spools. The dry fibers are impregnated at the job site during the winding process. The wrapping machines are used for automated wrapping on existing concrete columns.

Precured systems like shells of FRP composites, strips, and open grid forms are typically installed with an adhesive. The adhesive should be applied uniformly to the prepared surface and at a rate recommended by the FRP manufacturer to ensure full bonding of the laminate. The laminate surface should be cleaned and prepared in accordance with the manufacturer's recommendation.

23.3 STRENGTHENING OF MASONRY WALLS

The failure of masonry load bearing walls and the infill walls in framed buildings during earthquakes is due to the low tensile and shear resistance of the walls. For masonry load bearing walls, the modes of failure under in-plane forces are sliding shear cracking along a bed joint, diagonal cracking through several bed joints and flexural crushing (for tall walls). The walls can also fail due to out-of-plane bending. The different failure modes of masonry infill walls are local crushing at the corners and shear cracking along the bed joints. These are shown in Fig. 23.5. For slender infill walls, there can be diagonal compression failure due to the effect of buckling. Retrofitting

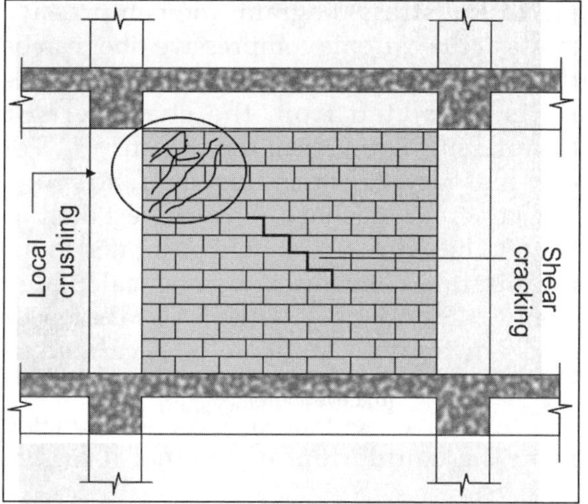

Fig. 23.5: Failure modes of masonry infill walls

existing masonry walls using FRP composites is a cost effective means to prevent the loss of life and damage to property during a future earthquake. The possible schemes of layout of FRP wraps are shown in Fig. 23.6.

23.3.1 Strengthening for In-plane Shear

The shear strength of a masonry wall (V_{UR}) is calculated by superposing the strengths of unreinforced masonry (V_m) and the vertical FRP laminates (V_f). The detailed aspects of bond characteristics, anchoring and stress concentration effects are neglected in the design provisions. A simple procedure for the design of strengthening a masonry wall is given below.

$$V_{UR} = V_m + V_f \qquad ...(23.1)$$

The shear strength of masonry is calculated from the permissible shear stress. As per IS:1905-1987, the permissible shear stress (f_s) is the least of: (a) 0.5 MPa, (b) $0.1 + 0.2f_d$ and (c) $0.125f_m$. Here, f_d is the direct compressive stress in the wall segment and f_m is the crushing strength of the bricks.

$$V_m = f_s bt \qquad ...(23.2)$$

Here, b = width of the wall

t = thickness of the wall

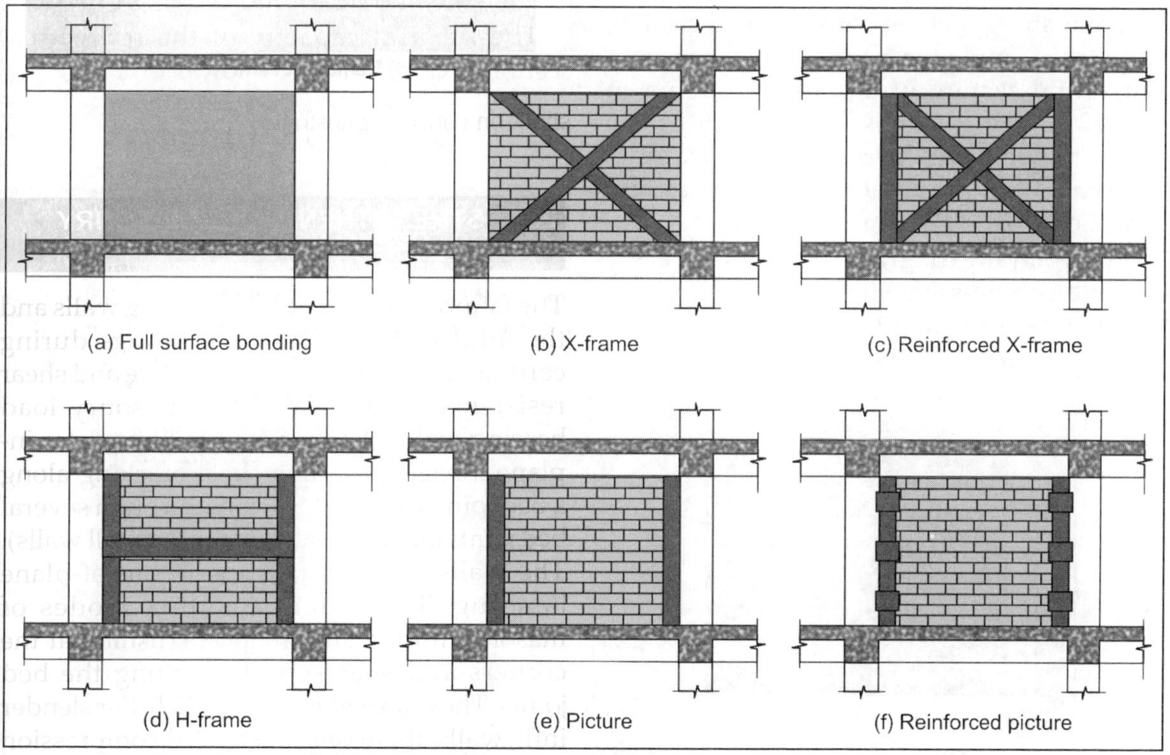

(a) Full surface bonding   (b) X-frame   (c) Reinforced X-frame

(d) H-frame   (e) Picture   (f) Reinforced picture

Fig. 23.6: Configurations of FRP laminates for retrofitting of masonry walls

The shear contribution of the FRP laminate is calculated as follows.

$$V_f = E_f\, e_{fe}\, A_{fv} \qquad \text{...(23.3)}$$

Here, A_{fv} = area of vertical FRP laminate

e_{fe} = effective usable strain in the FRP laminate

$$= 0.5 e_{fu} C_E$$

The effective usable strain in FRP can be taken as a factor of the half of the ultimate tensile strain (e_{fu}). C_E is an environmental reduction factor.

23.3.2 Strengthening for Out-of-Plane Bending

FRP laminates are bonded to masonry walls to check the failure due to out-of-plane bending. They also reduce the crack width and the lateral deflection of the walls. The design guidelines ensure compression failure by designing over-reinforced sections. This mode of failure gives warning due to plastic deformation of the masonry. Based on the working stress method for a cracked section, the strain and stress diagrams across a horizontal section of a wall are shown in Fig. 23.7. The thickness and width of the masonry wall are represented as t and b, respectively. The width b can be taken as 1 m. A_f is the area of FRP laminate.

In the strain diagram, e_m and e_f are the extreme compressive strain in masonry and tensile strain in the FRP, respectively. The strain e_f is limited to the effective usable value E_{fe}. In the stress diagram, the compressive stress at the extreme compressive fiber can be determined as $f_m = e_m\, xE_m$, where the modulus E_m is estimated from the characteristic compressive strength of the masonry f_m. For clay masonry, E_m varies from $550 f_m$ to $750 f_m$. For concrete masonry, E_m can be taken as $900 f_m$. The value of f_m should be limited to the permissible compressive stress calculated from f_m (IS:1905-1987, Clause 5.4.1). The tensile stress in the FRP laminate is calculated as $f_f = E_f\, xE_f$.

The depth of neutral axis kt can be estimated from the equilibrium of internal forces as follows.

$$1/2\, f_m\, ktb = A_f f_f \qquad \text{...(23.4)}$$

When the FRP reinforcement ratio p_f is defined as $A_f bt$, the coefficient k can be obtained in terms of p_f by the following equation.

$$k = \sqrt{(mp_f)^2 + (mp_f)} - mp_f \qquad \text{...(23.5)}$$

Here, the modular ratio $m = E_f / E_m$.

The flexural capacity of the retrofitted section is calculated as follows.

$$M_{UR} = p_f b f_f t^2 \left(1 - \frac{k}{3}\right) \qquad \text{...(23.6)}$$

23.4 STRENGTHENING OF REINFORCED CONCRETE BEAMS

The flexural capacity of a concrete beam can be increased by bonding unidirectional cloth FRP fabrics/sheets to the tension face, with

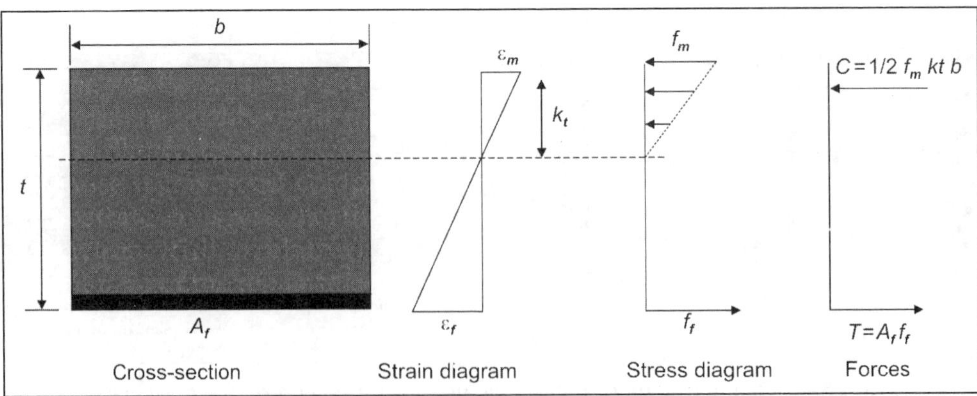

Fig. 23.7: Analysis of a wall section bonded with FRP laminate

the fibers oriented along the span of the member. The shear capacity of a beam can be enhanced by bonding fabrics at 45° to the side faces. The capacities can be increased up to about 40 percent.

23.4.1 Strengthening for Flexural Capacity

The flexural capacity of a concrete beam strengthened with FRP laminate is calculated based on the equilibrium of forces, compatibility of strains and the constitutive relationships. Based on the limit state method of analysis of IS:456-2000, Fig. 23.8 shows the ultimate limit state of a singly reinforced rectangular section with the strain and stress diagrams across the depth and the internal forces. It is assumed that the beam is propped during strengthening, so that under ultimate load the full section is effective. The section should remain under-reinforced, that is the steel should yield before the crushing of the concrete. The ultimate flexural capacity of a member must exceed the flexural demand [Eq. (23.7)]. The demand under factored loads is obtained from the analysis of the building.

$$M_{UR} \geq M_u \qquad \ldots(23.7)$$

The notations of the variables are as follows:

A_f = area of the FRP composite
A_s = area of the steel reinforcement
B = breadth of beam
C = compression in concrete
D = effective depth of the steel reinforcement
F_f = stress in the FRP composite
F_y = yield stress of the steel reinforcement
H = depth of the beam
M_u = ultimate moment demand
M_{UR} = ultimate moment capacity (resistance)
X_u = depth of the netural axis
E_s = strain at the level of steel reinforcement
E_f = strain at the level of FRP composite

A reduction factor $Y_f = 0.85$ is recommended for the strength from the FRP. The depth of the neutral axis (x_u) and the stress levels in the steel and FRP reinforcement can be obtained using force equilibrium and strain compatibility conditions. The yielding of steel under tension followed by the crushing of concrete under compression is the desired failure mode. Based on this, the thickness of FRP fabric should be calculated.

23.4.2 Strengthening for Shear

The ultimate shear strength of a concrete member strengthened with FRP composite (V_{UR}) must exceed the shear demand (V_u) as shown by following equation.

$$V_{UR} \geq V_u$$

The shear capacity of a concrete member strengthened by FRP composite can be determined by adding the contribution of the FRP reinforcing (V_f) to the contributions from

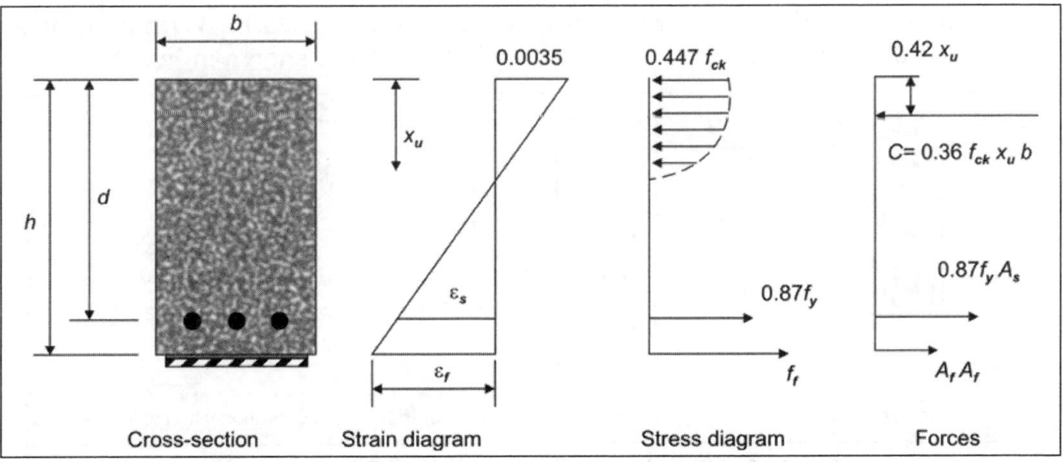

Fig. 23.8: Analysis of beam section bonded with FRP laminate

the reinforcing steel (V_s) and the concrete (V_c). As before, a reduction factor Y_f is applied to the contribution of the FRP composite.

$$V_{UR} = V_c + V_s + Y_f V_f \qquad ...(23.8)$$

The reduction factor, Y_f should be selected based on the known characteristics of the application, but should not exceed 0.85. Typical schemes of layout of FRP wraps for strengthening for shear are shown in Fig. 23.9.

The contribution of shear strength provided to a member by the FRP should be based on the fiber orientation and an assumed crack pattern. The shear strength provided by the FRP reinforcement is the force resulting from the tensile stress in the FRP along the assumed crack pattern. For discrete strips of FRP and cracking inclined at 45°, the shear strength is calculated using Eqs (23.9) to (23.11).

$$V_f = \frac{A_{fv} f_f (\sin x + \cos x) d_f}{S_f} \qquad ...(23.9)$$

$$A_{fv} = 2nt_f wt \qquad ...(23.10)$$

$$f_f = Ef_e E_f \qquad ...(23.11)$$

Here,

A_{fv} = total vertical area of the FRP wrap

d_f = effective depth of the FRP wrap

E_f = elastic modulus of FRP

f_f = stress in the FRP wrap

N = number of wraps

S_f = spacing of the FRP wrap

W_f = width of the FRP wrap

X = inclination of the FRP wrap with respect to the beam axis

E_{fe} = effective usable strain in the FRP wrap.

For continuous FRP sheet, the expression of V_f is modified based on the length of the sheet intercepted by an inclined crack.

23.5 STRENGTHENING OF REINFORCED CONCRETE COLUMNS

The failure of a column may be due to a single or a combination of the following deficiencies.

- Inadequate shear capacity
- Inadequate confinement of column core
- Inadequate splicing of rebar
- Inadequate capacity for sustaining combined load effects, such as biaxial bending.

The first three inadequacies can be improved by FRP wrapping. For increasing the shear strength or the confinement of concrete or the clamping force at a splice location, the fiber should be oriented horizontally perpendicular to the axis of a column. The wrapping of a column not only increases the strength of the concrete by confinement but also improves the ductility (energy absorption capacity) that is an essential requirement for resisting seismic forces. The increase in thickness of the wrapping improves the strength and ductility, but they are limited by the ultimate strain of the FRP. A typical wrapping of columns using FRP composite is shown in Fig. 23.10. Wrapping of beam–column joints is shown in Fig. 23.11.

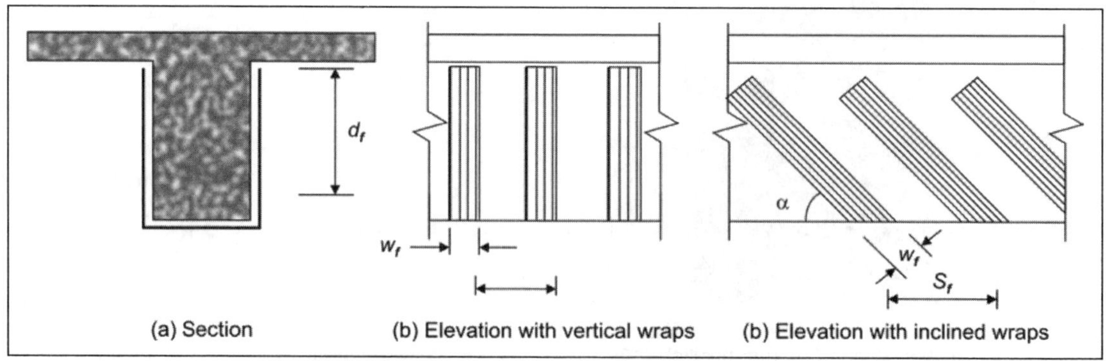

Fig. 23.9: Strengthening for shear using FRP composites

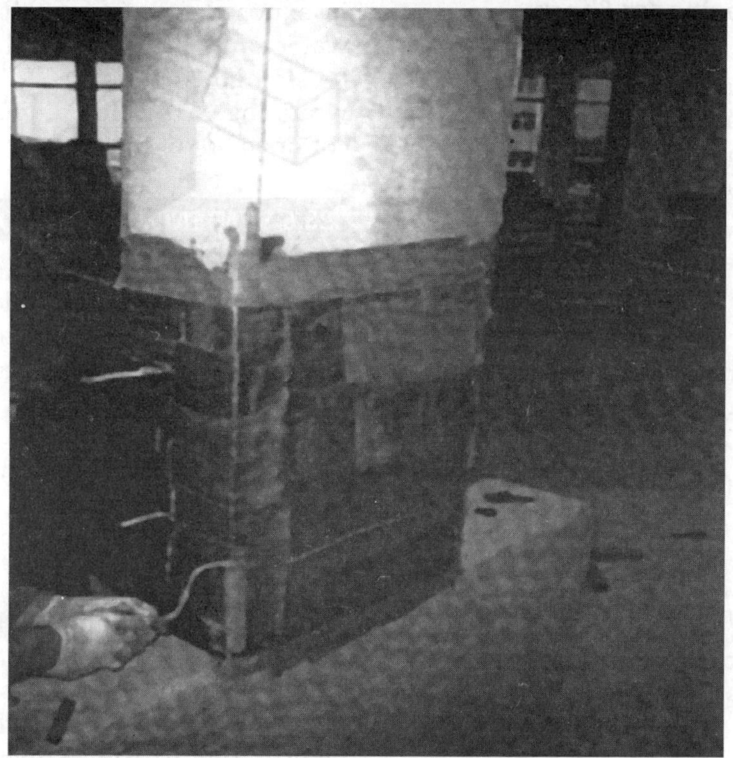

Fig. 23.10: Wrapping of column using FRP composites

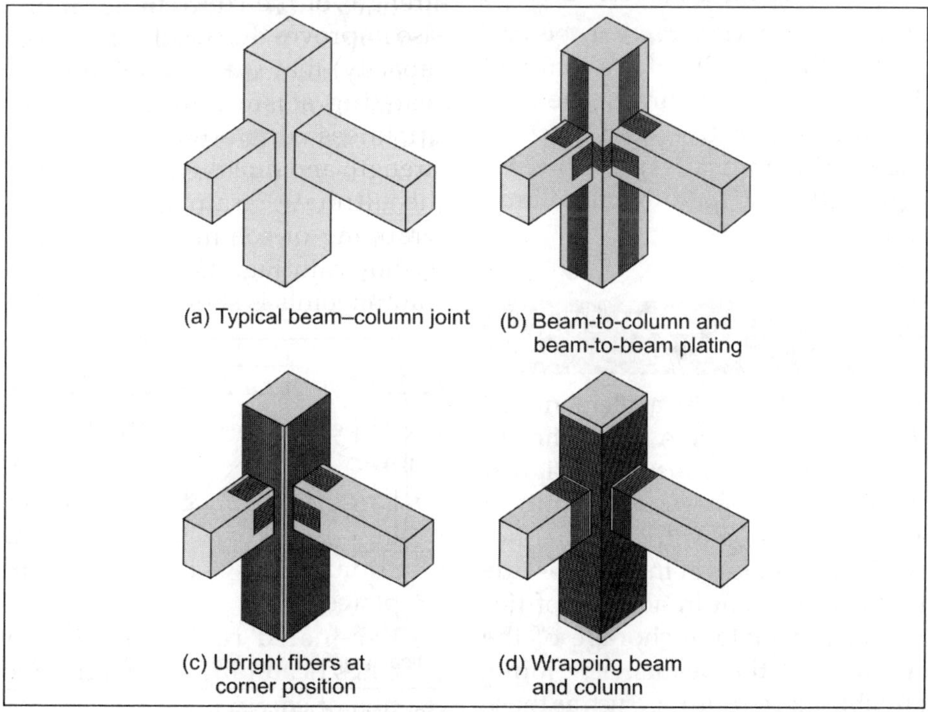

Fig. 23.11: Wrapping of beam–column joint

23.5.1 Strengthening for Shear

The shear failure is the most prevalent failure mode in columns under seismic forces. Assuming 45° cracking, the thickness of the FRP jacket (t_j) for shear retrofit can be determined using the following equation.

$$t_j = \frac{V_u - (V_c + V_s)}{2e_{fe} E_j D}$$

Here

D = the column dimension in the direction of loading

E_j = elastic modulus of composite

V_u = shear demand based on the flexural capacity in the potential hinge locations

V_c = shear capacity of concrete (IS:456-2000, Clause 40.2)

V_s = shear capacity from steel ties

E_{fe} = effective usable strain in the FRP wrap.

23.5.2 Strengthening for Confinement

The thickness of the FRP wrap can be calculated based on the capacity of the confined concrete. For noncircular columns, the confinement is not as effective as a circular column. There are empirical expressions for thickness of the FRP wrap for different column cross-sections.

23.6 STRENGTHENING OF BEAM–COLUMN JOINTS

A beam–column joint is a critical element in a reinforced concrete frame. A joint should maintain its integrity and must be designed stronger than the members framing into it. Quite often, the detailing at the joint is not taken care of. The weakness at the joints is due to the lack of confinement in absence of ties and the lack of adequate anchorage of the longitudinal bars of the beams. The joints which are suitable for retrofitting, such as those in exposed frames in industrial buildings, water tanks, bridges and other structures, can be strengthened using FRP composites. The wrapping of the FRP laminates enhances the confinement of the joint.

23.7 SUMMARY OF USE OF FRP

This chapter explains the retrofitting of masonry and concrete buildings using fiber reinforced polymer (FRP) composites. Brief descriptions on the constituent materials of FRP, the types and installation of FRP composite are provided. Guidelines for calculating the thickness of the FRP laminates for strengthening masonry walls, reinforced concrete beams and columns are given.

23.8 BASE ISOLATION

23.8.1 Base Isolation

The technique of base isolation involves introduction of devices above the foundation level of a building to increase the flexibility in the horizontal plane. In addition, damping elements are introduced in the superstructure, so as to reduce the relative displacements of the floor levels during earthquake. The base isolation devices decouple the building from the ground motion and also add substantial damping. As a result, the structures behaves very much like a rigid body sliding on a moving platform, as shown in Fig. 23.12.

23.8.2 Types of Base Isolation Devices

The types of base isolation devices are as follows:

- Laminated rubber (or elastomeric) bearings (layers of rubber alternated with steel shims sandwiched between two steel plates)
- Laminated rubber bearings with lead cores (lead plug at the centre for dissipation of energy)

- Sliding bearings (stainless steel and teflon coating layers)
- Friction pendulum devices (bearing)— one or two concave surfaces with sliding hemispherical bob.

Schematic sketches of different types of base isolation devices are shown in Fig. 23.13 and elevation of connection of base isolation devices is shown in Fig. 23.14.

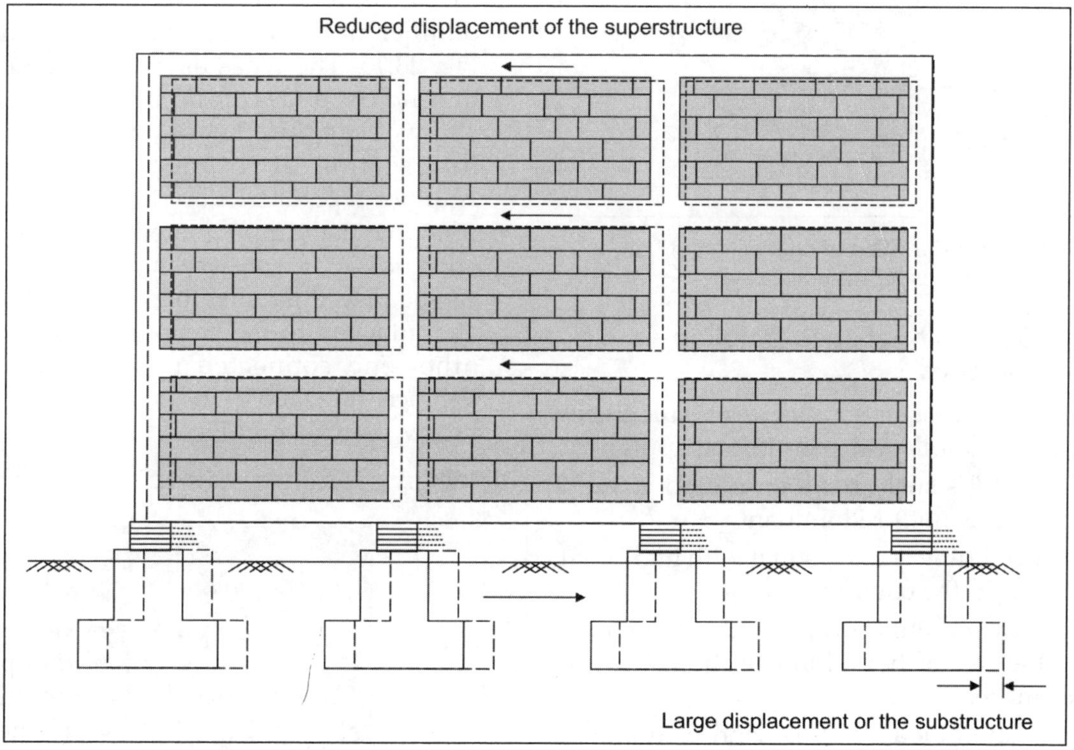

Fig. 23.12: Schematic presentation of the effect of base isolation devices

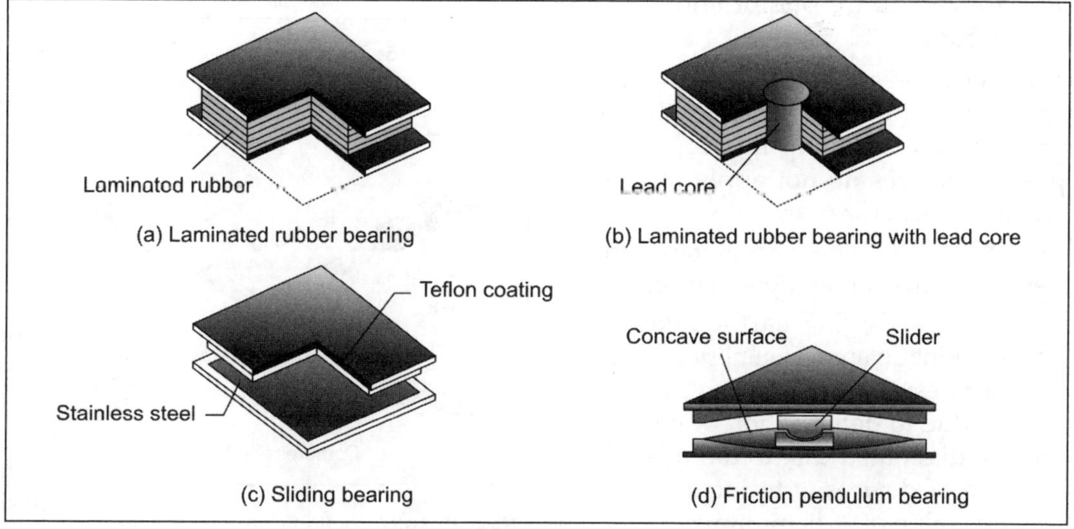

(a) Laminated rubber bearing

(b) Laminated rubber bearing with lead core

(c) Sliding bearing

(d) Friction pendulum bearing

Fig. 23.13: Different types of base isolation devices

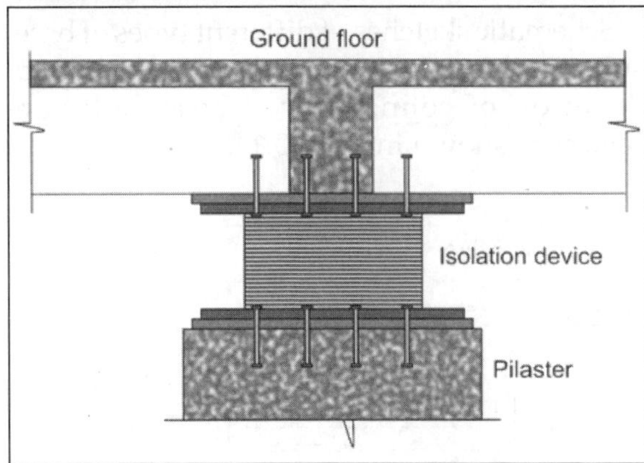

Fig. 23.14: Elevation of connection of base isolation device

23.8.3 Suitability of Base Isolation Devices

The use of base isolation devices is suitable in a building, if following conditions apply:

- The soil is such that long period ground motion is not predominant.
- The building is fairly squat with high axial forces in the columns.
- The site permits horizontal displacements of the base of the building of about 200 mm or more.
- The lateral load due to wind is less than 10% (approx.) of the weight of the building.

23.8.4 Passive Energy Dissipating Devices

The passive energy dissipating devices dissipate energy during the seismic vibrations by increasing the damping in the structure. The passive devices do not apply any force directly to the structure.

Types of passive energy devices can be represented by the following equation:

$E_i = E_k + E_s + E_d$; E_i = earthquake input energy; E_k = kinetic energy generated due to motion of masses; E_s = strain energy generated in structure due to deformation of members; E_d = energy dissipated due to damping. The object of use of devices is to increase E_d for given E_i, the value of E_s is minimized.

The types of passive energy devices are as follows:

i. Friction dampers: Dissipates energy through friction among metals along sliding surfaces. The typical force versus displacement behavior of a friction damper under cyclic load is shown in Fig. 23.15. The types of friction dampers depend on modes of sliding. Figure 23.16 shows a device that is located at the intersection of X-braces in a frame. Another type of friction damper consists of copper pads impregnated with graphite and encased in a steel casing. The damper can be placed parallel to a floor, with one end attached to the beam above and the other end connected to stiff inverted V-braces attached to the beam in the floor below, as shown in Fig. 23.17.

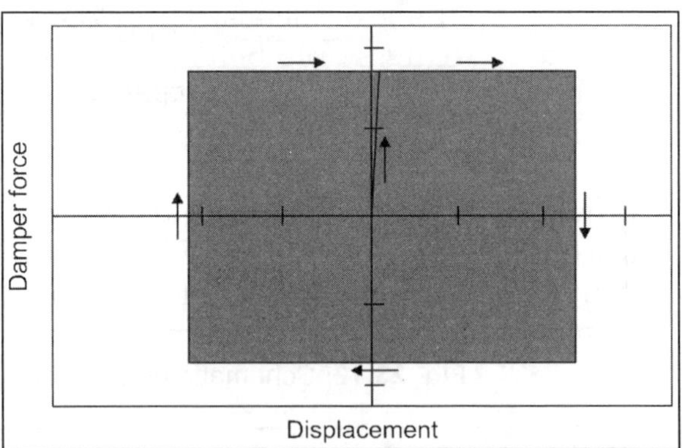

Fig. 23.15: Typical force versus displacement behavior of a friction damper

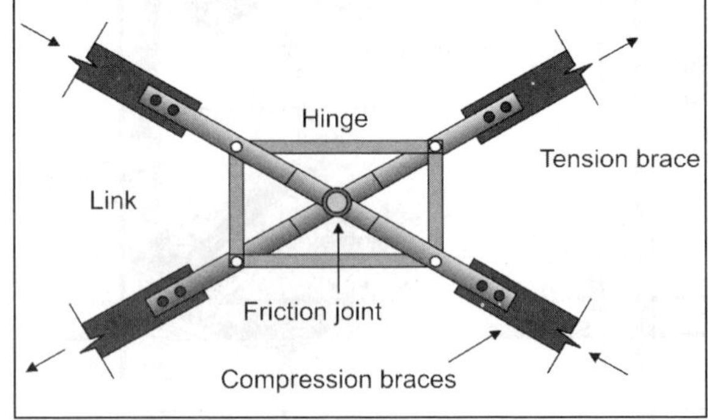

Fig. 23.16: A friction damper installed in X-braces

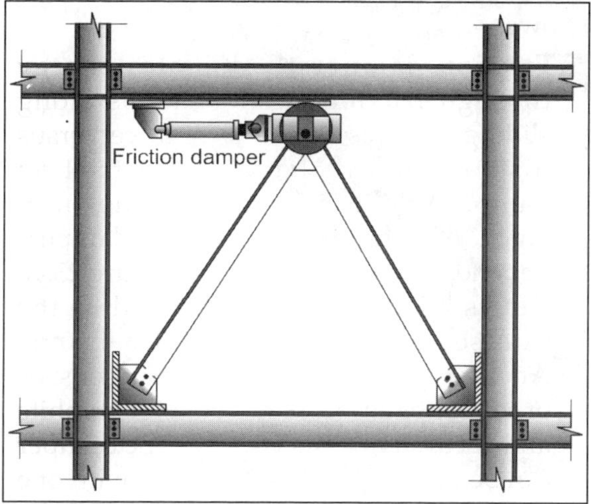

Fig. 23.17: A friction damper installed in a building frame

ii. Metallic dampers: Dissipate energy through yielding of metals. These dampers are having stable behavior, long-term reliability, and good resistance to environmental and thermal conditions. The metallic dampers are also capable of providing increased stiffness and strength to a building. Figure 23.18 shows a metallic damper installed in X-braces of a building frame. In this device, the compression brace loses contact from the rectangular steel frame to prevent buckling. Energy is dissipated by inelastic deformation of the steel frame in the direction of the tension brace. Figure 23.19 shows a metallic devices based on extrusion of load. Here, the process of extrusion consists of forcing a material through a hole or orifice, thereby altering its shape.

iii. Viscoelastic dampers: Viscoelastic materials exhibit combined features of an elastic solid and a viscous liquid, when deformed. The materials return to their original shape after each cycle of deformation and dissipate a certain amount of energy as heat. A typical device consists of layers of viscoelastic material bonded to a plate and sandwiched between two other plates, as shown in Fig. 23.20.

iv. Viscous dampers: Utilize viscosity of fluid to generate damping and dissipate energy. A device consists of a highly viscous fluid in a steel cylinder and a piston with an orifice. The section of viscous damper is shown in Fig. 23.21.

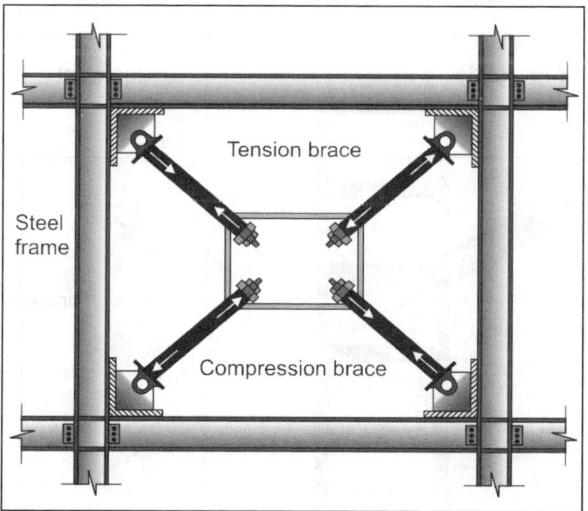

Fig. 23.18: Metallic damper based on yielding of steel

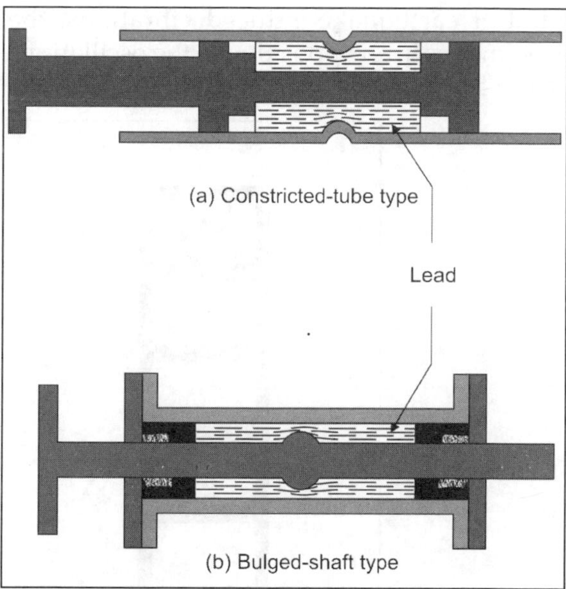

Fig. 23.19: Longitudinal sections of lead extrusion dampers

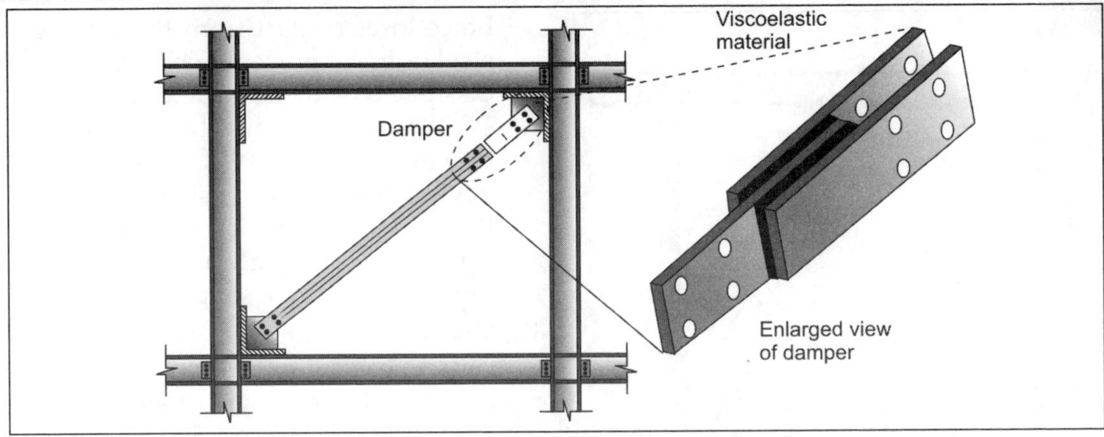

Fig. 23.20: Installation and schematic representation of a viscoelastic damper

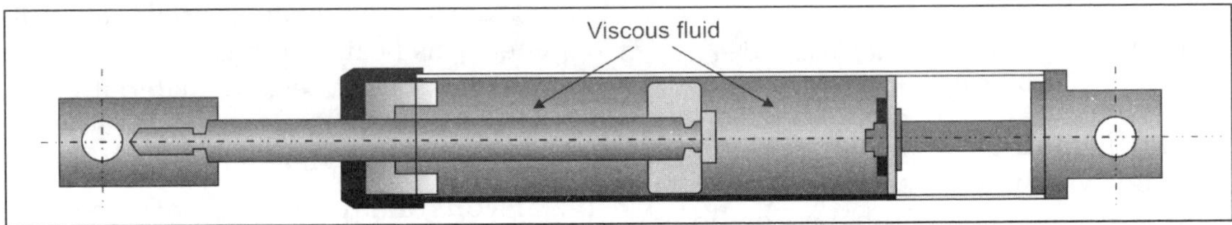

Fig. 23.21: Section of a viscous damper

Tuned Devices

Tuned devices consist of a heavy mass of solid or liquid attached with springs and dampers atop a building to reduce vibrations. Under wind or earthquake induced vibrations, the mass moves in an opposition to the oscillations of the building. Energy is dissipated by the dampers or the sloshing of the liquid. Because the natural frequencies of these devices are equal or close to those of the buildings to which they are attached, they are called tuned devices. Two types of devices are "tuned mass dampers" and "tuned liquid dampers". These are shown in Fig. 23.22.

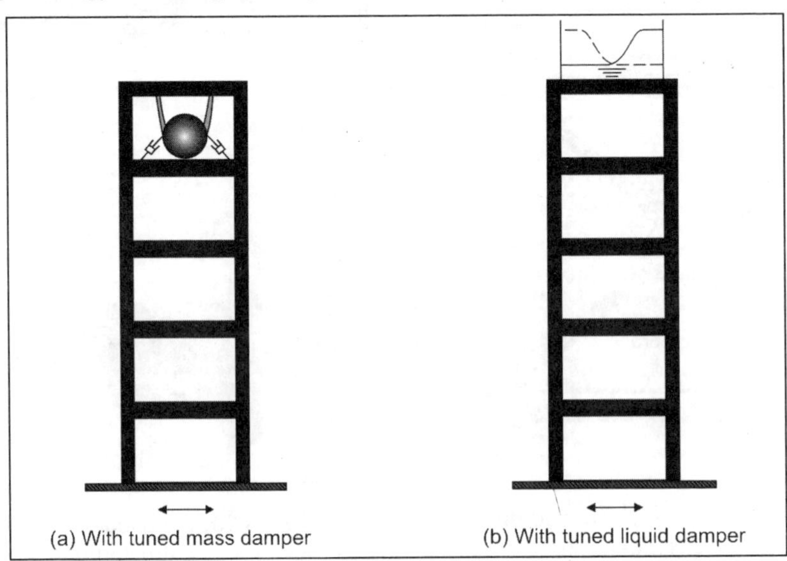

Fig. 23.22: Schematic representations of building with tuned devices

THEORY QUESTIONS

1. What are the various techniques for strengthening/retrofitting of structures?
2. Describe how repair/retrofit of non-engineered structures is carried out?
3. Describe in detail how various techniques and materials are used for retrofit of historical and heritage structures.
4. How retrofit strategies are used in case of RCC structures?
5. What are the methods for condition assessment of foundations and what precautions should be taken for overcoming deficiencies of foundations, while executing the job on ground?
6. How retrofitting is done by using FRP composites?
7. What is base isolation technique and how is it used for new construction of building?
8. How passive energy dissipating devices work? Explain with diagrams.
9. What are the various types of base isolation devices? Explain in detail.

Part 6
Case Studies

24. Retrofit of Buildings and Other Structures
25. Repair/Rehabilitation of Bridges
26. Repair/Rehabilitation of Marine Structures
27. Repair/Rehabilitation of Monuments
28. Repair/Rehabilitation of Irrigation Structures

24

Retrofit of Buildings and Other Structures

The building is a five storeyed residential RCC building located in zone v. Ground storey is an open ground storey to accommodate car parking. Various steps involved are:

Rapid visual screening (RVS) done as per procedure. RVS data is brought out. As per results, there is a requirement of detailed analysis.

Data Collection

The building is a RC framed structure. There are infill walls in all storeys, except ground storey (where there are three infill walls only around the staircase). The data collection as per procedure taught earlier is made in a tabular form and then preliminary evaluation, as per procedure is done. Then evaluation statements are prepared in a tabular form. Then based on preliminary evaluation certain structural deficiencies are established and thereby detailed evaluation is resorted to.

Observations from Detailed Evaluation

The observations from the detailed evaluation are summarized below:

- The equivalent static analysis shows that a number of beams and columns are deficient in flexure.
- However, all the beams and column sections have adequate shear capacity.
- The building complies with the drift requirement of IS:1893-2002.
- The pushover analysis in either direction failed to give a performance point before the collapse. Thereby the building needs to be retrofitted.

Retrofit

The retrofit scheme is to be now prepared. As a solution to the problem, 230 mm thick shear walls were proposed at a few locations throughout the height of building. This is global retrofit strategy. This will cause least intervention in the functional requirement of car parking.

Summary

The evaluation includes rapid visual screening, data collection, preliminary evaluation and the detailed evaluation. It was not possible to have a condition assessment of the building. The structural model of detailed evaluation was carried. The equivalent static analysis results show number of beams and columns are deficient in flexure. The pushover analysis in either direction failed to give a performance point before the collapse. So the buildings needed to be retrofitted.

24.2 A DOUBLE STOREYED LOAD BEARING RESIDENTIAL BUILDING AT MUMBAI

Salient Features

- Year of construction: 1965
- Investigation done: 1998
- Type of structures: Double storeyed load bearing structures
- Type of foundations: Spread footing
- Number of quarters: 44

Visual Observations

- Extensive cracking and spalling of concrete in sunshades, chajja, staircase and beams, etc.
- At some places reinforcement was exposed and corroded heavily.
- Major cracking observed on plastering on all faces.
- Severe seepage seen in most of the roof slab and external walls.
- Waist slab of staircase and soffit of beams exhibited delamination over 50%.

In Situ Evaluation and Nondestructive Tests

Delam Survey

Every column and beam was tapped by three different types of hammer. Most effective was the medium hammer, which gave delams for 15 to 25 mm depth. The hollow sound was recorded as hollow and the results of each structural member recorded on to observation sheet and that area evaluated for remedial measures.

Rebound Hammer Test

The rebound numbers are measured on concrete surfaces.

Ultrasonic Pulse Velocity Test

Values of the pulse velocity varied in the range 3.0–4.8 km/sec.

Half-cell Potential Test

Electrical potential values indicate that severe corrosion had taken place in some areas, as observed during the visual survey.

Carbonation Test

Carbonation has taken place beyond the reinforcement levels.

Conclusion

The main cause for distress to RCC element is:

- Inadequate thickness of cover concrete
- Highly permeable and porous concrete
- Carbonation of concrete
- Most distressed portions were chajjas along with lintel/beam, which is directly exposed to marine atmosphere and frequented by alternate wetting and drying.
- Seepage observed in roof slab was due to leakage from pipe line.

Recommended Repair Methods

- All sunshades/chajjas including balcony wall, side wall of steps and slab (partly) to be dismantled.
- In the joints between old and new concrete, epoxy based bonding materials to be applied on old concrete surfaces.
- Before removing affected RCC portion, temporary support to be provided properly.
- The use of lightweight precast lintels/chajjas in lieu of cast *in situ* was suggested. Accordingly corrugated sheets were provided.
- For RCC members, where overall integrity found good, cracks to be filled by putting epoxy grouting under pressure. For spalled concrete, polymer mortar/concrete with epoxy bond coat to be provided.
- The painting to complete surface to be applied with cement based paint with addition of polymeric compound.

The complete repair work got completed in 2000 and the performance after about 14 years is very satisfactory. The author was directly involved in complete assessment and planning of repair scheme, etc.

The part views of quarters before and after repair are shown in Figs 24.1 to 24.5.

Fig. 24.1: Part view before repair

Fig. 24.2: Part view before repair

Fig. 24.3: Part view after repair

Fig. 24.4: Part view after repair

Fig. 24.5: Part view after review

24.3 REHABILITATION OF RCC OVERHEAD RESERVOIR AT SILIGURI, WB

Synopsis

This paper deals with the reasons for failure/ distress of RCC staging of overhead reservoirs and suggests remedial measures for rehabilitation of this tank. The paper illustrates through case study how the rehabilitation work of staging of overhead tank of 50000 gallon capacity resting on RCC columns was proposed for restoration of structure.

Introduction

Reinforced cement concrete is most widely used material for construction of overhead water reservoirs, as it can be given any size or shape as desired. Since these structures are elevated, they are more prone to the effect of atmospheric weathering action, which in turn affect the longevity of such structures.

As it is well known, corrosion of embedded steel is the prime cause for damage of any RCC structures. The structure normally performs satisfactorily till such time the steel in reinforced concrete does not get corroded. Corrosion may even lead to catastrophic failure, in the absence of timely remedial measures. Steel corrodes due to atmospheric effects and impurities in various constituents used in concrete. The depth of concrete cover and concrete quality, i.e. density, impermeability, cement content and freedom from chemical pollutants are some of the factors which influence the corrosion of reinforced steel. The present study deals with the reasons for failure/distress of RCC staging of overhead reservoir and suggests remedial measures for rehabilitation of these tanks and precaution to be taken while maintaining such tanks in future.

Observations

A study was carried out in March–April 96 for an overhead tank of 50,000 gallons capacity resting on staging of 16 RCC columns braced together at different levels. The tank is located in North Bengal. The tank was constructed in the year 1977–78. The tank having dimension of 9.3 m × 9.3 m × 3.4 m is resting on columns of size 300 mm × 300 mm. The columns are interconnected at 3 different levels through bracings of size 250 mm × 250 mm; record drgs/CA drgs were not available. The distress in staging was noticed in the beginning of 1994. The distress was manifested in the form of cracks, spalling of concrete, rusting of steel in the bracings and columns. The cover to reinforcement of column and bracing was grossly inadequate.

In most of the cases, the spalling of concrete occurred from the bottom of bracings. Horizontal cracks were also observed at several places of a number of bracings; exposed

reinforcements were badly rusted. At the zone of lapping of reinforcement, the quantum of reinforcement was very high and there was inadequate bonding between concrete and reinforcement at that zone. Similarly spalling of concrete has occurred from the corner of column along with vertical cracks.

There was no sign of tilting or settlement of foundations. However, some seepage from the bottom of tank was observed. The water container was otherwise found in sound condition, as there was no visible sign of distress on external or internal surface except top slab cover of tank (from inside), which was having exposed rusted steel. The process of rusting of reinforcement and deterioration of concrete was observed to be progressing at a rapid speed, by comparing with the distresses noticed first time in the beginning of 1994, due to severe weather condition in that area. Any further delay in repair might cause sudden collapse of the structure due to the external forces like heavy wind and earthquake, etc. The overall view of overhead tank is shown in Fig. 24.6. The spalling of concrete and corrosion of steel in column and bottom/top side of bracing is shown in Fig. 24.7.

Fig. 24.6: Condition of water tank before repair

Fig. 24.7: Water tank after repair

Discussion

The tank was having major problem with bracings. The concrete of almost all the bracings had deteriorated extensively. As brought out the spalling of concrete had occurred mainly from the bottom of the bracings due to inadequate cover. The rusting/spalling started at the initial stage from the zone of lapping of reinforcement, (where quantum of reinforcement was very high), which subsequently progressed in zones where there was no lapping of reinforcement. Thereby, horizontal cracks have developed, even in those zones causing spalling of concrete from there. Similarly, due to insufficient cover to the reinforcements of columns, there was development of small vertical cracks at the initial stage. Subsequent rusting of reinforcement has caused spalling of concrete and development of vertical cracks in columns along the reinforcement.

Adequate precaution in terms of additional supports, careful dismantling of old concrete with modern equipment was an essential requirement in this case, to avoid any permanent damage to main structure. The major rehabilitation of this nature could be avoided, if system of proper periodical inspection and regular maintenance can be implemented on ground. Any seepage in water tank particularly through shaft type staging needs special attention, as this might result into sudden collapse. Here, in this particular case also the deterioration would have been much faster, if there had been added seepage through main structure. Here effect is mainly restricted to adverse humid atmosphere on RCC supporting structure having inadequate reinforcement cover.

Remedial Measures

The following remedial/rehabilitation measures were suggested:

- All unsound concrete to be removed by electric cutter/pneumatic chipper and reinforcement to be exposed all round.
- Existing steel to be derusted properly by brushing and applying rust removers, etc.
- Wherever existing reinforcement is reduced by 20% or more, additional reinforcement duly cleaned to be provided and welded on both sides.

- Epoxy based bonding material to be applied on old concrete surface (including old steel surface), after removing rust from the surface of concrete. The surface to be dried properly before applying bond epoxy coat, as per manufacture's instruction.

- Polymer modified concrete mortar to be applied. The materials to be used for mortar are cement, quartz sand, polymer and water in the proportion of 100, 400, 15 and 30 parts by weight. For damaged portion of member having depth more than 75 mm cement, quartz sand, aggregate (20 mm and down grade), polymer and water in the proportion of 100, 200, 400, 15 and 35 parts by weight to be used.

- 20 mm thick plaster in cement mortar 1:4 to be applied to overcome the deficiency of reinforcement cover.

The subsequence of repair is very important in this whole repair scheme, because of critical condition of staging. The repair of all the bracings in appropriate groups initially was suggested, and then the columns to be repaired in a group of maximum four at a time, after transferring load through strong steel prop to the bracings (below)/ground. Further, it was recommended that minimum three years performance guarantee clause be included in the tender for repair works. Though presently number of companies are manufacturing construction materials/chemicals, the products of following companies or equivalent were suggested for use:

- Hindustan Ciba-Geigy Ltd, Bombay
- Sika Qualcrete Pvt Ltd, Calcutta
- Krishna Conchem Products Pvt Ltd, Bombay
- Fosroc Chemicals (India) Ltd, Bangalore
- MC-Bauchemie (India) Pvt Ltd, Bombay.

Conclusion

The staging of water retaining structure is always exposed to the action of moisture/rain and as such there is a very strong need to provide well compacted dense concrete with adequate cover to avoid rusting of reinforcement and weathering action on concrete members/staging.

The overhead tanks need to be periodically inspected by the maintenance staff of GE and its soundness should be reported on 6 monthly basis through a regular report. This will function as an early warning system and avert the type of deterioration, observed in the tanks. Normal/special repairs of all the affected tanks need to be undertaken in a group. The present system of carrying out only periodical services of giving one coat of snowcem/colour wash every year, over the concrete surface even with cracks, spalling of concrete, etc. needs to be dispensed with. No periodical services should be done/allowed, unless structural soundness/stability is established. Presently numbers of companies are manufacturing various polymer based construction chemicals, which can be utilized effectively for repair/rehabilitation of this type of structure, as soon as any deterioration is observed in the structure.

The author was involved in planning and formulating rehabilitation scheme. The work was subsequently executed by the chief engineer of Siliguri zone successfully, under the guidance of the author.

Repair/Rehabilitation of Bridges

25.1.1 Background

- For thousands of years, masonry has been proved to be an extremely important construction material for a wide range of structures.
- A substantial number of rail and road bridges built-in masonry have lived a life over 100 years and some of them are still functional in spite of being over 150 years old.
- In recent past, a good number of these bridges have been identified as distressed, leading to the imposition of speed or load restrictions.
- These bridges require either rebuilding or rehabilitation to restore their original strength and to increase their residual service life.
- Since India is a developing country with limited resources, it has to utilize available fund judiciously.
- It is in the national interest to undertake restorative/rehabilitative measures instead of reconstruction/rebuilding, thus saving the limited resources.
- On Indian railway network alone there are 19,600 masonry arch bridges.

- Similarly on Indian road network, a substantial number of existing bridges comprise masonry arch type bridge.
- The design geometric of large number of such bridges is not known, yet these are catering to present day functional requirements.
- These are able to carry such high loads due to the inherent strength of structural form of arch and considerable factors of safety by way of conservative values for stresses in the materials used at the time of the design.
- However, most of the arch bridges are overstressed and show various signs of distress due to increased loads and material deterioration.
- Hence, there is a need to investigate the load carrying capacity of the existing masonry arch bridges and upgrade their strength to meet the present day as well as growing future traffic requirements, wherever possible.
- The assessment of the carrying capacity or load rating of the existing bridges can be done using theoretical methods, in conjunction with field tests.
- While theoretical methods can be used to investigate many variables quickly and cheaply, such analyses are limited in use due to idealization required.

- On the other hand evaluation based fully upon field tests provide realistic data pertaining to health of structure, but are enormously expensive.
- Model tests are generally used to investigate effect of parameters and are comparatively economical on cost and time as well as for the validation of theoretical procedures used.

To address some of the above issues, studies were carried out with the following objectives:

- To find out the total carrying capacity of masonry arch model experimentally and to study their collapse mechanism.
- To devise a technique for strengthening the test arches and determine the enhanced strength to see effectiveness of the strengthening.
- To develop theoretical method of local rating analysis for arches in elastic and post-elastic stages.

25.1.2 Strengthening Techniques

Strengthening can be categorized in three parts as follows:

- **Repairing:** Refers to superficial or cosmetic treatment involving racking and filling of joints and cracks with mortar and replastering the surface. Hence, the lost strength is not restored, but the surface deterioration is checked.
- **Restoration:** Refers to treatment by which original strength can be restored. It includes the grouting of the cracks with cement slurry with admixtures or epoxy materials to restore the compressive strength and enhance the tensile strength.
- **Retrofitting or upgrading:** Refers to increase in strength and load carrying capacity to desired level using different types of technique.

25.1.3 Defects in Arch Bridges

The major problem encountered in masonry arch bridges is due to the crack developed in the arch ring due to wide ranging reasons. The defects encountered are:

- Longitudinal crack in the center of barrel: These are produced due to eccentric loading because of traffic passing in opposite directions. This is called flexing.
- Cracks on the face of arch: Sometimes cracks appear on the junction of the spandrel walls and the extra-dos of the arch due to lack of bonding. These can appear due to bulging and sliding of spandrel walls, excessive rib shortening and distortion of arch rib under loads.
- Longitudinal cracks on intra-dos under spandrel wall: Where the spandrel wall is constructed monolithically with the arch barrel, longitudinal cracks on the inside edge on infra-dos appears due to difference in stiffness of spandrels, which act as deep beam and flexible arch barrel.

Cracking and crushing of masonry: The trapped water in the fill cause deterioration of the mortar and in some cases leads to complete loss of jointing material. The excessive impact load due to poor surfacing of roadway, such as presence of potholes, badly maintained rail joints, further increase the severity of cracking and leads to the crushing of the masonry.

25.1.4 Methods of Strengthening

The commonly used methods of strengthening of masonry arch ring are listed out below in brief:

- Pressure pointing
- Pressure grouting with cement slurry
- Grouting with epoxy resins
- Relieving arch
- Local strengthening
- Replacement by box section
- Introducing relieving girder/slab
- Tension ties
- Reducing the span of the arch
- Stitching.

25.1.5 Experimental Investigation

With the objective to determine the load carrying capacity of the arch rings and to understand the behavior of the masonry arches, some work was carried out at IIT, Roorkee. This has enabled to understand the behavior of the arches and their collapse mechanism under the gradually increasing static loads. Some of the strengthening techniques listed earlier have been tried upon these tested arches to determine the suitability and the efficacy of the strengthening techniques.

25.1.6 Assessment of Strength

The strength of masonry arches can be determined by the following methods:
- Pippard's elastic method
- Heyman's plastic method
- MEXE/MOT method
- Nonlinear finite element method.

In the theoretical analysis, the masonry arches are treated as fixed arches. The stresses at various cross-sections are determined and the critical section is located. The load that causes stresses at the critical sections to exceed permissible stresses is the load carrying capacity of the section. The fixed arch can be analyzed using strain energy approach to determine the net stresses at different cross-sections to establish the condition of plastic failure.

25.1.7 Load Carrying Capacity

Load rating/carrying of retrofitted arches can be calculated by assuming that there is perfect bond at the interface of old arch and new repair. Further, the repair by grouting is fully effective and homogeneity of the section/material is restored. The load carrying capacity of masonry arches much depends upon axial thrust. The theoretical loads predicted for the retrofitted arches can be analyzed by step-by-step analysis.

By carrying out experiment, it was observed that the collapse load values obtained experimentally and theoretically before and after strengthening is quite close to each other. The load carrying capacity of unstrengthened arch after retrofitting by ferrocement was increased by 65%. The load carrying capacity of unstrengthened arch after retrofitting by providing RCC relieving arch was increased by 191%. It was also observed that there was no loss of ductility in the arch behavior due to the retrofitting employed.

25.1.8 Conclusions

- Various types of defects in arches and their retrofitting techniques have been studied in detail. These measures would not only help in strengthening a number of existing arch bridges on roads and railways, but also will prevent the reconstruction of old bridges, which would eventually result in economy of huge public fund and avoid inconvenience to traffic caused by construction of new bridges.
- Filling the cracks of distressed arch by epoxy, i.e. restoration of cracked arch by epoxy was found effective in all the arch models. This was found to be more effective that restoration by filling joints with cement slurry.
- The ferrocement patching method of retrofitting has been found to be effective, which gives clear indication that increasing the arch sections can increase the load carrying capacity. But it is observed to be more effective, when the lair is nailed or bonded to old arch.
- The method of providing RCC relieving arch beneath the old arch has been found to be very effective in order to increase the load carrying capacity. But providing proper anchoring of the old arch and relieving arch at interface is found to be necessary.
- From step-by-step analysis, it is observed that theoretical points of formation of hinges in unstrengthened arches are close to practical points of hinge formation. The actual sequence of formation of hinge is same as the theoretical one. The values of collapse load calculated theoretically and found experimentally, both unstrengthened and strengthened are found to be very close.

25.2 RETROFITTING/STRENGTHENING OF MASONRY BRIDGE STRUCTURE

25.2.1 Introduction

The first train in India was made operational on 14 April 1853. Thereafter, railway network was spread very fast to cover the whole country in the shortest period. By 1900, 60% of the present railway network was laid. Appreciable number of bridges of that vintage still exists and majority of them are in sound condition.

During those days, substructures were made either in brick masonry or in stone masonry with lime mortar. As far as superstructure is concerned, it was also tried to avoid steel superstructure, which had to be shipped from UK unless the same was unavoidable (like in case of longer spans). For similar to medium size spans, arch bridges got preference over other bridges. Here, discussion will be about various types of bridges and strengthening/retrofitting work done normally for masonry structures.

25.2.2 Retrofitting of Arch Bridges

Retrofitting of arch bridges is required on account of the following:

- Age factor
- Increase in axle load
- Increase in load due to gauge conversion
- Increase in traffic density.

In case of increase in axle load of the train due to introduction of heavier freight stock or in case of proposed gauge conversion from meter gauge to broad gauge, load on the bridge would increase. In such a situation, thorough investigation of the existing bridges is required. If the investigation reveals that the bridge is in sound condition, then its theoretical analysis is done. If the theoretical analysis reveals that the stresses are coming within limit and the bridge is fit for the increased load either on account of introduction of heavier freight stock or on account of gauge conversion, then there is no problem in retaining such a bridge without any modification. In case, the theoretical analysis reveals that the heavier axle load is creating overstressing and the bridge is apparently in sound condition, then its load test is required before taking any decision.

Load Testing of Arch Bridges

Provision of load test exists in arch bridge code of Indian railway. The relevant extracts of the same are reproduced below.

The criteria for arriving at the safe load shall be:

- Under the proposed load, the crown deflection and spread do not exceed 1.25 mm and 0.4 mm respectively.
- There is no residual deflection or spread after release of load.
- There is no crack appearing on the infrados of bridge.

The above criteria will be applicable to segmental and nonsegmental arches of span 4.5–15 m, provided span/rise ratio lies between 2 and 5. The load test shall be conducted only after complete pressure grouting of the masonry.

In order to gauge conversion, the certification for the safety of the entire arch bridges, whether sound or distressed shall be based on load test on representative type of bridges. The criteria for assessing the safe load shall be same as specified in the above para, provided the following conditions are satisfied.

- The condition of masonry and its behavior under test load are satisfactory.
- Type of foundation and nature of soil on which it is founded are suitable.

Reason for conducting load tests of the arch bridges, if over stressing is found in theoretical analysis. Due to configuration, arch bridges transfer load through arching action in which complete arch ring remains in compression. There is always difference between the theoretical analysis and the strength potential available. Furthermore, analysis of arch bridges are being done considering arch rib as independent unit. In actual practice, it is partly monolithic with abutments and piers. On this account also, some extra strength potential is available with the arch bridges, which needs to be explored and exploited. To get benefit of the strength potential available in the entire elastic range, after rigorous analysis and field experience, testing procedure for the arch bridges has been evolved and the same has been made part of the arch bridge code, relevant para of which has already been mentioned above.

Experience of the Load Test of Arch Bridges

Experience of the load test of arch bridges has been very positive. The results of some of the bridges show, the theoretical analysis reveals that the bridges are overstressed in case of introduction of heavier axle load. However, after carrying out the load test, the bridges were found very much safe for revised load. Nowadays, such a heavy factor of safety is not being taken. Hence, even in case of allowing 100% overstressing, at least factor of safety of 3 will be there.

Guidelines Regarding Strengthening of Arch Bridges

Some guidelines about strengthening of arch bridges are given as under:

- For strengthening weak/distressed arch bridges, the method of jacketing at the intra-dos is preferable, if the resultant reduction in the waterway is permissible.
- In case of strengthening over extra-dos of arch bridges, the new arch ring should be designed to take the entire load, viz. dead and live loads.
- In case of strengthening below infra-dos of arch, the new arch ring should be designed as under:
  i. To take entire load by itself, where existing arch has transverse crack(s).
  ii. To take the entire load by composite action with the existing arch ring, where the existing crack(s) is/are all longitudinal or there are no signs of distress

in the existing arch, provided effective bond could be ensured between the old and new arch ring.
- In case of strengthening of abutments and piers of arch bridges, the design should be on the basis of composite action of the new material acting along with the existing one. It should, however, be ensured that a proper bond is established between the existing masonry and new materials by suitable means, such as dowels and post grouting through grout holes to be left while casting the jacket.
- In all cases of cracked masonry, whether in arches or in abutments and piers, it should be grouted under pressure to plug all the cracks before the additional material is provided.
- The space between the new arches, in the case of strengthening below the intra-dos of the arch, should be grouted under pressure for which grout holes should be provided in the new arch ring.

Strengthening of Arch Bridges by New Techniques

Mostly, it is experienced that deterioration starts first in arch ring, and thereafter at other places. This might be due to arch ring bearing has more impact than other members of the arch and also it being more seepage prone. The system of strengthening of the arch bridges described earlier may change the architectural features of the arch bridges. To ensure architectural features even after strengthening of the arches, a new technology has been developed, wherein by drilling holes in the arch ring at different locations, special type of reinforced bars having proper grip are inserted. After insertion of such specially designed bars, proper grouting is done and thus, the structures get strengthened. Railways have tried this system on some of the bridges and the result is satisfactory. The following two firms are working in this field:

- M/s Helifix
- M/s Cintec

Comparatively, this system is costly. Economy may be achieved in mass scale adoption or by use of local products. Further, performance needs to be watched over a longer period before further adoption.

25.2.3 Strengthening of Masonry Substructures

Strengthening of substructures is required on account of the following:

- Strengthening required coping up with the increase in load standard
- Strengthening required coping up with weathering action.

Strengthening Required Coping up with Increase in Loading Standard

Over a period of time loading standards have increased drastically. To enable heavy haulage of freight stock, increase in attractive effort of locomotives was done from time to time for various loading standards, as given below:

Sl. No.	Loading standards	Year of introduction	Total tractive effort
1.	BGML	1926	47.6t
2.	RBG	1975	75.0t
3.	MBG	1987	100.0t

To cope up with the strength requirement, jacketing is preferred solution.

Alternative Solution to Cope up with the Increase in Longitudinal Forces

Some transmission units have been developed by different companies. These units are attached with superstructure at the support point of the substructure. By virtue of its design, it transfers a fair extent of longitudinal forces to the bridge approaches, so as to make the superstructure safe. Names of some of the firm, who are supplying such units, are given below:

- M/s Colebrand of UK
- M/s Maurer of Germany
- M/s Jarrett of France.

Jacketing with FRP Material

Increasing use of FRP is being exploited for strengthening of substructure of bridges also. For that, TYFO systems are available. It involved wrapping the structure with fibers and then making it a part the substructure with the help of epoxy. By this method, with very little addition of thickness, structures are jacketed to take additional vertical as well as longitudinal forces.

Strengthening on Account of Weathering Action

Weathering action may result in one of the following reasons either alone or together with others:

- Loss of strength of the mortar
- Spalling
- Decay in bricks
- Development of different types of cracks

In case of loss of strength of the mortar, loose material is completely removed. Then, after pointing, plastering is done. In case of spalling, after removal of the loose material either plastering or guniting is done. In case of decay in bricks, either guniting or plastering or jacketing is the appropriate solution according to the extent of decay and strengthening requirement. Cracks are properly treated either by cement grouting or epoxy grouting.

25.2.4 Conclusion

Strengthening measures for masonry arch bridges, as described in this chapter, can be used both for railways and highways.

25.3 REHABILITATION OF CONCRETE BRIDGES

25.3.1 Introduction

When a structure is not able to resist the forces (external/internal) exerted on it, the structure shows signs of distress or results in failure. Sometimes there is a single explanation for a failure, but usually it is a combination of several reasons responsible for failure. After a bridge is constructed, it is absolutely necessary to follow a regular inspection schedule. Based on the regular inspection reports, systematic maintenance is essential to ensure long-term serviceability of bridges. There are mainly four types of bridge maintenance:

- Routine
- Preventive
- Repairs
- Strengthening/replacement.

The engineer responsible for inspection of a bridge structure should be experienced enough to assess the condition of the structure and recommend the maintenance schedule. This is effectively possible only when the causes of deterioration can be identified and rehabilitation strategy is finalized based on the assessment. Therefore, the chapter discusses causes of concrete deterioration, various methods for condition evaluation and also some repair procedures.

25.3.2 Causes of Deterioration

The most common forms of defects and deterioration that occur in concrete are summarized below:

Corrosion of Reinforcement

Corrosion of reinforcement is the most common cause of deterioration of concrete. Corrosion occurs when oxygen and moisture are present and the passivity of the steel is destroyed. As the steel reinforcement corrodes, it expands and causes the cracking of concrete at or near the level of reinforcement. When the pressure increases (due to increase in reinforcement volume) spalling of concrete occurs. There are several mechanisms, which may be responsible for the start and continuation of corrosion of reinforcement, e.g. ingress of chloride and carbonation of concrete, etc.

Chloride Ingress

Intrusion of chlorides into a concrete bridge component is the beginning of the deterioration problem that ultimately results in spalling of the concrete. Presence of more than the threshold concentration of chloride ions at the steel surface results in loss of passivity and starts corrosion of reinforcement. There could be areas of anodic regions and pitting corrosion may occur as shown in Fig. 25.1. The cracks produced at the start of the corrosion, provide easier path for further chloride ingress. This may further enhance the rate of corrosion. Extensive deterioration is expected at an early age after the corrosion has started. In many parts of the world, chlorides used as deicing agent were found to be mainly responsible for corrosion in concrete bridges. When the deterioration has taken place due to chloride ingress, the patch repair should be carried out after careful estimation of chloride content in the uncracked concrete. After patch repairs, new anodic regions may be created (if the chloride content is high in the sound uncracked concrete). This may accelerate the corrosion in the sound concrete.

Carbonation of Concrete

When the carbonation front reaches the reinforcement its passivity is lost and corrosion can start if moisture and oxygen are present. The rate of carbonation is highest at about 60% relative humidity and very low in dry state.

Chemical Attacks

The most common form of chemical attack is by sulfates. Naturally occurring sulfates of sodium, potassium, calcium or magnesium are sometimes found in soil or dissolved in groundwater adjacent to concrete substructures and foundations. When evaporation can

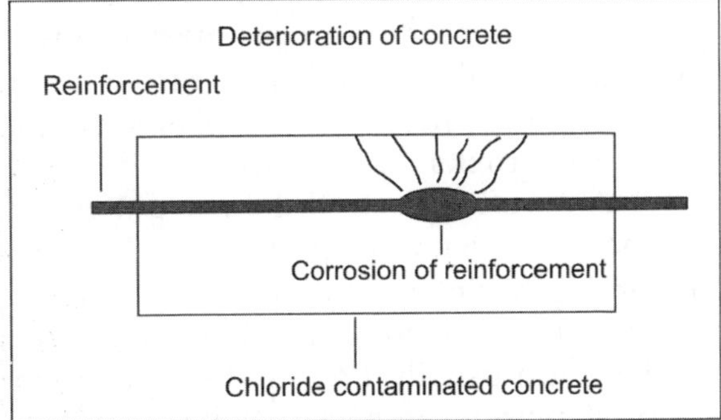

Fig. 25.1

take place from an exposed face, such as an abutment or retaining wall, the dissolved sulfates may accumulate at the face and increase the potential for deterioration.

The deterioration of concrete by acids is primarily the result of a reaction between chemicals and calcium hydroxide. In most case this reaction results in the formation of water soluble calcium compounds, which are leached away. Acid attack of concrete in bridge is uncommon, but it can occur as a combination of acid attack with scaling and even corrosion of embedded reinforcement.

Freeze–Thaw

At subzero temperature, the water inside pores freezes and due to increase in volume exerts pressure on concrete. The cracking and scaling is observed due to low tensile capacity of concrete. Due to freeze-thaw cycles, the microstructure of concrete changes substantially. Strength of concrete decreases and permeability increases. This is a harmful effect as far as durability of concrete is concerned. During this process coarse aggregate is exposed and eventually becomes loose.

Cracking

Cracking in structures may occur due to many reasons other than the corrosion of reinforcement. The cracking may be structural or nonstructural. However, even if they are not posing any problem to the structure, they may help in penetration of chemicals, oxygen, moisture, etc., and results in deterioration due to their harmful effects. The following are the types of cracks observed in concrete structures:

a. Plastic shrinkage cracks result from rapid drying of concrete in the plastic stage. These cracks are usually wide and shallow, often in a well-defined pattern or spaced at regular intervals.

b. Drying shrinkage follow the drying of restrained concrete that has hardened. They are usually finer and deeper than plastic shrinkage cracks and have random orientation.

c. Plastic settlement cracks occur due to upward bleeding of water and downward movement of solids. This downward movement is restricted by the layer of reinforcement and results in cracking in the cover zone of concrete. These cracks are nonstructural, but these are very harmful from the durability considerations.

d. Settlement cracks may be of any orientation and width, attributed to foundation settlement.

e. Structural cracks beyond the control of reinforcement, resulting from chemical reactions between the minerals aggregates and the cement paste. These cracks increase in numbers and width with time.

Miscellaneous

There could be several other reasons for deterioration, such as honeycombing and air pockets, excessive wear may appear on concrete surfaces exposed to traffic, erosion of concrete from the bridge piers and abutments by solid particles in rivers with high bed load, natural factors, design deficiencies, errors during construction, etc.

25.3.3 Inspection and Assessment

The objective of bridge inspection comprises monitoring and evaluation of the performance of each bridge component throughout its service life so that any deficiency in the performance could be detected and corrected early. For this purpose, a periodic or routine inspection followed by a detailed technical investigation is essential. Deterioration is initially assessed by visual inspection and the results are reported in standard forms, supplemented by photographs and sketches. After visual inspection, strategy for detailed inspection is decided. There are a number of techniques available for assessing the condition of concrete structure/bridge. Table 25.1 suggests some of the methods for condition assessment of concrete bridges.

Table 25.1

Parameter	Testing technique
Strength	1. Ultrasonic
	2. Rebound hammer
	3. Pull off test
	4. Penetration resistance
	5. Core testing
Chloride ingress	1. Mohr's method
	2. Ion selective electrodes method
	3. Photospectrometer
Carbonation depth	1. Phenolphthalein spray
Cover to the	1. Magnetic method
reinforcement	2. Thermographic method
Corrosion	1. Half cell potential measurements
	2. Concrete resistivity measurements
	3. Linear polarization resistance measurements
In situ absorption	1. Autoclam permeability system
	2. Initial surface absorption test (ISAT)

25.3.4 Repair Procedures

Among the factors to be considered in choosing the repair method to restore the structural capacity and service life are durability, anticipated future use of structure, cost and speed of repair, inconvenience to users, local labour market and availability of contractors, material availability, environmental priorities and aesthetics. Repair materials should respond similarly to changes in temperature and the applied loading, and they should blend in appearance. The following are examples of extensive repair/strengthening of concrete pier and reinforced concrete girders and smaller patch repairs.

Repair/Strengthening of Bridge Pier

A typical repair of deteriorated concrete pier by encasement in concrete is shown in Fig. 25.2. It can be used where much of the pier cross-section is lost or it needs to be strengthened to cater to increased loading or damage. The construction procedure involves the following steps:

- Remove deteriorated concrete and clean the surface of the column. The concrete surface should be shot blasted to provide a good rough surface for proper bonding of new concrete with old concrete.
- Use shot blasting/sand blasting to clean the exposed reinforcement in concrete pier above the water line and splice with new reinforcement cage around concrete pier as shown in Fig. 25.2. Use spacer to keep the forms in the proper position. A suitable coating material may be applied on the reinforcement.
- Place the forming jacket around the pier and seal the bottom of the form. Pump suitable concrete into the form through the opening at the top. The concrete used should have a low water–cement ratio and high workability, so that proper bonding can be ensured.

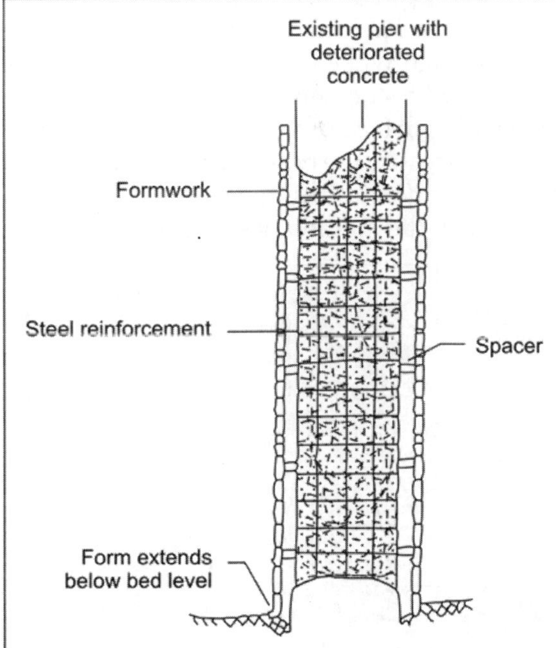

Fig. 25.2: Reinforcement around concrete pier

- Finish the top portion of the repaired section. The forms should extend above splash zone down to sound concrete. A suitable coating material may be applied on the surface to improve the permeation properties of the concrete.

Improving the structural capacity of a column/pier can also be achieved by encasing the column in concrete or steel jackets. The jacketing may be applied to the full length of the member or only to severely deteriorated section. The intent of this process is to increase the cross-sectional area of the column and also reduce its slenderness ratio. Partial encasement may also be particularly effective when an unbalanced moment acts on the column. Figure 25.3 illustrates two such concepts for increasing cross-sectional area and strengthening of concrete structures.

The complete encasement of an existing column in a concrete jacket has been used frequently for strengthening of columns/bridge piers. The reinforcement is placed around the existing pier perimeter inside the jacket. A difficulty most often encountered is the development of structural continuity between the old and new material. This is the critical stage where part of the load is to be transferred to the new material. The first step is usually surface preparation of the existing column because the bonding between the new and old concrete has been found to greatly depend on the method of surface preparation.

At this stage, it may also be desirable to jack the superstructure and place temporary supports on either side of the column. This may be necessary because shrinkage can cause compressive stresses on the column that will be reduced if the existing column is unloaded. Supports will also be required if the column shows signs of deterioration. Finally, this procedure will allow the new material to share equally both dead and live loads after the supports are removed. The other steps are similar to the steps explained above.

Jacketing techniques that have been used for seismic retrofitting of existing pier/columns are illustrated in Fig. 25.4. Figure 25.4a shows the addition of longitudinal reinforcement in the jacketed area around the existing column. The reinforcement is extended into the footing in predrilled holes to ensure the structural continuity by using dowels spliced with the new bars in the column. New ties are placed and concreting can be done as explained above. Shotcrete can also be used for applying a concrete layer on the existing pier/column. However, the shotcrete should only be preferred when an experienced person for doing the job is available.

A procedure used to improve the lateral capacity of a column is shown in Fig. 25.4b. There are two methods for adding lateral reinforcement: By wrapping the existing column with tensioned prestressing wire or by adding a series of hoops with a turnbuckle

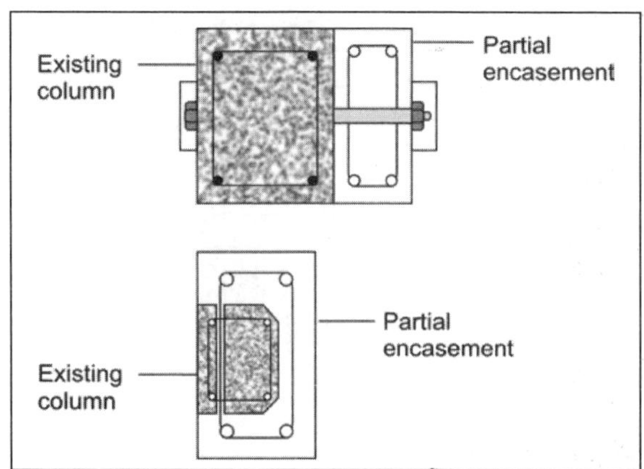

Fig. 25.3: Concept of strengthening

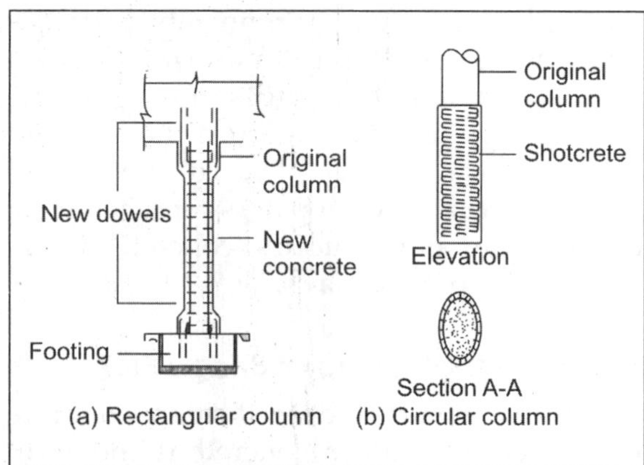

Fig. 25.4: Addition of reinforcement in jacketed area

included to pretension the two ends of hoops together. Both methods should include the application of a protective layer of shotcrete or cast-in-place concrete.

The jacketing of concrete columns with steel shapes is essentially similar. Three versions are shown in Fig. 25.5 and involve a primary load transfer between the steel and the column by shear friction. The column is first cleaned and treated. The steel shapes are treated with an epoxy coating and erected around the column, these steel shapes form the jacket, and new concrete is poured and compacted in the annular space.

Strengthening of RC Beams Using Epoxy Bonded FRP Composites

Strengthening of concrete members is usually accomplished by providing external reinforcement and concrete or concrete jackets, by epoxy bonding of steel plates to the tension faces of the members, or by extra post tensioning. A relatively new technique involves the replacement of steel plates by fiber reinforced polymers (FRP), or simply composites in the form of laminates of fabrics. These materials offer the engineer an outstanding combination of properties, such as low self-weight (making them such easier to handle on site), immunity to corrosion, excellent mechanical strength and stiffness and the ability of formation in very long lengths, thus eliminating the need for lapping at joints. The FRP strengthening technique has found wide attractiveness and acceptance among researcher and engineers in many parts of the world, and is no longer considered to be a new technique for certain types of strengthening jobs.

Typical FRP configurations for shear strengthening of concrete beams are shown in Fig. 25.6. The external reinforcement in Fig. 25.6a is in the form of epoxy-bonded laminates or fabrics. Another possibility is that of Fig. 25.6b, where the epoxy-bonded FRP fabric is wrapped around the beam. The effectiveness of the strengthening reinforcement depends on its failure mechanism,

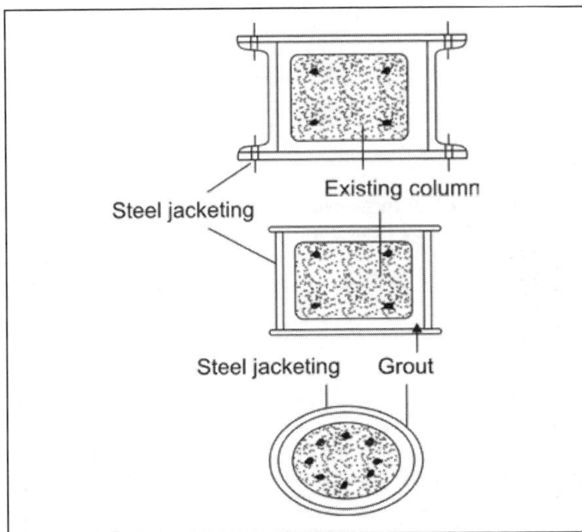

Fig. 25.5

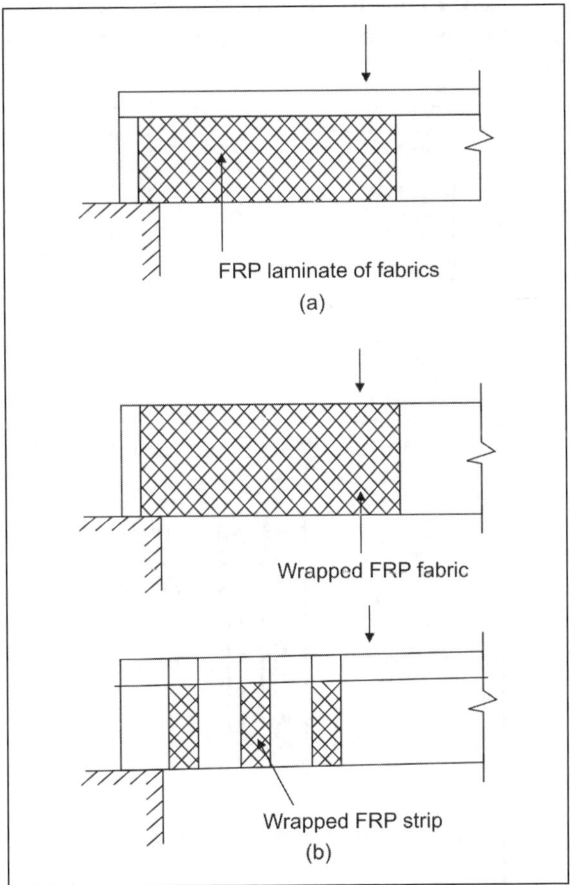

Fig. 25.6

which mainly depends on the bond between the FRP and concrete.

25.3.4.3 Patch Repair

When some part of the structure is deteriorated, then patch repair is a useful technique. First of all, edges of the patch to be repaired are made square. The surface is then prepared using a suitable surface preparation method. The bond between the old concrete and new repair material greatly depends on the surface preparation method. After cleaning and providing protecting cover to the existing reinforcement or new reinforcement, a bonding coat is provided on the prepared concrete surface. Subsequently, the repair mortar is placed and the surface should be smoothened. At the end, surface coating is provided to reduce further ingress of harmful substances. The procedure is explained in Fig. 25.7.

If deterioration has taken place due to ingress of chloride, evaluation of chloride content of concrete in the affected concrete is essential. In case of high concentration of chloride in the affected concrete, the patch repair may enhance corrosion in the area adjacent to patch repair due to formation of anodic region. However, at low concentrations it has not been found critical if proper surface treatment is used to reduce further ingress. The lower permeability and shrinkage of the repair materials has also been found to improve durability of repairs.

25.3.5 Conclusion

The most common defects in concrete bridges are cracking due to drying shrinkage; thermal cracking; spalling of concrete due to reinforcement corrosion. The corrosion of reinforcement occurs mainly due to lack of cover, chloride contamination, carbonation, water leakage, etc.

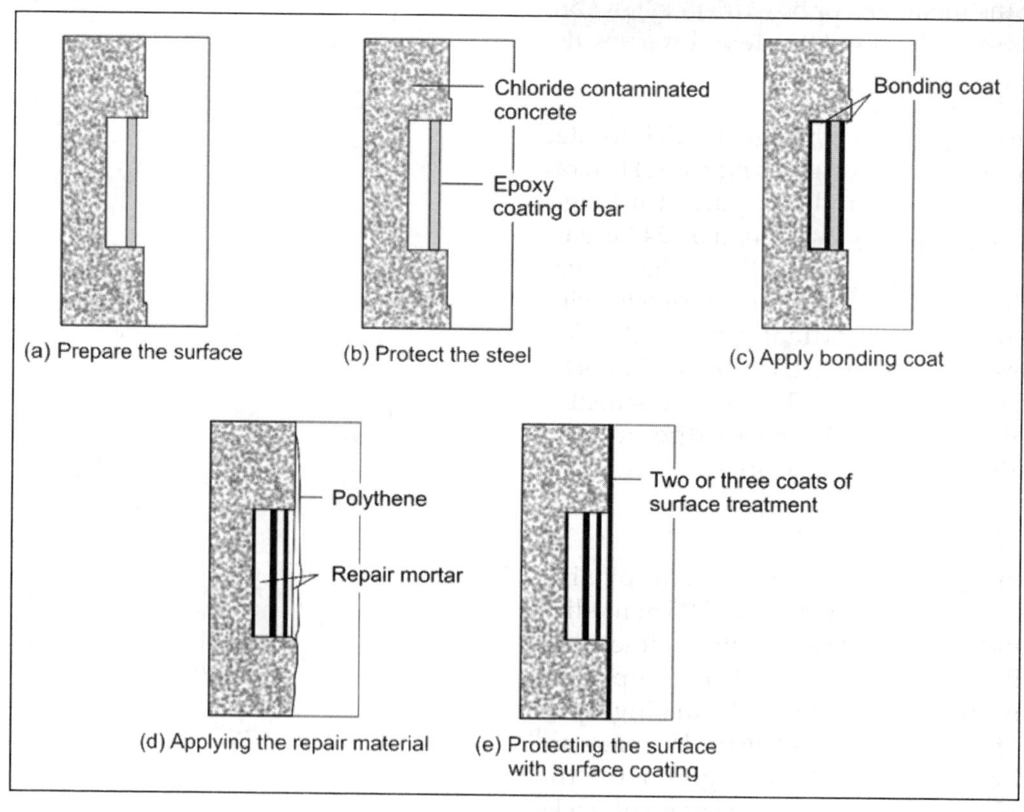

Fig. 25.7

The cause of deterioration must be diagnosed for an effective rehabilitation of concrete structures. Defects are commonly due to more than one cause. A detailed investigation is recommended before finalizing any repair strategy, e.g. chloride contamination requires removal of comparatively large areas of concrete, otherwise high concentration of chlorides in substrate concrete may create further problems in the patch repairs. In case of chloride-contaminated concrete, low permeability repair material is recommended and preferably a surface coating be applied after the repair, to reduce the chances of corrosion. Patch repair methods have a reasonable chance of success where corrosion is due to carbonation. FRP have been found to provide a practical and effective solution to rehabilitation and strengthening problems in columns and beams.

25.4 REHABILITATION OF A ROAD OVER-BRIDGE AT BANKIM SETU

25.4.1 Introduction

Bankim setu, road overbridge (ROB) is located over railway tracks originating from Howrah station at Kolkata and is a part of a longer flyover. The bridge was opened on 24 January 1981 and is in service since then. The railway authorities observed snapping of prestressing cables of the longitudinal girders at a few locations and appointed STUP consultants (P) limited (SCPL) in June 1999 for assessment of present condition of the bridge and for suggesting remedial measures.

25.4.2 Salient Features

The bridge is 26.0 m wide comprising 2 m × 9.9 m wide carriageway, 0.95 m median verge and 2.74 m wide footpath on either side. Overall length is about 294 m comprising 9-spans in ROB section, of varying span lengths from 21.22 m to 42.96 m. The piers are placed at different skew angles with the carriageway depending on the layout of tracks underneath. The skew angles are varying from about 4 to 25°. The superstructure in each span consists of 9 longitudinal prestressed concrete girders connected by diaphragms at suitable intervals. *Cast in situ* gap slab has been provided between top flanges of girders to serve as the bridge deck. Deck structures were designed for carrying the worst of the following IRC loading for vehicular loads:

i. 6 lanes class A
ii. 2 lanes of class 70 R or class AA and live load on footpath as per IRC:6.

The depth of the PSC girders was restricted by the available vertical clearance between the rail level and the level of the approach via ducts on both sides, and high grade of concrete of M42.5 and M45 with water–cement ratio of 0.4 were used to withstand the loading. Each cable consisted of 12 nos. HTS wires of 7 mm diameter of UTS 1520 MPa. Span no. 9 have the maximum span length of 42.5 m; and have maximum number of 23 cables in each girder, of which 11 cables have been anchored at the vertical face of girder (2.38 m deep) and the rest in the deck slab. The shape of girders is 'I' section with 1.8 m wide top flange, 0.2 m thick web and 0.61 m wide bottom flange. Galvanized metal sheathing duct of internal diameter 42 mm was used.

25.4.3 Investigation Strategy

Rehabilitation work for bridges involves study of original design and preliminary visual inspection prior to detailed field inspection. Localized testing is also carried out to establish thrust and direction of testing. The important part of bridge rehabilitation work constitute review of original design, establishing details of as-built structure and carrying out checks for the present day design loads using current design codes, considering present condition of structure. The entire forensic investigation and testing was to be planned in such a way that they provide information for carrying out these tasks. This procedure was adopted for bankim setu.

25.4.4 Detailed Inspections and Diagnostic Tests

The following special inspections and testing were conducted with specific purpose of yielding data as mentioned for evaluation of root causes of distress and extent of distress leading to appropriate repair methods and for carrying out analytical studies.

- Dimensional and geometric survey: To compare the 'as-built' dimensions and levels of the structural components for structural analysis to check if there is any settlement and tilting of foundations.

- Schmidt hammer test, ultrasonic pulse velocity test and core strength test: To compare variability of strength (in a comparative sense) and to serve as indicator of existing strength.

- Test pertaining to placement, present condition, potential of further corrosion and strength of reinforcing steel and pre-stressing cables: Following tests for the above purpose were conducted:
  - Cover meter study: To measure the cover to reinforcement.
  - Concrete resistivity test: To establish and quantify the quality and ability of concrete to protect steel from corrosion.
  - Carbonation test: To quantify the depth of penetration of carbon dioxide in concrete.
  - Half cell potentiometer study: To determine the electrical potential as measure of electrochemical activity between concrete and steel which is responsible for activating and progressing the corrosion process, and indicator for risk of corrosion.
  - Test on reinforcing and prestressing steel: To establish loss of section due to corrosion.

- Chemical analysis of concrete and petrography: Following tests for the above purpose were conducted.
  - pH
  - Alkalinity
  - Sulfate
  - Chloride content and gradient.
- Petrography related to alkali-aggregate reaction.

All these tests were carried out to assess the concentration/quantum of presence, compare with permissible limits and suggest remedial measures, if necessary.

- Air quality test: To assess the concentration of harmful gases to concrete like oxides of sulfur and carbon.

- Thorough visual inspection and defect identification: To assess the type and extent of distress in each structural member.

- Strain gauging of the girders: To measure the residual strain over a period of 10–15 days to find out if the structural behavior of the girder is within elastic range or not.

25.4.5 Extent of Major Distress Observed: Root Causes and Option of Repair

General Observations

- The alignment of the bridge and levels did not indicate any abnormal movement or settlement in construction.

- The environment is aggressive; the chloride penetration is atmospheric except one location of a girder, where it reduces its intensity from surface to inside. The chloride content is generally above the permissible limit for prestress concrete but within permissible limit for reinforced concrete. The excessively high chloride content in one location is an isolated case and might be due to particular batch of coarse/fine aggregate used in the construction. Potential of corrosion except in the soffit zone of longitudinal and cross girders are insignificant. The concentration of CO_2 and SO_2 in the air near span-2 and span-6 are high, due to fumes generated by diesel locomotive engine passing through this area. This has caused high content of sulfate in the concrete in this area.

- Core strength test results were comparable to specified design compressive strength of concrete. UTS of prestressing cables for span-9 matched with the design requirements. However, for span-8 this was close to jack end force. This could not be checked due to nonavailability of prestressing details during construction at site, and was therefore ignored.

- UPV test reflect that the concrete at bottom 1/3rd of longitudinal girder bulb and bottom 100 mm of cross girder is weak.

- At several locations the sheathing material has corroded and the cables were exposed. The cables have snapped at three girders for span 7 and 9. At most locations there are no gap between two adjacent sheathing. The material below the sheathing has been found to be basically cementing sand mortar. This is attributable to inadequate construction and/or detailing.

- Cover meter study indicated insufficient cover to reinforcement for cross-girder. This is also attributable to inadequate construction and/or detailing. Covers in many places are irregular.

- At locations where sheathing are exposed or made to expose and the sheathings were punctured to find out the condition of cables and extent of grouting inside the sheathings; it has been noticed that mostly the sheathings are void or the grouting was discontinuous. Rust formation was noticed at most locations.

- There are many other minor construction inadequacies like misformed joints, honeycomb in concrete, inadequate cover, etc.

Main Longitudinal Prestressed Girders

Designs checks for the prestressed girders have been conducted for the following alternative situations.

- Considering 50 mm reduced thickness of the bottom bulb due to bad concrete in this zone below the sheathings and all the cables are intact, the section was found to be within safe limits due to reserved strength in the original design. Hence, protective repair of the bottom concrete have been suggested together with application of protective coating of epoxy paint on the PSC girders.

- Considering the bottom layer of PSC cables to be snapped but no reduction in depth, the section was found unsafe. Retrofitting calculations were carried out for external prestressing to compensate the deficiencies and strengthening by external prestressing have been suggested for a few girders.

Repair to Weak Bottom Concrete of Cross-girders

It has been suggested to remove the weak bottom concrete of cross-girder and replace them with appropriate sound concrete and apply protective coating by epoxy paint.

- Repair to damaged sheathings and voids in sheathings.

  It has been suggested to expose the sheathings in weak concrete zone, replace the sheathings wherever corroded and regrout all the bottom sheathing to arrest further deterioration of prestressing cables.

- Other minor repair related to defects arising out of inadequate constructions, lack of proper maintenance has also been suggested.

- Repair options

  - For strengthening of PSC girders two options were proposed, namely strengthening with bonded carbon fiber reinforce polymer (CFRP) CarboDur plates and alternatively by external prestressing. The proposition of using bonded CFRP plate, although is being used widely over globe is a costly proposition, relatively new in the field with its efficacy yet to be proved in India. Therefore, external prestressing was favoured.

- For repair to bottom weak concrete to PSC longitudinal girders and RCC cross-girder, several repair materials were considered and finally polymer modified latex concrete has been used.
- Urgent action was suggested for the protective works mentioned above.

25.4.6 Repair Work

The report was accepted by the railway and the urgent work for the 3-girders was successfully carried out, under extremely strenuous conditions due to limited line block, presence of HT traction lines in very close proximity of working area, busy railway traffic under the bridge and heavy vehicular traffic on the bridge. The repair of the balance girders was completed subsequently.

25.4.7 Conclusion

The job has focused attention on:

- The aspects of quality control
- Importance of proper grouting and sheathing
- Importance of selection of suitable cross-section of structures prone to atmospheric pollution
- The need to select proper detailing of proposed structural components, so that concrete could flow properly to all parts
- The need for proper installation of bearings as per manufacturer's recommendation
- The need for regular periodic inspection of concrete bridges also.

Rehabilitation of bridges is a complex job and needs multidisciplinary approach for tackling various activities, such as investigations, interpretation, analytical studies, sensitivity analysis, formulation of repair plan and meticulous implementation during actual execution. Diverse situations are encountered and continuous interaction between design consultant and executing agency is required.

26 Repair/Rehabilitation of Marine Structures

Utilization of superplasticizers and other chemical admixtures in rehabilitation of marine structure along with a case study.

26.1.1 Introduction

Although cement concrete has been used for over 70 years, some of its inherent disadvantages like time lapsed in setting, low tensile strength, large drying shrinkage and low chemical resistance particularly in marine environment leave the civil engineer and concrete technologist in a dilemma. Each passing day is bringing more and more understanding about concrete and its behavior. It has now been more or less established that the deterioration of concrete and RCC structure along with its mechanism largely revolves around corrosion of steel, carbonation of concrete and its porosity and permeability.

With the tremendous advancement made in the construction technology and ease in manufacture of concrete at site, the use of cement concrete assumed a lot of importance. Concrete has become a material of choice, where strength, durability, impact, abrasion and fire resistance is required. To modify the properties of cement, mortar and concrete for diversified use, mixing of additives is one of the easiest options. Varieties of chemical admixtures and superplasticizers are now available as additives for altering the properties of both wet and hardened concrete. They are being used to modify the properties of concrete in such a way so as to make it more suitable for the work in hand, achieving economy and for saving energy. This paper brings out the necessity of using plasticizers and other chemical admixtures, for the repair and rehabilitation work with particular reference to marine structure. This has been explained through a case study.

26.1.2 Purpose for Use of Admixtures

The admixtures are mainly used for modification of fresh and or hardened cement concrete, mortar and grout. Among various types of admixtures, plasticizers/superplasticizers are used specifically for improving the workability of concrete and its flow characteristics without addition of extra water. In the process the following properties also get improved:

- Concrete becomes dense, strong and durable due to reduction in water to cement ratio for a given workability.
- Concrete can be made to flow for placement by tremie or by pump at reduced cement content.

- Concrete can be transported over a long distance without much loss of slump.
- Concrete becomes more volumetrically stable so that the spacing of construction joints can be increased as they are a vulnerable part of a structure.

26.1.3 Classification of Admixtures

The admixtures can be classified according to type of material constituting the admixtures or to the characteristic effects of their use.

26.1.4 Utilization of Admixtures in Repair/Rehabilitation of Marine Structure

The use of construction chemicals for repair is desirable for following areas of repair of marine facilities:

- Preventing corrosion of embedded steel, which is the prime cause for damage of any structure.
- Application of corrosion resistant barrier film on the reinforcement steel (which inhibit further corrosion).
- Application of useful bond coat which assures good bonding between old and new concrete.
- Rendering a strong, passive carbonation resistant polymer modified/polymer concrete cover of proper generic, wherever necessary.
- Applying a protective seal coat on the entire surface to avoid further aggression of chemicals.

The ordinary classical methods of repair/rehabilitation like replastering, jacketing and guniting, etc. are often seen not to offer satisfactory results only because of the fact that it does not adequately take care of the above aspects. It is often found that in traditional repairs, the same problem may soon recur. Investigations have brought out that the repair/rehabilitation measures in such cases failed due to two reasons:

- Corrosion of steel not being arrested
- Bonding between old and new concrete being inadequate.

For repair/rehabilitation of any structure, uses of construction chemicals are essential for obtaining long-term results. The various monomers and polymers are available in the market. Corrosion of embedded steel is the prime cause for damages to any marine structure. It is like a cancer, which progress at a slow pace and if neglected or not attended to in time, may spread over a large area and cause extensive deterioration of structural elements. It may even lead to catastrophic structural failure, in absence of timely remedial measures. The causes which create conducive conditions to accelerate the rate of corrosion in a marine atmosphere are as follows:

- Wave action (alternate wetting and drying processes)
- Presence of harmful gases in the surrounding/atmosphere
- Contact with acids/fumes
- Salts in the air
- High humidity (70%)
- Cracking due to applied loads

The methodology to be adopted for repair depends on the extent of damage to the particular structure. When corrosion of the steel has not started but carbonation of concrete has taken place to the depth of the reinforcement, coating can be applied to prevent carbonation. Depending upon the severity of carbonation, coating of polymer or epoxy resins or polymer modified mortar concrete can provide adequate protection. Such coating also stops penetration of chlorides and other deleterious elements. Whenever corrosion has started, the restoration techniques employed should depend on the extent of damage to the concrete and or steel. In this case the following guidelines are recommended:

- Removing all unsound concrete and exposing the reinforcing steel
- Removing the rust by appropriate methods, such as sand blasting, brushing or by applying liquid rust removers
- Anchorages of reinforcement by providing shear connectors

- Applying a polymer coating or an epoxy based coating to the steel reinforcement where exposed
- Restoration of concrete to the original surface level
- Injection of polymer modified slurry or epoxy of suitable grade to fill up the pores, internal cracks or segregation
- Applying a protective coating

High strength, resilient materials which have high resistance to attack from chlorides and sulfates are normally used for repair materials. The polymer modified concrete (PMC) is one such material, which is commonly used for repair/rehabilitation. It has the following advantages:

- High compressive strength at an early age
- High flexural and tensile strength
- Good water tightness
- High chemical resistance
- Good adhesion

The filling of crack is normally done by epoxy grouting. The epoxy grouting systems have high mechanical strength; they obtain strength in only a few hours and are resilient in nature. The epoxy systems are immune to sulfate and chloride attacks and are impermeable. They have got high compressive and tensile strength also. Since epoxy grouting system can be injected into even hairline cracks, effective repairs can be carried out with them.

For enhancing the life of structure, a coating of polymer based paint is recommended. The coating will provide additional resistance against ingression of air and harmful chlorides, sulfates into the concrete. It shall thus protect the entire structure from corrosion and spalling. However, these coatings need to be reapplied periodically, normally every five years or so, depending on type of product being used.

26.1.5 Precautions in the Use of Admixtures

Presently in India there are no standard codes of practice for utilization of various construction chemicals and their use is restricted by supplier to a select few who are acquainted with the product. In the absence of proper codes, careful attention must be given to following the manufacturer's instructions. An admixture should be employed only after appropriate evaluation of its effects, preferably by use with the particular materials at the intended site. Such an evaluation is particularly important in a tropical climate, as the chemicals have mainly been developed for use in colder regions. They therefore require long, term testing under tropical climate. Besides, such an evaluation is also needed when special types of cement are specified and where more than one admixture to be used.

Admixtures that modify the properties of fresh concrete may cause problems through early stiffening or delayed setting times. The cause of abnormal setting behavior should be determined through studies of how much the hydration of cement is effected by the admixtures. Early stiffening is often caused by changes in the rate of reaction between tricalcium aluminates and sulfate phases. Undue retardation can be caused by an overdose of admixtures, which can adversely affect the hydration of tricalcium silicate.

26.1.6 Case Study

The author has successfully used various admixtures for major rehabilitation of wharves along with other repair works of caissons, dry dock, etc. costing ₹ 50 million (approx.) at Bombay (Mumbai). The two wharves (contiguous to each other) having width of 16.45 m and total length of 500 m was constructed during 1957 and 1967 (one having length of 350 m and other of 150 m). The damages observed were excessive spalling of concrete, formation of wide cracks, excessive corrosion, falling of fenders and shearing off of piles, etc. In general, it was observed that various structural elements after these have been exposed/chipped off showed much more deterioration than was observed at the time of the site survey.

The two wharves were composed of fender portals, bollard portals, curtain wall (on one side of wharf only), wailings (connecting piles at the bottom of the bracing level), deck slab (37.5 cm thick) and wearing coat (17.5 cm thick) over deck slab. The general arrangement of the fenders and bollards in the wharf is shown in Fig. 26.1. Apart from a visual and dimensional survey, various other tests were carried out to diagnose distresses. Cover meter test, half-cell potential survey, ultrasonic pulse velocity test, petrography test, core test, water permeability test, chloride sulfate test and porosity test were carried out. In addition, analytical assessment of residual strength was also made. The structure being very old, design calculation/details were not available. Using a computer, a structural analysis was carried out that showed that all structural members were adequately designed based on earlier codal provisions. Because of the extent of deterioration repairs were required to wharf portals frames, curtain wall, wailings, deck slab, wearing coat and fenders. The repair methodology adopted for different types of defect are as follows:

All cracks at various locations were filled with epoxy grouting under pressure. The stages involved in grouting are:

- Cutting 'V' grooves with pneumatic cutter
- Drilling of holes at intervals
- Fixing of entry ports
- Application of epoxy sealant
- Cleaning of grooves with compressed air
- Injection of epoxy under pressure
- Cutting of entry ports
- Grinding of sealant

Spalling of concrete was repaired by various methods depending on locations. For soffit of slabs and fenders where the depth of deterioration was less than 75 mm, epoxy mortars were used. The sequences of activities involved were as follows:

- Square cutting the boundary of damaged concrete.

- Chipping the spalled surface by pneumatic chipper.
- Removing rust from existing reinforcement by wire brushing.
- Welding of additional reinforcement after carrying out anticorrosive treatment and an epoxy coating.
- Removing dust from surface of concrete and reinforcement by compressed air, drying of the surface before applying coating of epoxy.
- Applying an epoxy coat within an interval of 15 to 30 min after mixing of resin and hardener and quartz sand, in the proportion of 100, 50 and 800 by weight.
- Air curing, for 48 hrs.

For piles, beams, wailings, bracings and fenders/soffit of slabs (having a depth of deterioration greater than 75 mm), polymer modified concrete (PMC) were used. The stages involved in application of PMC is similar to that of epoxy mortar, except for mix of PMC and curing requirement. In PMC the materials used were cement, quartz sand, polymer and water in the proportion of 100, 400, 15 and 30 by weight, for the damaged portion of structural member having shallow depth up to 75 mm. However, where damaged portion is considerable and having thickness more than 75 mm, cement, quartz sand, polymer water and aggregate (20 mm and down grade) was used, in the proportion of 100, 200, 15, 35 and 400 by weight. Further, in case of PMC, water curing is also required. One pile which was sheared off (because of which the jetty was declared nonoperational) was replaced by two piles after cutting the deck slab, lifting the slab by jacks and casting two new piles under water by bored *cast in situ* method and thereby restoring the structure to its original position as shown in Fig. 26.3. The condition of a typical fender before and after repair is shown in Figs 26.1 and 26.2. The classification of admixures used is shown in Table 26.1.

Fig. 26.1: Condition of typical fender before repair

Fig. 26.2: Typical fender (placed over deck slab) after repair

Fig. 26.3: Repaired/rehabilitated structure

Table 26.1: Classification of admixtures

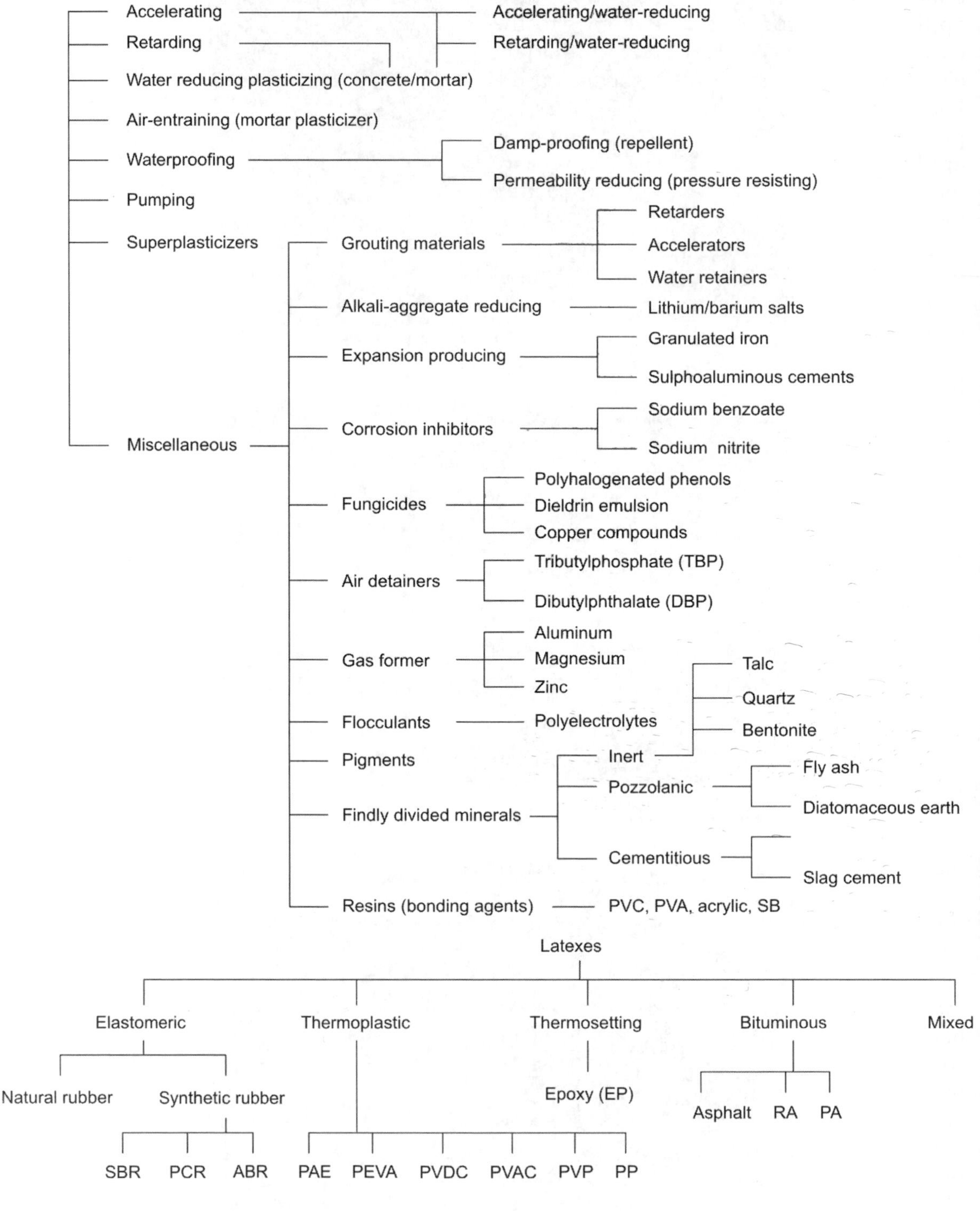

In other locations only a wearing coat, bonding coat of epoxy resin needed to be applied to the concrete. For repairs to the curtain wall (which is not a very important element of the structure) guniting was used. For enhancing the life of structure, a special penetrating thermosetting polymer was applied over the entire concrete structure including fender, after cleaning the structure by sand blasting.

The repair/rehabilitation work of the two wharves costing approximately ₹ 30 million carried out during 1993–94, appears to have stopped the deterioration and no further deterioration has been observed (by visual inspection) after about two years since the repair work has been carried out. It is expected that this work will extend the service life of the facility for 15 to 20 years, with normal periodical maintenance.

26.1.7 Conclusion

Corrosion of reinforcement is the main cause of distress in reinforced concrete marine structures. In India reinforced concrete structures mainly came into being about 3 decades ago. These structures were constructed without the use of any admixture. As the time passes there will be many more structures located in Indian coastal areas like Bombay (Mumbai), Goa, Cochin, Madras (Chennai), Vishakhapatnam and Calcutta which will come up for major rehabilitation work. This case study reveals that the use of admixtures is essential for the restoration of deteriorated marine concrete structures.

To modify the properties of concrete or mortar, a large number of admixtures have been tried and extensively used in other countries, to improve the qualities of concrete. In India, such effective concrete admixtures have only recently become available. But now a number of internationally known and time tested concrete admixtures are available all over India, but they must be tested under Indian tropical climate. World over, admixtures have been in use for over 20 years and their long term behavior patterns are known. The superiority of polymer modified mortars/concretes over normal mortars/concretes over normal mortars/concretes in rehabilitation field is established beyond doubt. In India, though the use of admixtures in the construction industry particularly in the rehabilitation field is growing, they have to have a set of standard and codes developed for them which will guide both the specifier and customer in their proper use. The admixtures/polymers should form a permanent part of rehabilitation scheme in the coming years.

Note: This paper was published/presented in ACI international conference held at Rome, Italy on 7 October, 1997 by the author.

The complete jetty work was examined almost after 20 years on 1 February, 2016 and was observed that the facility rehabilitated is still in serveable condition and no major repair will be required in near future.

26.2 REHABILITATION OF OIL JETTY AT NEW MANGALORE PORT

26.2.1 Introduction

The new Mangalore port is located at Panambur on the west coast of India, about 170 nautical miles south of Mormugao port and 190 nautical miles north of Cochin port. The project establishment of the construction of the port was set up in the year 1962. The port was declared as the ninth major port of India on 4 May, 1974 and was formally inaugurated on 11 January, 1975. The port trust board was formed under the major port trust Act on 1 April, 1980. Since then the port has been functioning as tenth major port trust. The port has been developed in stages. The details of the jetty are as follows:

- Constructed in four stages
- Construction started in 1968 (1st stage of development) at a cost of 45 crores
- Construction completed in May 74 (1st stage)
- 2nd stage during 1977–1980, at a cost of ₹ 39 crores

- 3rd stage during 1994–1997, at a cost of ₹ 238 crores
- 4th stage development took place during 1999.
- In 4th stage construction of additional one crude oil jetty, multiuser jetty and a wharf wall was carried out.

26.2.2 The Existing Jetty Design Aspects

The details are as follows:

- RCC open jetty type structure with raker *cast in situ* bored piles.
- The design vessel of jetty was with dead wt. of 30,000 ton; length 200 m; breadth 12 m; draft 9.15 m; top level of deck slab 3.66 m.
- The structural details of jetty: Breasting dolphins 10 m × 13 m – 2 nos each supported on 16 nos of 1200 mm diameter RCC vertical piles. Mooring dolphins of 8 m × 8 m – 4 nos each supported on 10 nos of 800 mm diameter and 6 nos of 750 mm diameter RCC raker piles. Jetty head – 25 m × 12 m –18 nos of 750 mm diameter RCC vertical piles.

  Approach trestle 100 m × 406 m – 7 bents, each bent supported by 2 nos 800 mm diameter RCC vertical piles.
- Concrete was of M25 for piles, beam/slab M20 and for pile cap – M35.

26.2.3 Planning Aspects of Upgrading and Rehabilitation

The planning aspects are as follows:

- The main jetty was constructed in 1975; over the years, shown signs of deterioration in the piles of breasting dolphins and service platform.
- On investigation, it was observed that the concrete had deteriorated mainly for about 6 m length below the pile cap, in most of the piles.
- For deteriorated piles, it was decided to provide four piles on either side of the breasting dolphins, i.e. eight piles of 1200 mm diameter integrated into existing breasting dolphin, by extension of breasting dolphin cap.

- Existing fenders were replaced by rubber fenders of 1000 H × 1000 L with a 1.5 m × 1.5 m steel frontal frame.
- Existing piles were rehabilitated by removal of damaged concrete; replacement of reinforcement wherever necessary and rebuilding of piles to original section by use of PMC/PMM (polymer modified concrete/polymer modified mortar).
- Rehabilitation of superstructure by use of epoxy mortar, wherever necessary. The cracks in the concrete were repaired by epoxy injection grouting.

26.2.4 Construction Aspects of Upgradation of the Jetty

The following are the major constructional aspects during upgradation:

- A water-tight specially fabricated mild steel chamber in two halves was used, which had outside diameter of 3 m.
- Total height of shell was 6.5 m.
- Two semicircular halves were moved to the position of pile to be repaired with help of floating pontoon.
- Steel shell was kept in position with 25 mm steel ropes hung from the breasting dolphin/service platform with assistance of turnbuckles and chain pulley blocks. Two halves were brought together and joined with bolt and nuts from top and bottom.
- Neoprene rubber seal was used in between two halves of working chamber and around pile, to prevent entry of sea water inside chamber.
- The fitting and fixing of chambers was carried out with help of divers.
- After making chamber water tight, sea water inside chamber was pumped out.
- Any entry of water into chamber was arrested by use of epoxy seal/M-seal, etc.
- Dewatering process continued till the time repair continued.
- After dewatering, surface of pile was inspected and surface preparation was

carried out by using high pressure water jet. Epoxy resin injection grouting used for minor cracks.

- Spalled concrete was removed with the help of pneumatic hammer.
- Wherever reinforcement was corroded, the same was replaced by new reinforcement.
- Steel primer was applied on steel surfaces, after cleaning the reinforcement to give protective coating.
- Deteriorated concrete of the piles was replaced with PMM/PMC.
- Surface of pile was cleaned with degreasing solution followed by water.

Similarly, superstructure repair was carried out by using epoxy mortar/polymer modified concrete. The epoxy mortar proportion was of 1:0.55:9.3 of resin, hardener and quartz sand.

26.2.5 Summary

The original jetty was constructed during 1975 and the piles and superstructure were showing signs of deterioration.

- The bearing dolphins were upgraded by installation of the additional three piles on either side.
- The service platform was extended by the installation of one additional pile on either side.
- The existing structure of the breasting dolphin and service platform was integrated with the new construction by exposing the reinforcement of the existing structure and provision of additional steel reinforcement.
- The piles were repaired underwater in dry by the use of watertight MS chambers assisted by divers.
- Wherever reinforcement was severely corroded, the same were replaced by new reinforcement steel by welding.
- Polymer modified mortar was used for replacing the damaged concrete in the piles.
- The superstructure concrete was repaired using epoxy mortar.

- Fine cracks in the superstructure were repaired by epoxy resin grout.
- The rehabilitation and the repair works were carried out successfully.

26.3 REPAIR/REHABILITATION OF MARINE STRUCTURES: KANDLA PORT

26.3.1 Introduction

The Kandla port was hit by a cyclone on 9 June, 1998 with wind velocity exceeding 180 kmph. The tidal waves up to 11.0 metres hit the Kandla port area, leaving over 1500 people dead and causing instant destruction of crores of rupees worth of property. The cyclone also caused devastation in the entire Kandla port area, massive annihilation, destruction of structures and communication network, power line and extensive damage to the port and its buildings and structures. The extensive damage caused to the port and the structures was initially estimated at around 76 crores.

The entire Kutch district especially Kandla was cut off from the rest of the world. The psyche of the local people was rendered numb due to the shock and devastation caused by the cyclone. Clouds of gloom and despair prevailed everywhere. The extensive damages and consequent decommissioning of the Kandla port (KPT), the no. 1 port of India, was shocking news to the entire world.

The electronic media including international agencies like BBC, CNN, etc., flooded into Kandla and it was a challenge to the authorities to handle them in this small port town. More difficult was to arrange and manage the visits of VIPs and experts who included Prime Minister, Home Minister, Chief Minister, Governor and others. The day temperature was above 46 degree and there was no electricity and drop of water to drink.

26.3.2 Restoration Work

The following activities were carried out:

Crisis management: KPT authorities took immediate action of forming worked out the action plan for relief operations. A project monitoring cell, which was set up under the overall supervision of deputy chairman, worked incessantly towards monitoring and speeding up of the restoration process. The entire Kandla area was disinfected to prevent outbreak of epidemic. The damages to the infrastructure facilities were extensive. Due to tremendous velocity of the wind, as sheets had blown off and most of the cargo stored in go-downs had been destroyed. KPT undertook the repairs on a war footing in order to resume the cargo operations. The preliminary repair works were carried out on a war footing and the godowns and transit sheds were reallotted to port users within a short period. The walls and roof of the port buildings were severely damaged and so its repair works were taken out in a phased manner.

Cargo handling was finally resumed on the 25 June with the berthing of a vessel at one birth, which was least damaged.

The repair of cranes, oil jetties, oil pipelines, railway lines, marine operations and relief measures were undertaken simultaneously. The support of government and other agencies was taken. Further inspection by team of experts constituted had inspected the various marine structures damaged due to earthquake. Summary of their observations is as below:

- Three berth numbers (6, 7 and 8) have practically no damages and hence, can be commissioned for operation. The cracks/spalling of concrete noticed in some of the piles of these berths, which are minor and can be repaired as part of normal maintenance.
- Five berths (1 to 5) are very old and their circular cracks noticed in a number of hollow piles underneath and beams in the rear. These cracks are due to flexural rather than by shear. Further tests were suggested before recommending any details repairs. There are horizontal separations seen across the apron at the expansion joint locations. There is a longitudinal separation which seems to be running along the junction of the edge of the top slab of the quay and the transit shed. These are considered to be very serious.
- At the oil jetty (no. 1) cracks have been noticed in some of the piles and beams, which require extensive studies before taking up the repair works.
- The new berthing jetty (under construction) seen to have moved towards the water side by about 15 cm. The remedial measures needed for the jetty repairs, further analysis before finalizing the scheme suggested.
- Maintenance jetty suffered extensive damages to the piles, decks beams and fender blocks. The structure needs critical analysis before deciding on its future use.
- The bunder basin, the southern wharf suffered substantial damages beyond economic repair.
- Some warehouse in the cargo jetties, the exterior columns have suffered extensive damages between the lintel and first floor beam level. A proper repairing scheme has to be worked out, giving the construction sequence and methodology to support the load from the above.
- In respect of many buildings, which suffered extensive damages, the foundation condition would need to be checked up carefully before working out suitable remedial measures.
- The mudflow has created large subsidence and cracks in the roads, pavements and stack yard. As a remedial measure, it was suggested to fill the cracks with clean sand mixed with water to the maximum extent.

26.3.3 Technical Aspect of Damages

Liquefaction damage may take several different failure modes according to differences in structure and geotechnical condition, as brought out below:

- In pile foundations and caisson foundations, structure itself is damaged due to dynamic effect of surrounding liquefied soil and superstructures. They are sometimes greatly displaced by a lateral flow movement of liquefied sand. These foundation behaviors may invite severe superstructural damage like collapse of bridge girders or break joints between connected structures.

- In shallow foundation, superstructures resting on them tend to settle into liquefied sand layer, often unevenly and tumble down in the extreme, deteriorating their structural members, if they are not strong enough. This foundation slabs tend to break in the bending mode during the settlement, allowing the intrusion of liquefied soil inside the structures.
- In quay wall retaining structures, such as diaphragm walls and caisson walls are displaced offshore due to seismic inertia force and presumably due to liquefied solid pressure. This directly harms port structures like loading/unloading equipments. Liquefied backfill soils inevitably suffer large lateral deformation as well as vertical settlement, inflicting various structural damages to deep foundations and shallow foundations nearby.
- Embankments resting on liquefied soils settle, deform and fail in shear mode, deteriorating road pavements and railway tracks on them and endangering river banks. In road or railway embankments, differential settlements tend to concentrate at joint sections between different foundation conditions, such bridge abutments, concrete ducts, etc. Sometimes, embankment soil itself liquefies, if saturated, inflicting large deformations and even flow failure. Earth fill dams of cohesionless soils with insufficient compaction may suffer such failure mode, leading to disastrous dam failure during or after strong earthquakes.
- Underground or buried structures are often displaced upward relative to ambient liquefied soil, resulting in uplift of light structures and failures in their joints, if liquefied soil is displaced due to lateral flow, buried pipes or conduits in it are deformed in compression, extension or bending mode, leading to failure in joints or in other parts.

Methodology of remediation may be widely varied due to difference of structures, difference in failure modes, degree of importance of structures, degree of damages, site condition, etc.

26.3.4 Action Plan for Repair/Rehabilitation and Reconstruction

Various steps involved are as follows:
- Estimation of damages (nature and magnitudewise)
  - Using in-house expertise
  - Taking help of external expertise as required
- Prioritization (of the repairing, strengthening and retrofitting works)
  - Financial controls and availability of funds
  - Nature of damage and magnitude
- Coordination
  - Through reports, meetings, etc.
- In-principle approval for the restoration and refurbishment
- Estimating
  - With reference to cost
  - With reference to time frame
- Preparation of tender documents and invitation of tenders
- Award of works
- Monitoring of the implementation
- Completion of the work
- Inspection of the structures before commissioning.

26.3.5 Procedures for Repair of Piles, Pile Caps and Beams in the Jetty

The following procedures are followed:
- **Scaffolding:** Suitable scaffolding around the beams and the piles are to be erected to facilitate working around the beams over the piles of the jetty.

- **Surface preparation:** Cracked loosened concrete adhering to the reinforcements are removed by gentle chipping. The sound core concrete is thus exposed. Rust sticking to the reinforcement and concrete are removed by sand blasting or any other established methods.
- **Anchoring:** Holes with 20 mm diameter and 300 mm length are drilled in the concrete piles beams using electrically operated hammer-drilling machines without causing much vibration so as not to disturb the sound concrete. Clean the holes, wash the same and the concrete surface with potable water and allow it to dry. 10–12 mm diameter MS rods 300 mm long are cut to required length are kept ready for anchoring. They are inserted into the holes partially (250 mm) and anchored using polyester anchor resin. These rods projecting out of the holes are used for trying fresh reinforcement.
- **Placing of reinforcement:** After examining carefully the loss of diameter due to corrosion suitable mild steel (Fe 250) reinforcement is replaced and properly tied to the existing reinforcement.
- **Priming the reinforcement:** Reinforcement steel should be cleaned to a bright condition, paying particular attention to the back of the exposed steel bars making sure that the surface is totally free from any rust. An unbroken coat anticorrosive zinc reach primer is applied using a suitable brush over the reinforcing bars. If there is any doubt to having achieved an unbroken coating, a second application should be made as soon as first coat is fully dry.
- **Application of microconcrete:** Apply microconcrete consisting of 12 mm down-graded granite stone chips and approved nonshrink cement based grout (rendroc RG or equivalent product) material in the ratio of 1:1 by weight and correct quantity of potable water for an average thickness of 100 mm into a formwork/shuttering suitably positioned around the damaged area.
- **Protective coating:** If required, the finished surface can be given a protective brush coating of epoxy paint.
- **Curing:** Immediately after removal of the formwork, the surface should be cured thoroughly or be spread with curing membrane.

26.3.6 Repair Scheme for Overhead Water Tank Columns

The following are the procedures followed:

- Chip the surface of the column to remove all plaster.
- Hack the surface of the column well, if not already done for plastering.
- Clean the surface of the column using wire brush to remove all the loose materials. Wash the surface with potable water.
- The new bars have to be anchored by drilling holes in the raft foundation of the depth of 300 mm.
- At beam and column intersection where full strips cannot be used, "U" hooks to be used appropriately.
- Additional bars introduced are bonded with concrete using polymer based epoxy.
- New longitudinal bars have to be tied using new column ties.
- The additional bars are provided only at the corners with the spacing of the corner bar and adjacent bar are less than 75 mm.
- Leak proof formwork should not deform or leak to pressure of microconcrete. The formwork shall be well oiled for easy removal. Proper supporting arrangements are to be made for keeping the shutter in correct line and length.
- Encasement is done using microconcrete rendroc RG(S) or equivalent product with 75% aggregate (washed and cleaned) by weight of size 6.4 mm and down size.
- It should be ensured that the clear cover to the new steel is 50 mm.
- The concrete shall be well cured for 10 days.

26.4 STRUCTURAL REASSESSMENT OF OFFSHORE PLATFORMS

26.4.1 Introduction

Offshore oil and gas production started at India in 1976 from Bombay high. Today there are about 190 platforms on Indian offshore, contributing about 70% of indigenous production of oil and gas of the country. Some of these platforms installed in late seventies, have already completed their design lives, but they are continued to be operated and are expected to serve for another 10–20 years. Most of these platforms are supported by 4-legged or 8-legged steel jacket structures, secured to sea bed by steel piles. During the service, these platform suffer from operational hazards, extreme storms, fatigue, corrosion, possible damages due to fire, explosion, dropped objects, ship collision, etc. Many a time, platforms are also subjected to additional loads due to operational requirements, such as additional clamp on wells, deck extension, etc. Further, the design criteria themselves have changed significantly in the last about 35 years. The structures need to be safe not only against collapse due to this additional loads or damages, but also against gross distortion, which could jeopardize safety equipment and escape route. In view of this, assessments of the integrity of the structure as well as mitigation measures are essential for the safety of the platform.

This chapter deals with preassessment of a typical unmanned platform in western offshore of India, suffering from minor damage like holes and dents in tabular members and suitable remedial measures suggested for rehabilitation of the platform. Ultimate strength analysis using pushover collapse method has been performed to determine the platform capacity in terms of reserve strength ratio (RSR). A reliability analysis has also been carried out accounting for uncertainties in wave height, current model, etc. thereby reliability index (a) has been ascertained and integration measures suggested.

26.4.2 Reassessment of Offshore Structures

Bombay high platforms are in operation since 1976. The earlier platforms have already over their design service lives. In spite of periodic maintenance of platforms, the operation of existing platforms beyond their intended service spans requires special attention. The need has arisen to look closely into various aspects of extending the service lives of platforms in accordance with operational requirements. In engineering practice, it is widely recognized that if an existing structure does not meet present day design standards, it does not mean that the structure is inadequate or not serviceable. The application of reduced criteria for reassessing existing facilities is also recognized in risk management literature and is justified on both cost-benefit and societal grounds. Analytical techniques provide a powerful tool in assessment of offshore platforms.

API-RP-2A guidelines stipulate four levels of analysis, viz. screening analysis/design basis check, design level or elastic level analysis, ultimate strength analysis, and probabilistic analysis in order of increasing complexity, for reassessment of the fitness-for-purpose of the platform. Ultimate strength analysis and probabilistic analysis are more complex in nature and carried out when the structure is not found to be safe in design level analysis.

26.4.3 Ultimate Strength Analysis

While carrying out the linear elastic level analysis, if any member or joint in the structure is found to be overstressed as compared to a specified laid down norms in the applicable design code, the entire structure should not be considered as unfit for purpose. The ultimate strength analysis in such cases may be carried out for assessment of actual capacity of platform with the aim to understand the true behavior of the structure/mechanism of failure.

The ultimate capacity of a platform is expressed in terms of ratio known as the reserve strength ratio (RSR).

RSR = Ultimate capacity of platform R_u/Min. reference level wave load S_r.

R_u is the capacity of the platform at the ultimate limit state (ULS). The ULS capacity thus represents the limit state of platform beyond which any increase in the loads would result in collapse of the platform. S_r is the nominal global lateral load on the platform, which will normally be the best estimate of the base shear on current accepted guidelines or requirements for design, such as given in the relevant section of AP-RP-2A. The RSR is thus an important index that attempts to reflect the platform's overload capacity compared to minimum current reference level force, AP-RP-2A, permits a RSR value as low as 0.8 for an unmanned and low consequence platform.

26.4.4 Probabilistic Analysis Level

In a probabilistic approach, the purpose of design is to ensure that the probability of a structure becoming unfit for use within its intended lifetime is "acceptable small". The uncertainties involved in analyzing the performance of a structure are accounted for in estimating the probability of failure or, its complement, the reliability of the structure, which is now widely accepted as a more rational measure of structural safety. The reliability of a structure is the ability to fulfill the design purpose for some specified time, which in case of offshore platforms is the design life of the jacket platform.

Complex structural systems, such as offshore jacket structures consist of a large number of structural elements, each of which can fail in a number of different ways. Because of the considerable degree of redundancy there are usually large numbers of possible combinations of element failures that can result in the failure of such structures. Depending on failure behavior of structural elements a number of limit states is identified for individual element and reliability analysis is first carried out for each of these limit states.

These element limit states are sometimes referred to as "failure elements". In a redundant structure, a sequence of element failures leading to the collapse of the structure constitutes a "failure mode of the structure". If the system probability of all failure modes are identified it will be computationally very expensive to include all these "cut sets" in system reliability evaluation. Therefore, a suitable strategy is required to identify only "stochastically dominant" failure modes that contribute most to the system probability of failure. The probability of failure of a structural element can be mathematically expressed by using standard formula.

Two methods of reliability analysis, viz. component or simplified reliability analysis, and system reliability analysis are commonly used. While in the former method, the whole structure is treated as one component for performing analysis, the latter method involves the identification of all the significant modes in which structure can fail taking due accounting of the redundancy and ultimate strength of the structure. Considerable research effort has been made in recent years on the application of system reliability methods to fixed offshore structures. These efforts have mainly concentrated on the reliability analysis of the structure under extreme.

26.4.5 Case Study On Reassessment of a Typical Process Cum Production Platform in Western Offshore of India

The assessed platform is the one of the oldest platforms (commissioned in 1976) located in the Western offshore of India. It is a process cum production platform. The platform is supported by a tabular 4-legged jacket structure and is secured to sea bed by 4 main piles and 2 skirt piles of 1.2 meter diameter each with a penetration of about 99 meter below the sea bed. Platform model details are as below:

Details of the Platform Model

Details	Case study platform
Type of platform	Production cum process
Water depth	72.54 m
Year of commissioning	1976
Number of joint nodes	254
Number of element beams	540
Total weight of decks	1000 MT
Total weight of jacket steel	1250 MT
Total weight of piles	282 MT
Number of legs	4 + 2
Height of jackets	78.18 m
Mud level	(–) 72.54 m
Number of well conductors	4
Number of risers	9
No of electrical conduit pipe	1

The assessment initiator in case of the above platform is following:

- The main damages reported in the inspection report, i.e. a hole in the member MA75 and one incomplete weld in member MG76.
- Change in marine growth criteria (earlier 38 mm thickness throughout the depth, presently –100 mm thickness from (+) 3 m to (–) 30 m and 50 mm thickness from (–) 30 m to mud line.
- Change in hydrodynamic coefficients, such as drag and inertia coefficients (Cd and Cm) values in the new addition of API code (earlier drag and inertia coefficients were 0.6 and 2 respectively. Present values of drag coefficient are 1.05 and 0.65 and inertia coefficient are 1.02 and 1.6 for rough and smooth members respectively).

In-place static analysis was performed for the platform by considering 5 wave directions typical to Indian offshore and the results clearly indicated that the platform does not conform to current API codal provisions as unity checks are exceeded in case of few members and joints. Based on this, it was concluded that the changes are significant and a formal reassessment was necessary. As a part of the reassessment, the following steps were carried out:

- Ultimate strength analysis
- Simplified reliability analysis.

After taking into account input parameters, modeling, and progressive collapse analysis and system reliability analysis were carried out. The results obtained are brought out in next section.

26.4.6 Results

Ultimate strength analysis has been performed for the platform structure where in it was seen that the jacket strength is not fully mobilized due to pile failure on account of lateral loads. The reserve strength ratio (RSR) obtained for the case of 32 mm thickness of pile is 1.16. Although this value is low, when compared with the values for modern day platforms (wherein PSR of 2 and above is expected), yet this can be considered to be acceptable when compared with API-RP2A recommended value of 0.8 for low environmental consequence unmanned platforms. The main reason for the low value is due to the base shear increasing from 9.28 to 15.45 Mn (66% increase) as a result of change in design criteria over the years.

The reliability analysis gave a reliability index and probability of failure of 2.27 and 0.011 for the case of 32 mm pile thickness. The probability of failure value is due to low PSR values obtained as mentioned above and the uncertainty in the estimation of ultimate capacity and wave load on the structure. Due to lack of statistical data on wave heights, the coefficient of variation of 0.15 was assumed.

26.4.7 Conclusions

The platform does not meet the current design criteria by performing linear elastic analysis. However, it is established that based on advanced levels of structural analysis, viz. ultimate strength analysis and reliability analysis, the platform has been assessed to be fit for continued use, till further inspection. It was recommended to carry out grouted clamp repair for the top horizontal framing member (MA75), which has a large hole, as a remedial measure to ensure the member strength in this particular case.

27
Repair/Rehabilitation of Monuments

27.1.1 Introduction

The contribution of structural engineering in the study and design of restoration has been very important in the last few decades. This is supplemented by new investigation instruments; most sophisticated monitoring networks, information systems to mathematical models, from special devices for use on site, new technology and techniques for repair. However, results of this large back up have been only partially successful. This is due to the unmethodical and often casual advances at the forefront of progress, the lack of interdisciplinary vision and insufficient cultural awareness. Thereby result has been investigations seldom inserted within a coherent global programme. The illusion of understanding the real behavior on the basis of mathematical models which were not completely reliable, interventions carried out using new technology and materials that were not only insufficiently tested to simulate the real conditions and thus ensure durability, but that also deeply altered the original conception.

Therefore, critical review of all matter relating to restoration of historic building and monuments should be focused to determine the following preliminary objectives:

- Safety level of monuments
- Whether or not interventions are needed
- Criteria of interventions and the appropriate technology to use, taking account of double requirement to alter as little as possible the original conception and to ensure safety and durability.

The evaluation of the safety levels corresponding to various possible interventions gives not only a measure of the improvement of the behavior, but also helps in the choice of the most appropriate criteria. The safety evaluation, however, is a very difficult task and cannot unfortunately, always be obtained following mathematical analysis. As a rule the process of arriving at a judgment is very complex and is achieved by an interlacing of objective and subjective aspects. There are three main routes to follow, i.e. observation of the reality, mathematical analysis and historical survey.

Observation of reality lies in the survey of the monument as it stands today, through the observation of the quality of materials, the crack and failure patterns, the foundation system, the ground morphology, etc. This knowledge can be supported by chemical and mechanical tests and by data recorded on a monitoring system in order to highlight the evolution of various phenomena. The mathematical analysis helps in evaluation of stress levels and deformations corresponding in different kinds of external and internal force systems. The selection of any effective measures should be based on a careful analysis of the structure and its materials, the exposure conditions and the mechanism.

Historical survey of the monument is an indispensable process, since history provides an experimental laboratory on real scale that we have yet to discover and decode by research, review and interpretation of historical documents. The main difficulty is that history is not written for structural engineering purposes and thus objectivity of the facts must be partially rebuilt through the subjective reinterpretation of the research.

In deciding criteria and technology for interventions, the historical values and the associated risks must be taken into account, because monuments are precious objects that must be respected and altered as little as possible. This statement can lead, however, to some contradictions as higher risks must sometime be accepted, to avoid or limit modifications to the original conception. These risks depend on one hand, on the minimum level of interventions that related to the incertitude of the safety judgment, can become insufficient, and on the other hand on a delay of the decision in order to carry out deeper and deeper investigations.

The possible reversibility of the intervention techniques is also a key factor on taking decision of intervention and the time limit for their realization. Nowadays, the cultural trend is for interventions to be reversible, that is to allow for the possibility of their removal. In principle this philosophy is correct, taking account of the fact that judgments are not always sufficiently reliable and it therefore seems useful to allow for the possibility of applying better techniques and materials that will become available. This type of philosophy was proposed in the restoration of Lord Jagannath temple at Puri, Orissa, where it was revealed that forces resulting due to corrosion of wrought iron steel dowels created cracks in the wall and corbels, which support the roof. Reinforcement was designed in order to limit deformability and thus damages that periodically affects the corbelled roof. This reinforcement consists of a stainless steel space frame with bolt jointed purlins placed below the corbels to give a secondary support to these cantilevered stones. It is important to note, however, that this philosophy must be accepted as a guideline rather than a compulsory method; situations where reversibility is neither possible or convenient are not frequent, as for example, the reinforcements of floors, the connections between walls and the strengthening of deteriorated masonry with appropriate grout injections.

Some of these above issues are critically addressed in this article with a case study of restoration of one of the oldest and finest monument of our country— Lord Jagannath temple at Puri.

27.1.2 Historical Background of Jagannath Temple

Lord Jagannath temple at Puri was constructed by King Chodaganga Deva in the first quarter of the 12th century on the remains of an earlier temple. It was dedicated to Lord Jagannath (Purushottama). The monument is standing on a high platform connected with the ground level by a flight of 22 steps (believed to be a part of its foundation). The edifice is quite massive and is a product of accumulated experience of the Orissa temple architecture and stands very close to the sea, only a mile away from the coast. To protect the temple from invasions by Sultans of Bengal, who had earlier attempted to loot the jewellery of Lord Jagannath, two massive prakaras (compound walls) capped with battlements were provided making the temple look like a fort. The temple complex inside the outer boundary wall (known as "Meghananda Pachira") measures roughly 202 m × 196 m and is approachable through four gates from the four sides. Each of these four gateways is an architectural marvel with pyramidal terraced roof embellished with figures from Hindu pantheon, dancers, drummers, etc. The inner boundary wall measures roughly 122 m × 85 m. The height of temple is about 70 m.

27.1.3 Structural Problems of the Main Temple and Need for Intervention

The construction of temple was done in ashlar stone masonry with blocks of khondalite (a local sandstone) laid in courses. For the construction, no mortar was used. Instead the stones were joined with the help of wrought iron U-shaped clamps or dowels and were supported one over another resulting in fascinating wall and corbelled roof, in shape of frustum of pyramid. The geometrical arrangement of the stones perfectly matches with the proven thesis of arches, where all elements are primarily subjected to compressive forces. The wall thickness of the main temple is about 6 m and the main temple has three floors, i.e. three corbelled roofs inside, which are being supported by huge wrought iron beams (about 4 cm × 4 cm solid section), spanning over about 10 m of length. The wall face externally has been plastered with about 7.5 cm thick coat of lime plaster, applied in nine distinct layers, while inside wall has thin coat of plaster up to a height of about 3 m.

The temple was declared as a centrally protected monument by the Archeological Survey of India in 1975. The deplastering of the main temple, which started from 1975, has brought to light an array of sculptures and other decorative details, minute designs and figures carved and cut in stone with such consummate skill that they create an illusion of wood or ivory carving. During the deplastering work, the conservation measures adopted by the ASI were resting of bulging masonry, replacement of damaged and missing architectural members, grouting of voids in the masonry, and repairing and strengthening the damaged beki and amalaka.

A calamitous event took place on June 14, 1990, while deplastering and subsequent conservation was in progress by the ASI. A piece of the stone from amalaka weighing about six tons fell from the western side of the shikhara. On close examination it was found that amalaka comprises several members bound to one another by iron clamps or dowels fixed in grooves in the stone surface. Plastering over these iron clamps has led to their deterioration and finally to the fall of stone block. With great difficulty the amalaka was restored and the problem rectified.

The investigation carried out inside the shikhara of the temple in 1991 brought to light the existence of two more floors inside the garbhagriha. The ground floor and first floor are connected by a flight of steps, but no passage is there to connect the second floor with first. Wide vents (45 cm × 50 cm) are there on the floor sides below the beki and second floor is approachable only through these. Each floor rests on the corbelled arches.

It was found that the condition of the inner walls and corbels required attention. Similarly, on the southern side of the temple major structural weakness were detected in the form of multiple cracks, missing architectural pieces, etc. Another unfortunate happening took place on 13 August 1992, when two pieces of corbelled stone from inside the southwest corner of the garbhagriha collapsed, inciting fresh panic about safety and preservation of the temple. Close examination revealed missing of 10 more pieces of corbel stones and the main reason to above was the lateral thrust from the 900 years old rusted wrought iron clamps, which were used in the temple construction for joining the stones together. In the second storey, broken corbels stones were found lying on the floor. Close examination revealed that 26 corbels had got detached and fallen. To prevent damage of the floor by further fall, a cushion of paddy was provisionally provided within the 10 m × 10 m (approx.) space of the second floor.

27.1.4 Intervention Schemes

The expert committee formed made certain recommendations. The salient features of the scheme are as follows:

- Replacement of missing corbels
- Replacement of rusted wrought iron clamps with stainless steel clamps
- Sealing of stone joints by a mortar comprising stone dust, cement, polymer (acrylic type) and nonshrink additives
- Grouting the inner core with polymer modified flexible, nonshrink cement grouts
- Anchoring the loose cantilevers and corbels stones with the help of 1.5–2.5 m long thread steel anchors, grouted with resins
- Providing stainless steel anchors with low viscous epoxy resins
- Providing a stainless steel space frame as a secondary defence to support the ground floor corbelled roof
- Lateral confinement of the entrance corner walls of each floor by stainless steel flats
- Improvement to the existing ventilation system by drilling appropriate holes through the ceiling of the first and second floor and provision of suitable ventilation duct in the top floor
- Provision of temporary support to the ceiling of the first and second floor in the form of tabular scaffolding system to monitor corbel movements and any associated deformation in those floors
- Desalination of the external facia stones by paper pulp technique
- Application of suitable chemical preservative to the facia stones (in place of methyl acrylate) and suitable biocide treatments.

To complete the work, a task force was formed to assist and guide the existing staff, who supervised the work. The work was executed in various phases keeping in mind the sanctity of original work.

27.1.5 Consolidation of Temple Wall and Corbels

The walls and corbels of the main temple had shown gross deformations with weakening of interconnections of structural elements, thereby endangering the very stability of the structures. This is primarily due to the ingress of rainwater through the weathered joints into the thick dry stone walls thereby leading to the rusting of iron dowels/crams. It has been proved with success that injection of polymeric grout into the pore structure of the masonry diminishes the splitting forces and at the same time increases the adhesion between the stones. Therefore, for structural stability and proper load distribution of the space frame and to prevent the water ingress into the core of the structure, it was felt necessary to take adequate conservation measures by sealing the joints, surface cracks and grouting the walls with a material which besides meeting the general requirement, would also be compatible with the structural behavior of the ashlar masonry structures. Considering the condition of the temple structure, its structural design and inherent weakness, results of the analytical study regarding the stability of the temple against earthquake forces and requirement for uniform load distribution from proposed steel space frame, it was essential that the injection grouts should meet the following main criteria:

- Should not be excessively brittle.
- Should be elastic having resilience and permit minor movements without fracturing or cracking.
- Should firmly adhere to the stone surface.
- Should be able to set without curing, without any deformation.

Polymer modified concrete (PMC) grout was experimentally found effective, meeting above requirements and was used accordingly.

27.1.6 Seismic Safety Assessment of the Temple

Puri region falls in seismic zone III, according to the available quantitative seismicity maps.

Instances of occurrence of earthquakes of magnitude around 5 on Richter scale have been recorded at about 100 to 150 km southeast of Puri. The analytical study has been carried out by the civil engineering department of IIT, Kharagpur to check the stability of the block structure of the main temple against earthquake forces. A mathematical model using two dimensional finite element analysis procedure of the block structure (as a fair assumption) was carried out, where main temple has been discretised into suitable block elements. The results of the analysis indicate that the temple structure will safely withstand forces developed due to seismic excitation up to 5 on Richter scale. The soil structure interaction effect could not be included in the above analysis due to nonavailability of the foundation details and soils bore-logs below the temple foundation. Future research work should focus on this aspect. Since the loosely joint stone blocks as in the case of massive structures built in ashlar masonry are less vulnerable to the earthquake induced vibrations than the solid structures, it was essential to see that the grout materials using injection techniques will not solidify the structure fully and alter its structural behavior completely. For this purpose, the use of PMC grouts was considered favorable, which do not alter the flexibility of a structural system to a great extent.

27.1.7 Treatment of the Distressed Outer Fascia

As already explained, the outer and inner facia of the temple core are made from khondalite stones. Khondalite rocks being of metamorphic sedimentary origin, with evidence of post magnetic action, is essentially heterogeneous with several types of inclusions. Subjected to the exposure of salt laden air, moisture, airborne and sand particles, etc., both in isolation and combination, the surface of the temple structure and the sculptural features have shown extensive distress. The anxiety is two-fold structural strength and sculptural

features. There is evidence of large leaching from the surface layer and ingress of moistures of deep inside the walls, not only along the unmortared joints, but also within the main (individual) rocks bodies. The processes have been further accelerated due to high humidity and temperature. This has significantly increased the porosity of the stones, thereby hastening the process of deterioration. Evidently, the formation of the water soluble salts has taken place. These soluble salts have subsequently leached out from the interior of the stone structures enhancing their porosity to several folds. All this appears to explain a typical poker faced features on the surface of the stone. Evidently there is loss of strength of the rocks progressively with time aging. As the some of the choice (khondalite) is not very uniform, samples collected from different locations exhibited different stages of distress. It is highly relevant and appropriate to emphasize here that this nonuniformity will dictate the need for difference in treatment mode for different stones even when the basic chemical system is the same. Moreover, it should be also recognized that this process of progressive deterioration gets accelerated with time. The objectives of the chemical treatment to outer surface were therefore planned considering all these aspects.

27.1.8 Conclusion

In case of historical buildings and monuments made of masonry, it is important to have sound assessment and diagnosis of structural behavior prior to any permanent intervention. Since masonry is a nonelastic, anisotropic and nonhomogeneous material, it is generally impossible to predict the behavior of masonry buildings from the mechanical properties of the constituent materials as the input parameters in calculation procedures which are based on the assumptions of the theory of elasticity. It is especially so in case of ashlar masonry buildings having thick dry stone walls where the core has relatively loose stone blocks and the outer veneering structural members are held together by iron clamps/ dowels. These dowels in due course of time have rusted into powder forms leaving voids in between stone blocks where they once were. These types of structural problems are mainly tackled by structural consolidation with the help of grouting as was done for the Puri temple. When dealing with the repair by grouting of masonry structures, the improvement of their carrying capacity has certainly to be considered as one of the most important aims of the intervention. Nevertheless the long-term performance and durability of repairs has also to be taken into account; repairing should also ensure an increase of the service life of the structure. The effectiveness of the grout from the point of view of bond strength, improvement of the mechanical strength of decayed masonry and durability of injected grout to thermal cycles should be experimentally verified before application. The main problem associated with grouting is determinations of pressure distribution as well as the analysis of stress applied to the mass of masonry during injection. These are in general difficult to be solved theoretically. The difficulty stands on the inability of geometrical presentation of voids system, the particularity of the system of voids itself, as a flow duct, the great variability of flow conditions and further more on the number of parameters involved (kind and composition of grout, kind and geometry of masonry, etc.). The control of applied stress is made through practical limit state checks in accordance with the type and geometry of masonry in the point of pressure application. In case of Puri temple restoration work, two checks of blowing stone out of masonry face and check of horizontal swelling of masonry were made to control the grouting process.

28

Repair/Rehabilitation of Irrigation Structures

Hirakud dam is one of the first major projects taken up after independence in India. The work started in 1948 and was completed in 1956. It is a composite structure of earth, and concrete and masonry, built across river Mahanadi about 15 km upstream of Sambalpur town as shown in Fig. 28.1. The main dam has an overall length of 4.8 km, and is the longest among major projects. Besides the main dam, there are earthern dykes on both sides on plug low saddles. The main dam is a combination of concrete gravity masonry and earth fill sections. The two concrete sections incorporate the spillways and are 402 m and 256 m along the left and right arm of Mahanadi river separated by 2,300 m long earth dam in the middle. On the left bank there is earth dam on 1,353 m length and nonoverflow section is 110 m long and there is a power dam of 149 m length, with transition of 134 m in between. The maximum height of the dam above deepest foundation is 59 m. The average height is about 40 m as shown in Fig. 28.2. The dam is founded on granite rock with banks of schistose rocks. The gross and live storages are 8.1 km^3 and 5.8 km^3 respectively. The left spillway (402.3 m long) has 21 bays of ogee spillway (51 ft wide × 20 ft high) controlled by radial gates and 40 sluices (12 ft wide × 20.34 ft high) controlled by vertical gates as shown in Fig. 28.3. The right spillway has 13 bays and 24 sluices of the same size. The total spillway capacity is 42,450 cu m/sec out of which 40% can pass over the spillway and 60% through sluice gates. Such large sluice capacity is a unique feature of Hirakud dam. The sluices have to be operated first before opening the crest gates.

The first filling of the dam took place in July 1956. Some horizontal cracks were noticed in the right operation gallery as well as in the foundation gallery. Subsequently cracks also appeared in the sluice gate shafts and sluice barrels. The cracks were predominantly horizontal. The cracks were treated off and on

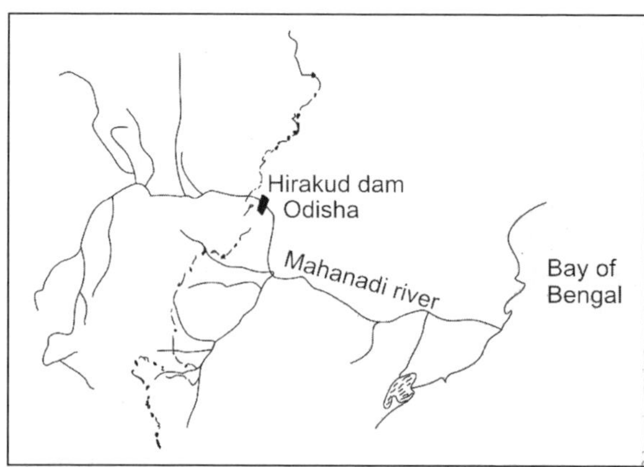

Fig. 28.1: Location of Hirakud dam

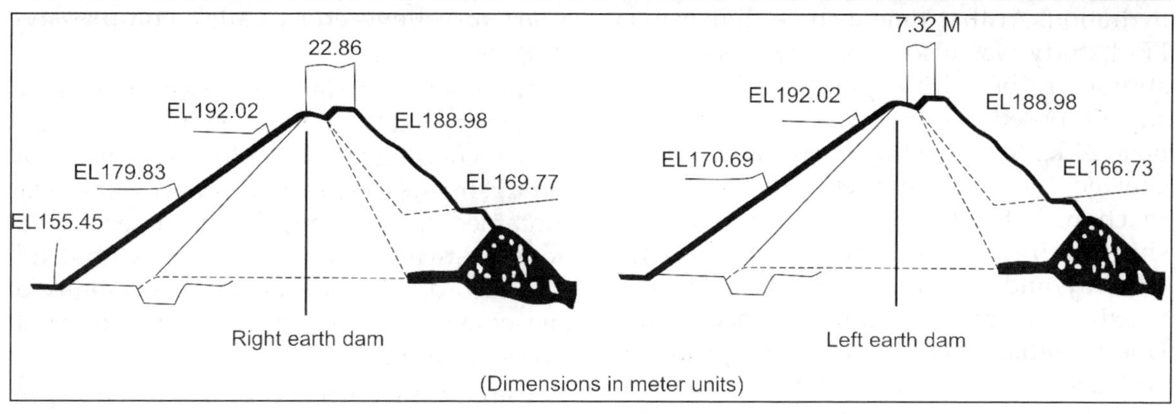

Fig. 28.2: Cross-sections of right and left earth dam

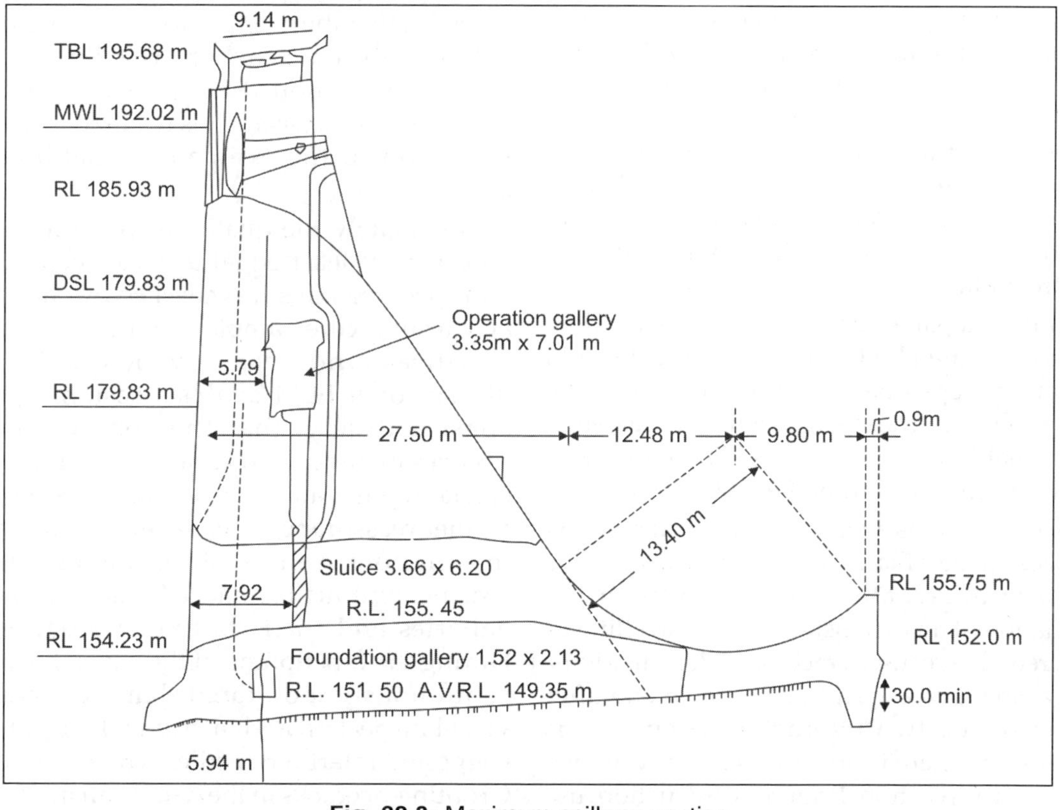

Fig. 28.3: Maximum spillway section

with cement grouting from inside. Practical manifestations of distress were experienced in 1974 in the form of difficulty in lowering hoist beams of two of the sluice gates and snapping of bolts along the guide rails in the gate shafts. In 1980, when two of the radial gates of the right spillway got jammed, it was found that the vertical face had deflected by 2 to 3 mm towards the bay. Other signs of distress were also observed.

In 1981, a committee chaired by Dr YK Murthy was appointed to study and advise on the causes of cracking and remedial actions. The committee concluded that the initial cause of cracking in the right spillway was high tensile stresses developed due to heat of hydration. At that time a three dimensional FEM study was also done which found the stresses to be within permissible stresses superimposed indicated a possible tension of up to 24 kg/cm². Testing of concrete samples showed conclusive evidence of alkali-silica reaction. Extensive use was made of river shingles in making concrete for the right spillway, and the shingles were found to have reactive minerals including strained quartz. Due to initial cracking, water found ingress and facilitated alkali-aggregate reaction with caused swelling of concrete resulting in cracks, jamming of gates, etc.

The committee recommended grouting of cracks, sealing of upstream face with epoxy mortar and instrumentation for monitoring. Subsequently underwater mapping of cracks on the upstream face of right spillway was done by the Indian Navy. In 1987 M/s Fressynet and a Netherlands company were allotted the work of underwater sealing of upstream face.

In 1990, a panel chaired by Sri GN Tandon, was constituted by Government of Odisha to review the repair being undertaken, access the safety of the dam and suggest further remedial and rehabilitation measures. The panel was active up to 1996 and had a last visit in 1999.

ASR activity is a result of chemical reaction between free alkali in cement and reactive silica in aggregates, which results in the formation of an expansive gel and swelling of concrete. This causes cracks and deformations. It is a slow process and comes to an end after either of the two interacting chemicals is exhausted. There is no known remedy, which can stop the reaction. Prior to construction, use of nonreactive aggregates, low alkali cement (> 0.60% equivalent Na_2O) and admixtures of pozzolan were suggested as precautions. The phenomenon has been noticed on several dams in USA and Europe as also at Rihand dam in India. However, instances of functional disability are rare in a few cases closed joints between blocks of concrete dams were saved

apart to relieve cross valley compressive stresses.

The cracks on right spillway of Hirakud dam exceeded 2,110 m in length in 1990 and were increasing at about 2 to 3% per year. On the upstream face, most of the cracking was contained in an 8 m high band between elevations 163 m and 171 m (Fig. 28.4). Construction records did indicate some possibility of fast concreting and possible high thermal stress in this bell.

The cracks were however shallow in-depth, and mostly less than 5 mm in width. In its very first meeting, the panel reassured Government of Orissa that the eventuality of a catastrophic failure of the dam could be ruled out and no risk to the continued useful performance of the dam was foreseen. This resulted in a sharp reduction in the estimated rehabilitation costs.

Fortunately the quality of concrete in the dam was at least equal to if not better than that specified. This was checked by taking out and testing core samples, and measurement of ultrasound wave velocity. A two-dimensional FEM analysis assuming very conservatively, that the concrete up to upstream face of operation gallery was cracked, gave adequate factor of safety and further reassurance. The panel recommended installation of more strain meters, stress meters, joint meters and tell take, etc. in the galleries and their continued monitoring. During the functioning of the panel, the panel continuously monitored dam behavior by visual inspection and instrument output, and suggested interim remedies where necessary. Grouting of cracks in the dam from inside was ruled out as it was considered unnecessary and could on the other hand prolong ASR activity by introduction of cement grout. The panel considered that the costly underwater sealing of surface cracks was a palliative which had no permanent bearing on the safety of the dam and suggested that this could be avoided.

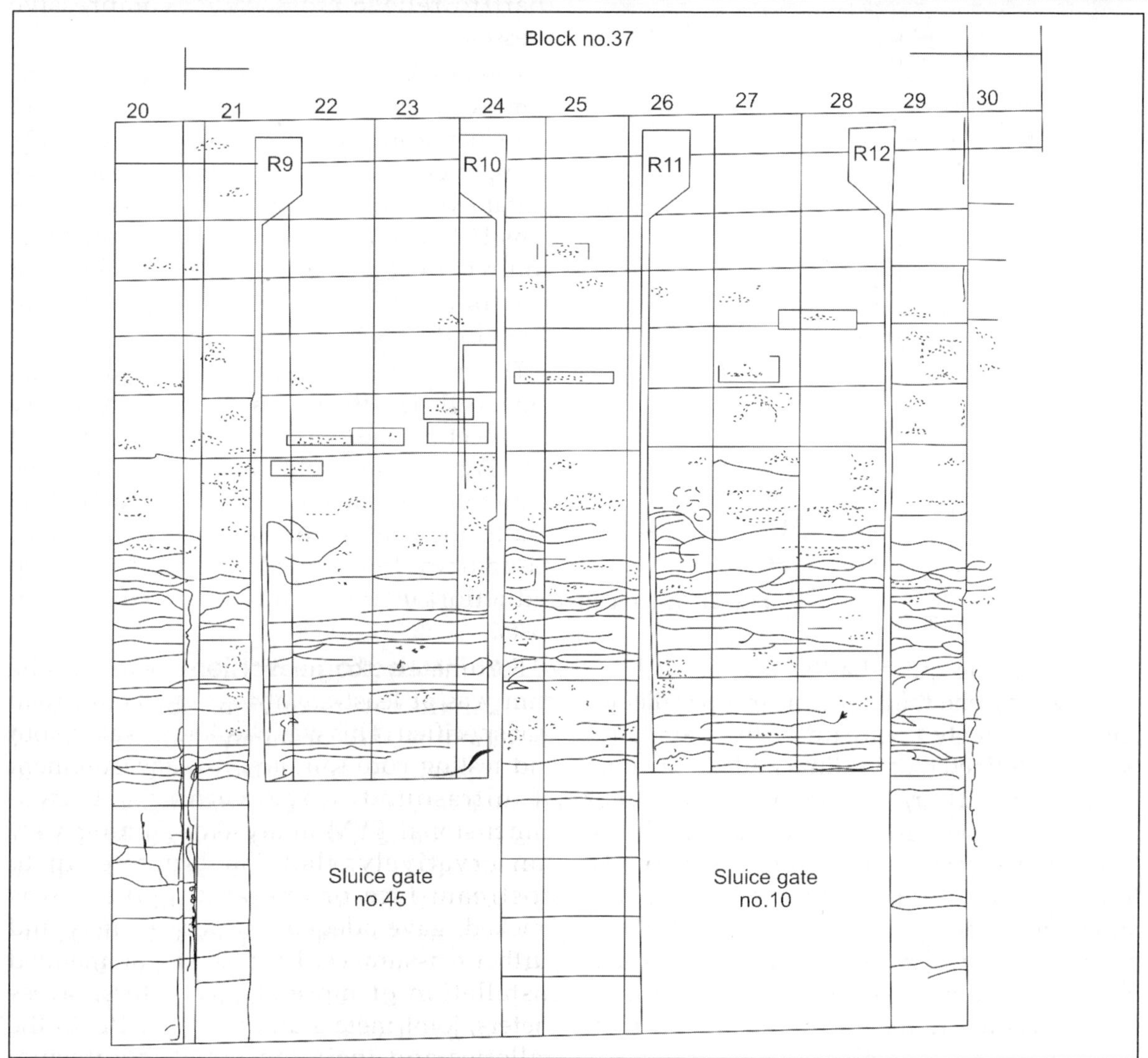

Fig. 28.4: Typical cracks on u/s face

28.2 LAR DAM (IRAN)

The problem in functioning was reported for about two years between 1985 and 1987, with the leakage problem of Lar dam as UNDP consultant.

Lar dam is located about 100 km north east of Tehran, immediately downstream of the confluence of Lar and Delichay rivers, which discharge into the Caspian sea. It is embankment dam 105 m high above river bed with a crest length of 1,170 m. The objectives of the dam included irrigation, hydropower and augmentation of water supply to greater Tehran. The consulting engineers for the project were Sir Alexander Gibbs and Partners (UK) and major contractor Impreglio SPA of Italy and Tessa Co. of Iran. The gross storage capacity at FRL of 2,531 m was 960 mm^3. Average annual inflow is 430 mm^3.

28.2.1 Geology

The right abutment is in Lar limestone formation which has been subjected to tectonic

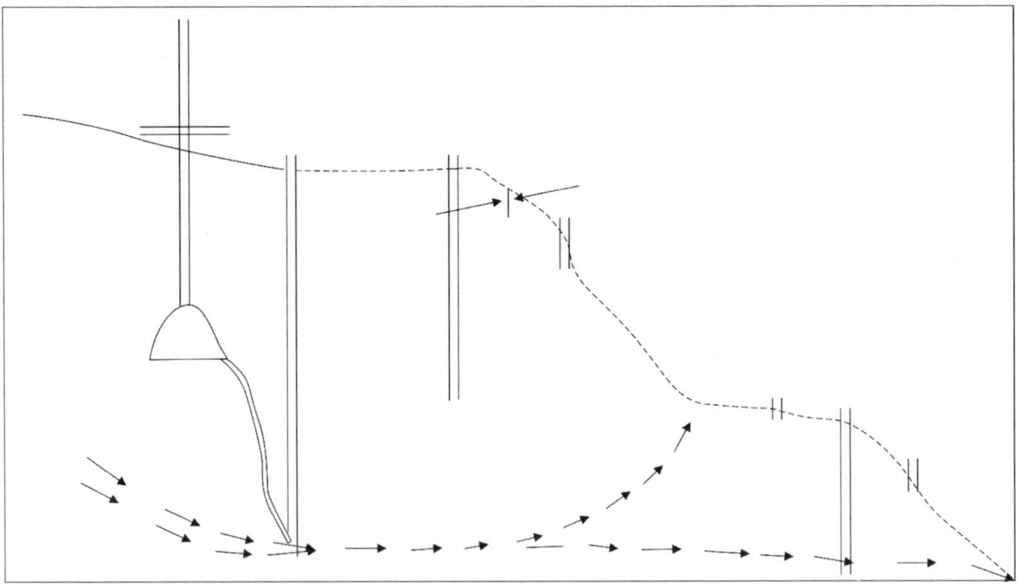

Fig. 28.5: Lar dam piezometric level

activity and is quite badly shattered. It is traversed by one fault zone nearly parallel to and immediately upstream of the dam, and two well defined longitudinal fault zones, one in the river bed buried under bed deposit and the other on the right abutment about 600 m from the valley face. The karstic (cavernous limestone) features are evidenced by cavities and springs and one huge cavern about 300 m below the full reservoir level. The name karst is derived from karst mountains in former Yugoslavia where these conditions were first studied by Karl Terzaghi.

The left abutment is composed by damav and (volcanic mountain at site) lavas, old take deposit and igneous detrital rock. In the bed, there are old lake deposits to a depth of almost 300 m. The permeabilities of the top sands and gravels are 10^{-2} to 10^{-4} cm/sec and lower silts and clays from 10^{-7} to 10^{-9} cm/sec. The upper sands and gravels are supposed to be cutoff by double diaphragm walls while grout curtain has been extended between and below the walls. Subsequent measurements of piezometric heads upstream and downstream of the curtain show almost total ineffectiveness of the partial vertical cutoff.

28.2.2 The Leakage Problem

The karstic nature of right abutment was identified by the consultants but they failed to estimate its gravity and to provide adequate counter measures. From 1980 when storage was first started, the reservoir could not fill beyond eln, 2500.31 m (up to 1987), 31 m below FRL with storage of about 380 mm^3.

This was due to heavy leakage from the reservoir, substantiated by water balance studies and direct evidence of increase in discharge of two springs, Lar and Haraz discharging into the river about 7 to 11 km downstream of the dam site. Their discharge prior to storage behind Lar dam was about 0.5 m^3/s which increased to a range of 5 to 10 m^3/s after construction of the dam. The correlation of their discharges to reservoir levels and, tracer observations clearly establish direct linkage (Fig. 28.6).

The problem was to try to establish the leakage paths from geological exploration and piezometric data and to devise grouting plans to curtail the leakage. The discovery of the huge cavern no.2 with a cross-sectional area of about 1,125 m^2 and observation of drilling mud introduced into the cavern emerging

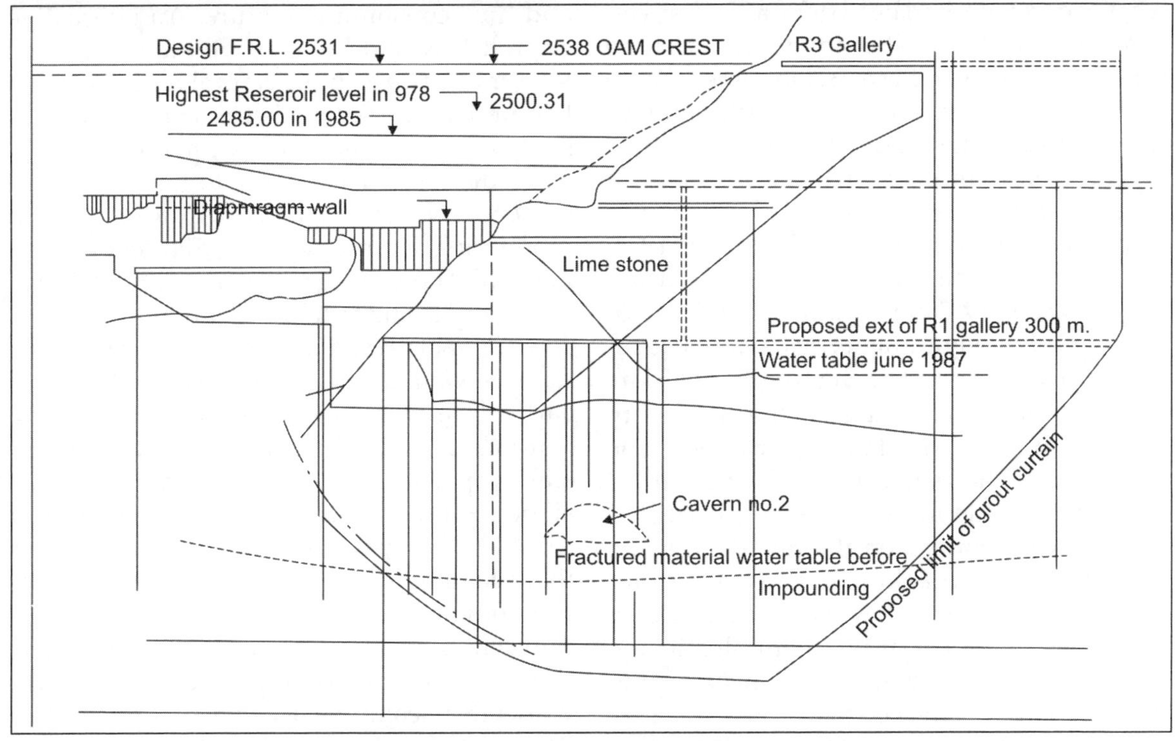

Fig. 28.6: Lar dam (Iran) leakage control measures

from Lar and Haraz springs had given hope that the main leakage was perhaps through this cavern and could be stopped once the cavern was plugged. Accordingly, elaborate work was taken up. First the cavern was filled with gravel poured through bore holes and then a planned grouting of the gravel was done. However, prior velocity measurements by diffusion methods and discovery of badly crushed zones with high grout intake further to the right of and lower than the cavern belied these hopes. In spite of very extensive grouting involving huge expenditure, there was only a small curtailment of leakage by summer of 1987. According to last available information, the remaining leakage had not been stopped. The reservoir fulfils only partially the first objective of irrigation.

28.3 WILLINGDON DAM (TAMIL NADU)

Willingdon dam is located on a small river Venganoor Odai in Cuddalore district of Tamil Nadu, about 240 km south of Chennai. It is fed by a supply channel from river Vellar besides intercepting the small run off from Venganoor and another small stream. The dam was constructed during 1913–1923 and has a command area of 11,200 ha.

The dam has a maximum height of 17.2 m, an average height of 13 m, and a length of 4,023 m. It has a storage capacity of 73 mM^3 of water. The dam rests on 4 to 5 deep soft clay below which there is hard stratum. The designer did not seem to have taken advantage to available experience at the time. There was, of course, no theoretical analysis method available. Even the construction supervision seems to have been lax. The entire section except for a 2 m thick cover of sandy clay on the downstream side was built of clayey soil of CH classification. There was no zoning and no drainage arrangement for the dam section. There is no record of methodical compaction. Even the slopes were not uniform the upstream slope was an average of 2H: 1V, but fluctuated somewhat and was flatter whenever repairs had been done after distress, from time to time. The downstream slope varied more, being as flat as 3:1 in some reaches as steep as 1.5:1 in others. The upstream face was pitched with boulder stones.

Old reports indicate that the problem of slippage of upstream slope started immediately after construction of the dam, from 1924. Subsequent slippages were reported during 1971, 1990 and 1993 mostly after drawdown of the reservoir. Longitudinal cracks were also reported at the dam crest in length of about 1.5 km.

The problem was referred to the dam safety review panel by the state government. The first visit of the panel in which the author joined was made in 1994. Several distressed reaches were noticed on the upstream slope mainly in the reaches 1509–1810 m and 2031–2313 m. On the downstream slope also there was evidence of incipient slip in the reach 2050–2500 m with subsidence varying from 15–45 cm at the crest.

The panel considered that as a matter of fact the dam section was unsafe in its entire length of over 4000 m, and a permanent upgradation would require substantial rehabilitation work for the entire dam to bring it to an acceptable standard of safety. This would be a very costly and time consuming venture. An alternative approach would be to rehabilitate the more vulnerable zones in the central reaches which had undergone severe distress in the past and still posed a risk. Such approach would upgrade the structure only to limited degree of security subject to continuous careful monitoring and prompt remedial action when needed. The state government opted to act according to the latter alternative.

The panel prescribed a series of tests for the dam as well as the foundation material, as earlier tests were inadequate and did not give consistent results. Reliable information about soil properties is necessary to devise stable slopes. For immediate relief for the distressed reach on the upstream side it was suggested that a loading rock toe to be provided with its top above lowest water level to be built by dumping stones under water. Above the rock, previous fill was to be placed on old earth work after cleaning it of all vegetation to bring it to a uniform slope of 2.5:1. A 1.5 m deep toe drain discharging downstream of spill section was also suggested. In order to improve drainage of the existing clay embankment vertical, sand filled trench drains, 10 m apart were also proposed Fig. 28.7.

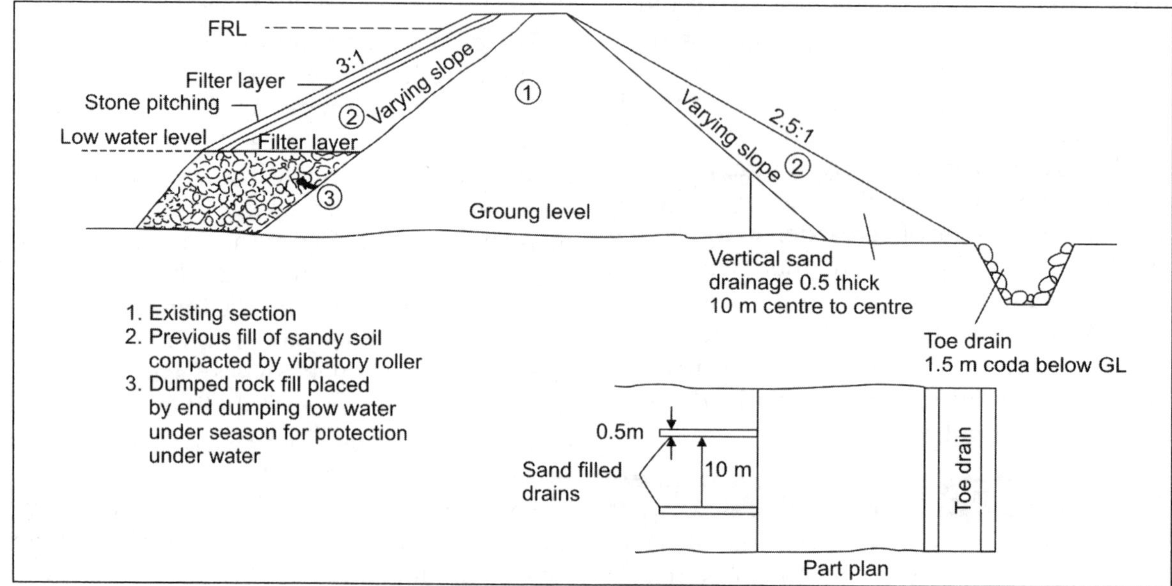

Fig. 28.7: Illustrative section after rehabilitation

These recommendations were implemented at site except for vertical trenches on the downstream slope. During 1996 and 1997 when reservoir was filled, there was further slippage of the downstream slope in the distressed reach resulting in depression of up to 2.5 m at the crest. Subsequently when the reservoir was lowered, there was distress in the upstream reach, from RD 2025 to 2113 (88 m long) with wide crack and pushing out of the dumped stone platform. Evidently the soil strength in the dam, and more so in the foundation was even lower than the adjudged by inspection and limited test results at the time of earlier recommendations. Restoration work was again recommended on both slopes. The downstream slope was further flattened to 3.5:1 with berms and provision was made for a rock toe, drainage filter and toe drain. Additional rock toe was recommended on the upstream slope (Fig. 28.8).

During execution of the work the upstream slope slipped further in a part of the reach but became stable after attaining an average slope of 4:1. It has remained more or less stable since then. During 1999 the project engineers proposed construction of a parallel dam, 50 m downstream of the existing dam in a length of about 450 m. The panel, however, did not favour this costly proposal and opined that with the measures recommended earlier, the existing section could be stabilized. A further setback occurred on the downstream side during September 2001, when excavation near the downstream toe was done for placement of the rock toe. A slip with a vertical face of almost 2 m development involving almost the full crest width. Even after placement of rock toe, slow slippage and upheaval continued in a reach of about 85 m. On earlier recommendation of the panel some relief wells had been constructed. Most of them had got choked and some got tilted in 2001. Construction of additional and more reliable relief wells and a few clay quick lime columns was also recommended. It was recommended that restoration of crest of the dam be delayed and carried out very gradually with continuous monitoring. However, even with this measure, the gradual slippage continued even till March 2002, when it seemed to stop. The panel was of the view that further exploration and testing was necessary before a restoration design could be finalized.

28.4 MASSINGIR DAM (MOZAMBIQUE)

Massingir dam in south Mozambique is located on elephants river, a tributary of the Limpopo river system. The project was conceived as a multipurpose storage project

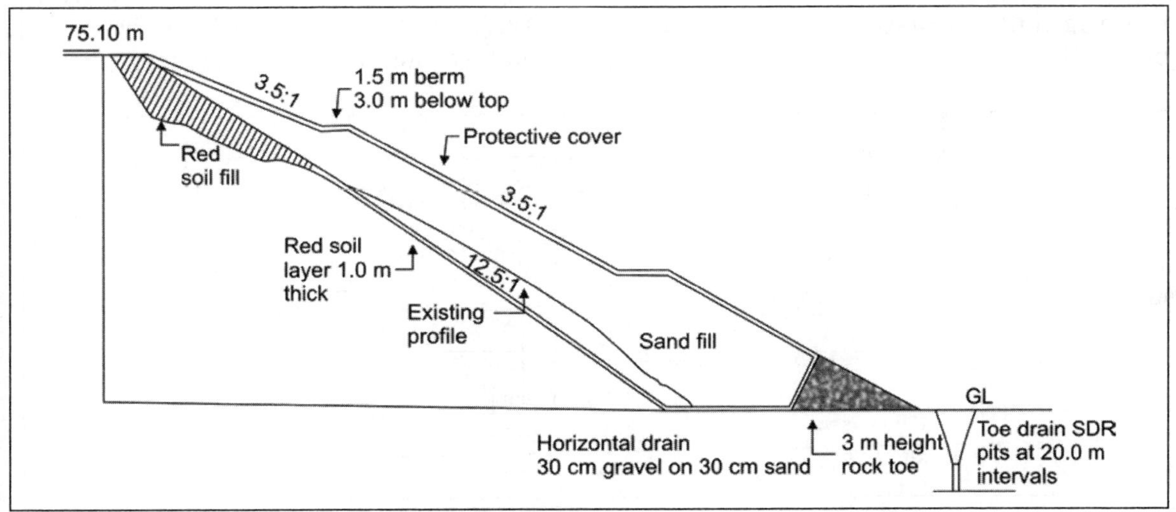

Fig. 28.8: Downstream slope treatment

with the primary objective of development of irrigation and additional benefits of hydropower and flood control.

The design was originally done by COBA, a Portuguese consulting firm. The main construction contractors were also portuguese except for RADIO who was awarded drilling and grouting work. The composite dam is 4.629 km long with a maximum height of 48 m in the deepest portion of the valley. The project was started in 1972 under portuguese administration. Independence was achieved in 1975 after which the work was resumed and completed in 1977. Problems of seepage and high residual water pressures in previous strata downstream of the dam particularly in the right embankment reach manifested themselves in the first year of filling itself and continued in subsequent years. Against the design full reservoir level of 125.0 m the reservoir could not be filled beyond 114.0 m due to distress and danger signals. These included boil formation between chainags 380 and 410 m, piping 500 m downstream of the toe from chainage 1,200 m and a worrying blowout across chainage 400 in 1979. There was noticeable seepage and piezometers showed unsustainable water pressure in the foundations.

Subsequently the filling was restricted to elevation 110.0 m. Thus only a fraction of the originally anticipated benefits could be realized. After a review by African Development Bank, WAPCOS (India) were entrusted with the work of preparing a feasibility report for the rehabilitation of the dam in October 1988. Sometime later expert from India was invited to advice on the problem. The right bank dam in which these phenomena were prominent rests on clay layer of an average thickness of 8 m underlain by highly pervious alluvium which in turn rests on relatively pervious blowout and piping failure (Fig. 28.9).

Studies indicated that at full level eln. 125.0 m the total seepage discharge through the dam foundation would be 2 m^3/s . This would not be a loss as it will be picked up at the diversion weir for irrigation canal located downstream. Further power generation to installed capacity of 60 MV could still be done. Thus, the solution has to aim at ensuring the safety of the dam, not curtailing seepage flow. Therefore, the more costly and cumbersome solution of providing a positive cut off was excluded after comparative study. The solution recommended consisted of providing 85 relief wells with an effective diameter of 50 cm at 25 m centre to centre and also counter berm at the downstream toe. The top elevation of the counter berm was at elevation 108 m sloping down at 5:1 to elevation 101.0 m at which relief wells followed by a toe drain were provided. With these remedial measures, it was analytically checked so that safety of dam could be ensured as shown in Fig. 28.10.

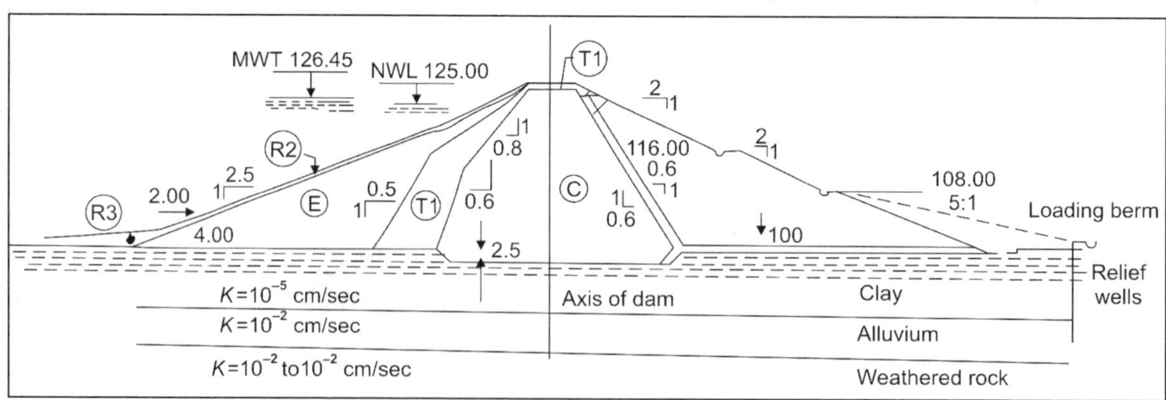

Fig. 28.9: Massingir dam

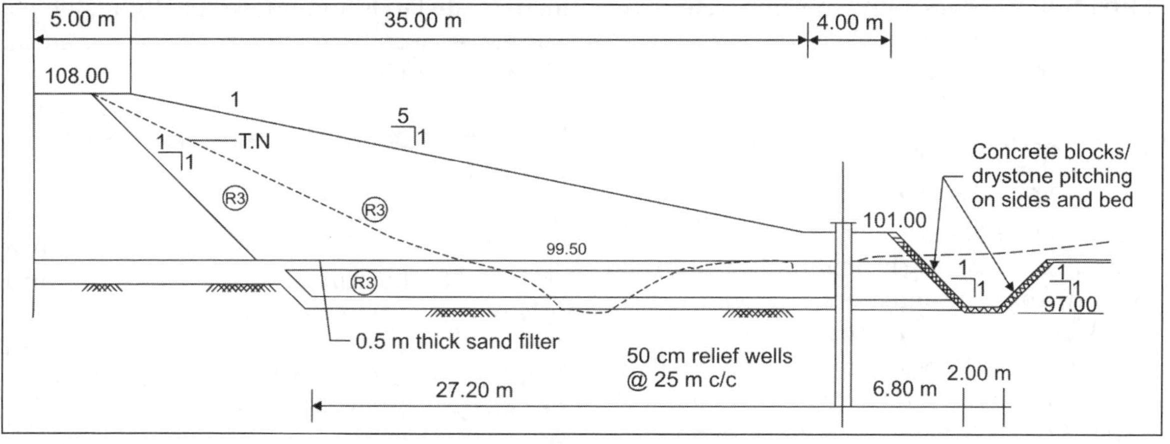

Fig. 28.10: Massingir dam remedial measures

28.5 UPPER INDRAVATI HYDRO-ELECTRIC PROJECT

28.5.1 The Project and the Problem

The Upper Indravati Hydroelectric Project (UIHEP) is a major multipurpose project in the State of Orissa. It has a reservoir of 110 sq. km spread formed by 4 dams and 8 dykes. The primary objectives of the project are: generation of 600 MW of electricity and provision of irrigation facilities for 1.28 lakh hectare of land in Kalahandi district. Water for power generation is taken from the reservoir by an intake structure near the upper reach of the reservoir. The water passes through a 7 m dia. and 3934 m long head race tunnel designed for 210 m^3/s discharges. At surge tank, the tunnel is bifurcated into two pressure tunnels of 5.25 m diameter each, leading to a valve house. Four steel penstocks of 3.5 m dia. each descend from the valve house to 4 turbines each of 150 MW capacities.

The four lines of penstocks (numbered from right to left) are encased in 7 RCC anchor blocks and supported on 38 saddle supports. Thereafter the four penstocks diverge at different angles and are encased in RCC anchor block no. 8 and thereafter meet 4 generating units. Out of the four penstock lines, line no. 1 and 2 were tested in 1998. It was noticed that a crack occurred at all the

anchor blocks tested, in longitudinal direction along the crown of the pipe where the concrete cover was thinnest. The test pressure was taken 1.33 times the hydrostatic pressure. It was also seen that the crack appeared even when the test pressure was less than the internal hydrostatic pressure. In some cases, the crack was not noticed initially but was subsequently notice at comparatively low pressures. Interestingly the crack occurred on all the anchor blocks including anchor number 8 and beyond it, along the axis of the penstock, such that the alignment of the penstock could be seen clearly though the penstock was buried in anchor block concrete.

28.5.2 Design of Anchor Blocks and Penstock Pipe

A check on the design of penstock liner showed that the plate thickness was sufficient to withstand the hoop stresses at different locations. The anchor blocks were designed by the designer for the conditions: (a) to prevent the pipeline from sliding down, (b) to control the direction of expansion, (c) to resist the unbalanced hydrostatic forces at the change of direction of the pipeline, and (d) to prevent movement on account of water hammer pressures. However, the supports were not tested on vibrations, bending stresses at bends and flexural stresses in lower portion at first filling. It was found that concrete cover was not checked for stresses on the basis of theory of elasticity.

28.5.3 Possible Causes of Cracking

Tension Around Penstock Periphery

Calculation on the basis of theory of elasticity assuming the opening in an infinite continuum showed, that the tensile stress valid from 144.44 kg/cm^2 at the crown of opening to 92.59 kg/cm^2 at the edge of the concrete cover; very high stresses. However, calculations on the basis of an opening as a sluice (USBR TM No. 457) showed tension of the order of 52 kg/cm^2 at the crown. In both these studies steel liner was not considered. Therefore, f.e.m studies were done with (a) steel pipe surrounded with concrete and (b) sans steel penstock. Tension due to temperature was only 12.6 kg/cm^2 and due to vertical loads only 0.2 kg/cm^2. Thus it was seen that tensile stress varied from 64.87 to 26.25 kg/cm^2 which could not be resisted by the concrete cover especially at the top where the cover was thinnest.

Inadequate and Wrongly Placed Reinforcement

It was seen that the reinforcement was inadequate to take tension in concrete of the order of 729 t and the concrete was bound to crack. Also the bars on the top were taken vertically on either side and anchored in concrete. Thus on any radial plane in the concrete around lower half of the penstock, there were any cracks in the lower portion. It was necessary in such a case to provide hold down straps, as is recommended by E Mosonyi.

Inadequate Lapping in the Reinforcement

Length of the reinforcement bars around the pipe was of the order of 12 m. Thus, the reinforcement had two bars with a lap which was only 45 d. Calculation as per IS:456-1978 showed that the lap should have been 70 d.

Inadequate lap is an undesirable thing, which can cause serious problem as happened at Rihand dam UP.

Distribution of Reinforcement Along the Vertical

Actually the whole tension zone should have been divided in four parts and separate bar provided for the full radial thickness; whereas only one bar was provided.

Fracture Mechanism Stresses at Crack Tip: Propagation of Crack

The penstock pipe was in direct contact of the concrete. The design drawing showed that no padding of any sort was provided between steel pipe and concrete. Had this been done the dilation in pipe due to high hydrostatic pressure would have been absorbed by the sealing around the pipe and there would have been less tension and consequently less chances of cracking. Since this was not there, the high tension was transmitted to the concrete mass, as was clearly shown by f.e.m studies.

Due to application of repeated loads or due to combination of loads and environmental attacks, a crack can grow with time. The longer the crack the higher is the stress concentration induced by it.

The study was carried out and based on this study, certain rehabilitation measures were suggested as brought out below.

Rehabilitation Measures

The following treatment was recommended by us and was carried out in the field.

- Clean about 1.2 m wide strip 0.6 m on either side of the crack by scrubbing and cleaning properly.
- Spread at least 2 mm thick epoxy adhesive (Lapex standard) on the clean surface.
- This painting treatment should also be given on vertical sides of Abs where crack was visible but in a width of about 0.5 m.
- Spread expanded metal sheet, about 25 mm above the paint, over the painted

surface with centre line along the crack. If expanded metal is not available 8 mm dia bars mesh placed 75 mm c/c be provided.

- Place 75 mm thick epoxy concrete to enter below the expanded metal sheet and over it. The ends of the concrete to be tapered at both ends.

  Resin concrete was to be made of sand and gravel mix with epoxy as binding material. The binder requirement is about 250 kg/m^3 (mix proportion 1:6 to 1:8). The concrete to be vibrated or pressed or vibro-pressed. This concrete would give flexural strength of about 100 kg/cm^2.

- Paint the top surface of resin epoxy concrete by epoxy paint again.

28.6 REMEDIAL MEASURES FOR THE GATES OF HIRAKUD DAM

28.6.1 Introduction

Hirakud dam project is built across river Mahanadi at about 15 km upstream of Sambalpur town in the state of Odisha. This happens to be the first post-independence multipurpose river valley project in India. Besides irrigation, power generation the project provided flood protection to 9500 sq km of delta area in the district of Cuttack and Puri.

Hirakud dam intercepts 83400 sq km of Mahanadi catchments. The reservoir has live storage of 5375 m. cum with gross storage of 7189 m. cum (as per revised figure of 1988).

Other silent features of the project are as follows:

A. Dam and Reservoir

Top dam level	RL 195.680 M	(R.L. 642 ft)
FRL/MWL	RL 192.024 M	(RL 630 ft)
Dead storage level	RL 179.830 M	(RL 590 ft)

Storage capacity in mecum (MSc ft.)	Original	Revised (1988)
Gross	8136 (6.60)	7189 (5.83)
Live	5818 (4.72)	5375 (4.36)
Dead	2318 (1.88)	1814 (1.47)

B. Spillway

Spillway capacity: 42459 cumecs (15 lac cusecs)

Crest level: RL 185.928 (RL 610 ft)

Size of sluice: 3.658 × 6.20 m (12 × 2034 ft)

No. of sluices: 64 (40 on left and 24 on right)

Sill of sluices: RL 155.448 m (RL 510 ft)

No. of crest gates: 15.54 × 6.10 m (51 × 20 ft)

Type-solid gravity with ogee profile and ski jump bucket.

28.6.2 Brief Description

The fixed wheel gates for the under sluice and the emergency gates were supplied by MAN of Germany including the Cranes to operate them. The radial crest gates were supplied by Jessop of India along with the rope drum hoists. 64 under sluice (40 on left and 24 on right spillway), of size (3.66 m) × (6.2 m) (height) are provided.

The gates are operated from an operating gallery at El 169.8 Mts. by using EOT Cranes (8 Nos) of 50 T capacity provided in this gallery (5 in operating gallery in left spillway and three in operating gallery in right spillway). The gates are designed for a water head of 33.53 Mtrs. At centre of gate (corresponding to FSL 192.07 m). The gates are upstream sealing type. Apparently this type of arrangement has been chosen as the operating gallery is located below the FSL, and also to reduce down pull and consequent reduction in hoist capacity.

Each gate is made in two parts which are connected by butt straps at ends. The units can be separated by removing the bolts from butt straps for maintenance and repairs and for taking off the gate through the gallery. Each gate is provided with 16 wheels of 620 mm dia. Plain cylindrical pin bearings are provided for the wheels.

The sealing is provided by bulb type stainless steel cladded seals in the sides and top and flat rubber seals at the bottom. Distress on the gates was also noticed which caused some operation and maintenance problems.

To carry out an overall assessment, the services of Mr Narendra Singh was taken in 1991. He submitted a report outlaying the procedure or revamping the gate systems. There was a second inspection by the German expert from MAN (original supplier) in 1994. He also stressed on the need of repairing the gates and laid emphasis on the repair of the cranes and hoists.

The major problems in the under sluice gates were as follows:

- Nonclosure of the gates with its own dead weight. The last part of the gate closure was affected by tacking the gates from the top in most of the gates.
- Pitting of skin plates, horizontal girders and other structural components of the gate leaf.
- Pitting of embedded parts and seal seats.
- Alignment of track guide and seal seats.

Nonclosure of the gate was attributed mainly due to the problems of the wheels. The wheel system of Hirakud was different from that of the wheels that are being adopted for the gates with the recent designs. The fixed wheel gates that are being currently designed are fitted with self-aligned roller bearings in the wheels. Wheels are accommodated inside the vertical girders. The wheel system at Hirakund has the following arrangement.

- The axles rested on two vertical girders and cantilevered out.
- The wheels were outside the vertical girder.
- The wheels rolled on cylindrical pins. The pins acted like needles of the needle bearing with the wheel inner race acting as the outer race of the bearing and axle surface acting as the inner race of the bearing.
- The axles had provision of eccentricity for imparting seal compression and a seat for provision of oil seals on the inner side of the wheel. The outer side had no oil seals or 'O' rings to check water ingress.

After the first gate was taken out of the slot, the following observations were made:

- The wheel inner bores were not concentric. Over the years of operation, the inner bores had become oval.
- The axles had become oval.
- The axles at the bottom of the gates had undergone severe pitting because of water ingress. The seal seats were severely pitted for which these axles had to be rejected. Any type of restoration by welding would have made the axle prone to cracking under load.
- Most of the bolts of wheel cover had been damaged. It was essential to replace them as it could not be opened by spanners.
- The greases in the wheel had been spoiled because of excessive water ingress. The seals were damaged. The grease nipples were damaged.

Two committees were set up to regulate the repairs of Hirakud dam. The Supervisory Committee comprising of SE Hirakud dam, EE main dam, E.E. HDRD, SE mechanical, EE mechanical and the high level technical committee (HLTC) comprising E-in-C (water resources), CE (Hirakud), CE (design) and CE (mechanical). As per the recommenda-tion of these committees the following works were taken up.

- The inner race of wheels were undercut and deposited with hard facing electrode, so as to achieve a BHN of 300. The inner race was ground finished so that least resistance to rolling of the pin was achieved.
- The axles were heated to a temperature of 800°C and water quenched to achieve a BHN of 300. The portion of the axle on which the pins rolled was ground finished. So that, it offered least rolling friction. Adequate tolerance was worked out so that the pins of inner race of the wheel and outer race of the axle did not interfere with each other to ensure smooth running of the wheel.

- The wheel outer race was built-up by undercutting and depositing the outer surface by submerged arc welding process. The BHN was maintained at 200. The hardness of the wheel out race was maintained below the harness of the embedded rails.
- The wheels were given a camber of 5 mm. This machining was done in CNC Lathe. This lathe was specially retrofitted by Kirloskar, Mysore for this purpose.
- The cylindrical pins used earlier had composition of ST-70-11. They were replaced with the new pins of composition 15 Cr 16 Ni 2. These pins had higher corrosion resistance because of high chromium content and higher hardness because of chromium and nickel. These pins were ground finished to the required tolerance.

Seal Friction

a. The other reason for nonclosure was seal friction. The seals used earlier for side seals and top seals were stainless steel cladded. The seal friction for the stainless cladded seals was to the tune of 5.3 tons. The weight of the gate was 20 tons. These seals imparted a seal friction of 20% of the gate weight. The seals were replaced by teflon cladded seals to reduce the seal friction.

b. The first act was to take the gates out of the slots and bring them out for repairs. It was decided to set up a central workshop in the island, almost midway between left and right spillway. The gates were to be brought to the central workshop from both the spillways. There were 40 gates in the left spillway and 24 gates in the right spillway. Each gate consisted of two parts weighting 10 MT each (approx).

The top pieces of the gate was taken out of the slot and dismantled in the slot. It was loaded to the trolley with the help of 50 T.E.O.T. Crane and the bottom part was held in the slot latching arrangements. Each segment of the gate was taken out by the EOT. Crane and placed in a trolley in a vertical position. There was provision of tracks for the trolley to cover the entire length of the gallery for catering to activity of each slot. The trolley was taken to the gallery exit. The direction of the trolley was changed by 90° by a rotary table located at the end of tracks. The face of the exit was perpendicular to the trolley track. These tracks terminated at a shorter distance out of the exit. So it was necessary to extend the tracks to accommodate the gantry systems.

The purpose of the gantry system was to unload the gate from the trolley, put the gate in flat position and load it on to trucks. One such gantry was set up in the left spillway and one in the right spillway. These gantries were also utilized to handle the repaired gates. In reverse operation the repaired gates were loaded to the trolley in a vertical position, taken to the respective slot and lowered by the 50 ton EOT crane.

In the island, a track network was done. eight number trolleys were made with swivelling wheels. Each trolley could take a half leaf. Each gate leaf had to go through series of activities. These activities were carried out in designated berths. The berths with sheds on top were at right angle to the main track and were earmarked to carry out specific activities like welding, wheel fixing, epoxy application, etc. The trolley with the gate leaf was positioned on the main track near the designated berth. The wheels of the trolley were swivelled to 90°.

The swivelling arrangement of the main track at the bottom of the wheel would swivelled that portion of the rail from the main track, so as to align it with the tracks of the berth. The trolley was then pushed sideways into the berth to carry out the designated activity. When the last wheels of the trolley crossed the main track, swivelling arrangement of the main track was swivelled back to the original position and aligned in the direction of main track to take other trolley movements. The main track had the swivelling arrangement in front of every berth.

The activities that were carried out in gates were as follows:

- **Wheel opening:** The first activity after bringing the gates to the Island, was opening of the wheels and axles. They were sent for the revamping as outlined earlier.

- **Sand blasting:** After opening of the wheels and axles gates were placed in vertical position in the trolley so that sand blasting could be carried out on both the sided of the gate leaf.

Over the years, several coats of coal tar epoxy had been applied on the gate leaf. Acting along with corrosion in the surface, it imparted a lot of resistance to cleaning by sand blasting. At places especially the corners, pneumatic chippers had to be utilized to remove coats of paint and rust.

- **Fixing of 8 mm plate in the web of horizontal girders and repair of horizontal and vertical girders:** The horizontal girders were strengthened by fixing of 8 mm plates on the web. Down hand welding was done by keeping the gate in vertical position.

Excessive corrosion in horizontal girders, vertical girders and skin plate was cause of concern. The horizontal girders co-acting with the skin plates are designed to transmit the water load to the vertical girders. The vertical girders through the wheel transfer the load to the embedded parts.

Under the water load this deflection of horizontal girders.

Under normal conditions, the deflection of the horizontal girder and the movement of the wheel are taken care of by the camber in the wheels. The wheels roll over the contact of the wheel face and the rail still remains as a line contact.

In case of unchecked corrosion, the section modulus of the horizontal girders goes on decreasing. This result in excessive deflection of the horizontal girders.

In case of excessive deflection the available camber in the wheel becomes insufficient.

This results in the wheel cutting into the rails surface. Additional resistance to lifting and seating of the gate is encountered.

If the gate is left to deteriorate, it may so happen that the excessive deflection of horizontal girders may result in jamming of the wheels on the rails and the gates may become defunct. So it was essential to carry out the strengthening of the horizontal girders and carry out welding in the skin plate and vertical girders.

- **Repair of skin plate by welding, grinding:** The next operation was to put the gate in flat position so that skin plate was exposed for restoring the pitted portions by welding and grinding. The gate leaf had to be moved under the gantry for this operation and then moved to the designated berth for welding work.

- **Fixing of axles and wheels:** The gate was turned upside down. In this condition, the balance pitting on horizontal girders, vertical were filled up by welding. Axles and wheels fitted in this position. A layer of coal tar epoxy was applied.

- **Application of epoxy and fixing of the seals:** The gates were tuned once more to put the skin plate up. A layer of epoxy mortar was applied. The materials used were supplied by M/s FOSROC (India) limited. The epoxy mortar provided a good resistance to corrosion. This added on to the dead weight of the gate. The gates were fitted with teflon cladded side seal and top seals and cleared for transpiration.

The gates were transported in trucks and unloaded by the gantries at the exit end of operation galleries and taken inside in trolleys to be lowered by cranes into the respective slots.

For completion of one half gate leaf, the gates had to be turned upside down several times. The entire operation of handling the gates in the Island was carried out by 8 nos. of Riggers. Each half gate leaf weighed approximately 10 MT. The gantries had manually operated winches which were effectively used for handling the gates.

- **Embedded parts:** The activity of repairing the pitting and checking the alignment of embedded parts was taken up as soon as the under sluice gate was taken out of the slot. Most of the top seal seat joints had been severely corroded. This was responsible for damaging the top seal and offered additional resistance to seating of the gates. Top seal plate was of stainless steel. Welding and grinding was done in this zone to ensure smooth movement of teflon cladded top seals.

- **Embedded parts of sluice gates:** It was observed by the dam authorities that misalignment in the embedded parts was also one of the reasons for the nonclosure of the gates. In some of the gates, the lifting beam could not be lowered and engaged to the gates. The guides had been misaligned and created obstruction for the lifting beam. In many places the bolts, which held the guide rails in position and were embedded in the second stage concreting, had snapped. These bolts had undergone shear failure by swelling of concrete.

It was suggested that the guide rails be welded to a base plate of adequate thickness. The studs be relocated by shifting its positions to the base plate. Additional holes be drilled in the concrete. The studs be anchored to the old concrete by dash fastener and the voids be filled with epoxy grout.

The gate grooves and sluice (about 1000 mm on upstream side and 1250 mm on downstream side) are lined with steel plate. The gate shaft is also lined in height of about 1000 mm above the top seal seat.

The steel claddings were sand blasted to clean the surface. The severely pitted portions were filled up by welding and grinding. Severe corrosion was observed at the joints of stainless steel seat and the steel cladding. These were repaired and a layer of epoxy was applied over the whole surface. However in some barrels these activities could not be carried out because of excessive seepage of water.

- **Emergency gate**

There are two nos. of emergency gates for each spillway. He overall size of the gate is 5.15 m (width) × 8.38 m (height). The gates are fixed wheel type having 16 wheels of 690 mm dia. The emergency gates are operated by 2 no.s of 100 m gantry cranes placed on both the spillway. The lifting speed of the crane is 1.2 m per minute and the travelling speed is 6.0 m per minute.

These emergency gates are operated by a lifting beam which lowers the gates on the bottom sills. As the gate on the seal the lifting beam gets unlatched by lever mechanism which is operated by an inclined fixture embedded into the concrete. This fixture is so positioned that it operated the lever only when the gate has reached the bottom. It was observed that operations through these fixtures were not dependable. So the operations of latching deglitching of the lifting beam were modified and were operated from the dam top by a string.

Underwater videography was done to assess the misalignment of the guide. A steel cage was fabricated and this was lowered from the dam top by a jjb-derrick arrangement. This was operated by electric winch and could be shifted from one emergency gate groove to the other by wheels. Sufficient counterweight arrangement was done for preventing over-toppling. The divers went down in the cage to assess the misalignment and obstructions. In the cage, a reference line was fixed from which the bulged concrete on the dam face was mapped. The alignment of embedded parts was taken with reference to a fixture. All the grooves were cleaned by underwater chipping. It was observed that, in the groove of 49 and 51 a few rods were protruding from the 2nd stage concreting. Underwater cutting of these rods was taken up to clean the groove. Underwater chipping of the dam face was also done to seat the emergency gate. No emergency gate had been lowered in these grooves in the last 44 years.

- **Gantry crane:** The emergency gates are lifted by gantry cranes of 100 MT capacity. Each spillway has one crane. The electrical contacts and components that were used in the panel boards are no more available in the market and need replacement. These cranes have hydraulic thruster brakes for its main hoist, auxiliary hoists and cross-travel. The brakes of longitudinal travel are foot operated hydraulic brakes. The operation of this brake is extremely essential considering the tall structure of the gantry crane. The revamping of the brake is extremely essential considering the tall structure of the gantry crane. The revamping of the brake system is essential. The present practice of braking is by reversing the direction of motor. This may result in severe damage of the gears, motors and the electrical controls. The other significant defect is in the gear box of the main hoist. During lifting, the motor transmits the torque to the gear train through a set of friction discs. At the time of lowering of the gates these discs get disengaged. These discs need urgent replacement. These can only be procured from the original equipment supplier. It was suggested to the dam authorities to modify the gear box so that the friction discs that are available in the market for earth moving applications be utilized in the crane. As the population of earthmoving equipment is high, the availability of spare parts is also good. Keeping the emergency gates and gantry crane in operational condition is absolutely necessary for the repair of the under sluice gates.

- **Radial crest gates:** The maximum discharge through the radial crest gate is 6 lacs cusecs. The skin plates were in good condition. This was scrapped and repaired with coal tar epoxy. The gates were fitted with hood plates to prevent spillage of water on the horizontal girders and end arms by wave action. All the components in the downstream of the gate were painted with enamel paint.

- **Suggestion:** The gates were taken out of these slots and made to lie in flat position for first time in its operation period of 48 years. The earlier maintenance was carried out by the dam Authorities. The maintenance consisted of greasing the wheels, replacement of cylindrical pins and grease nipples as and when required, replacement of side seals and top seals and painting with coal tar epoxy. These activities were carried out by taking out the gate partially from the slots and carrying out the repairs *in situ*. No major welding had been done for restoration of the pitted portions. It was not possible to carry out assessment and repairs *in situ* with gates partially hanging. All that could be done, in the existing circumstances, was being done. So it is essential now, that the gates be taken out of slots completely and put in horizontal position for inspection and proper maintenance. Periodical checking should be carried out by completely opening the wheel system and replacing the grease and worn out parts. Special emphasis should be given on the inspection and replacement of oil seal and 'O' rings. At least 13 gates should be done every year so that major overhauling can be done for each gate in a cycle of 5 years. This should be carried out over and above the regular maintenance schedule.

Note: The author is grateful to Mr KB Rajoria for guiding/helping in getting details of some part of the book (specially case studies).

THEORY QUESTIONS

1. How retrofit of a five storeyed building can be done explain through a case study?
2. How rehabilitation of double storied residential building at Mumbai was carried out? Explain with a case study.
3. Explain how rehabilitation of RCC overhead reservoir at Siliguri, WB was carried out.
4. Explain how assessment and retrofitting of masonry arch bridges can be carried out.
5. Explain how retrofitting/strengthening of typical masonry bridge can be carried out.
6. Explain how rehabilitation of road overbridge at bankim setu, WB was carried out.
7. Explain how repair/rehabilitation of jetty at Mumbai was carried out.
8. Explain how repair/rehabilitation of marine structures at Kandla port was executed.
9. Explain how structural reassessment of offshore platforms can be done.
10. Describe in detail how structural preservation of Lord Jagannath temple at Puri was carried out.
11. Explain how repair/rehabilitation of Hirakud dam structures was carried out.
12. Explain how repair/rehabilitation of upper Indrâvati hydroelectric project, Odisha was carried out.
13. What is the specialty of repair/rehabilitation of irrigation structures. How remedial measures for the gate of Hirakud dam was planned?
14. What was the problem of upper Indravati hydroelectric project and what rehabilitation measures were taken?

Bibliography

1. Aboutaha, R.S; Englehart, M.D; Jirsa, J.O. and Kreger, M.E. (1999), "Rehabilitation of Shear Critical Concrete Columns by Use of Rectangular Steel Jackets", *ACI Structural Journal*, American Concrete Institute, January-February, Vol. 96, No. 1, pp. 68–78.

2. ACI 440r–96, "Report on Fibre Reinforced Plastic Reinforcement for Concrete Structures", American Concrete Institute.

3. Alcocer, S.M. and Jirsa, J.O. (1993), "Strength of Reinforced Concrete Frame Connections Rehabilitated by Jacketing", *ACI Structural Journal*, American Concrete Institute, May-June Vol. 90, No. 3, pp. 249–261.

4. ATC 40 (1996), "Seismic Evaluation and Retrofit of Concrete Buildings: Vol. 1", Applied Technology Council, USA.

5. B.L.Gupta and Amita Gupta, *Concrete Technology*, Standard Publishers Distributors, Delhi, 2004.

6. B.L.Gupta and Amita Gupta, *Maintenance and Repair of Civil Structures*, Standard Publishers Distributors, Delhi.

7. B.S.Nayak, *A Manual on Maitenance Engineering*, Khanna Publishers.

8. Barnes, R.A, Baglin, P.S; Mays, G.C. and Subedi, N.K. (2001), "External Steel Plate Systems for the Shear Strengthening of RC Beams," *Engineering Structures*, Elsevier Publications, Vol. 23, pp. 1162–1176.

9. Basu, P.C. (2002), "Seismic Upgradation of Buildings: An Overview", *The Indian Concrete Journal*, The Associated Cement Companies Ltd; August, pp. 461–475.

10. Bowles, J.E. (2001), *Foundation Analysis and Design*, Fifth Edition, McGraw-Hill, Inc.

11. Dr.B.Vidivelli, *Rehabilitation of Concrete Structures*, Standard Publishers Distributors, 2007/2014.

12. FEMA 172 (1992), *NEHRP Handbook of Techniques for the Seismic Rehabilitation of Existing Buildings*, Building Seismic Safety Council, USA.

13. FIB Bulletin 24 (2003), *Seismic Assessment and Retrofit of Reinforced Concrete Buildings*, International Federation for Structural Concrete, Switzerland.

14. Ghobarah, A; Aziz, T.S. and Biddah, A. (1997), "Rehabilitation of Reinforced Concrete Frame Connections Using Corrugated Steel Jacketing," *ACI Structural Journal*, American Concrete Institute, May-June, Vol. 94, No. 3, pp. 283–294.

15. *Repairs and Rehabilitation of RCC Buildings*, CPWD, G.of I, New Delhi, 2002/2011.

16. *Seismic Retrofit of Buildings*, CPWD, IBC and IIT, Madras, by Narosa Publishing Pvt Ltd, 2008.

17. IS:13920-1993, "Indian Standard Code of Practice for Ductile Detailing of Reinforced Concrete Structures Subjected to Seismic Forces", Bureau of Indian Standards.

18. IS:13935-2009, "Guidelines on Seismic Evaluation, Repair and Strengthening of Masonry Buildings", Bureau of Indian Standards.

19. IS:13935-1993 (R-2003), "Guidelines for Repair and Seismic Strengthening of Buildings", Bureau of Indian Standards.

20. IS:1893-2002, "Criteria for Earthquake Resistant Design of Structures", Bureau of Indian Standards.

21. IS:1905-1987, "Code of Practice for Structural Use of Unreinforced Masonry," Bureau of Indian Standards.

22. IS:2911, "Indian Standard Code of Practice for Design and Construction of Pile Foundations", Parts 1 to 4, Bureau of Indian Standards.

23. IS:456-2000, "Plain and Reinforced Concrete Code of Practice", Bureau of Indian Standards.

24. IS:15988-2013, "Seismic Evaluation and Strengthening of Existing Reinforced Concrete Buildings Guidelines", Bureau of Indian Standards.

25. Jain, S.K. (2001), "Seismic Strengthening of Existing Reinforced Concrete Buildings, Repair and Rehabilitation", *Indian Concrete Journal*, The Associated Cement Companies Ltd; pp. 45–51.

26. Jhonson S.M., *Deterioration, Maintenance, and Repair of Structures*, McGraw- Hills Publishing Company Ltd., New York, 1965.

27. K.B. Rajoria, Ashok Basa, The Institute of Engineers(I), *Rehabilitation and Retrofitting of Structures*, Macmillan Publishers India Pvt Ltd, 2010.

28. M.S. Shetty, *Concrete Technology-Theory and Practice*, S. Chand and Company Ltd., New Delhi, 2005.

29. Manual for Seismic Evaluation and Retrofit of Multistoreyed RC Buildings (2005), Indian Institute of Technology Madras and Structural Engineering Research Centre, Chennai. Project sponsored by Department of Science and Technology.

30. Mukherjee A. and Joshi M.V. (2002), "Seismic Retrofitting Techniques using Fibre Composites", *The Indian Concrete Journal*, The Associated Cement Companies Ltd; August, Vol. 76, No. 8 pp. 496–502.

31. Mukherjee A; Kalyani A.R. and Joshi M.V. (2004), "Upgradation of RCC Frames with FRC-I, Design of Elements", *The Indian Concrete Journal*, The Associated Cement Companies Ltd; October, Vol. 78, No. 10, pp. 15–19.

32. Mukherjee A; Kalyani A.R. and Joshi M.V. (2004), "Upgradation of RCC Frames with FRC-II, Design of Structures", *The Indian Concrete Journal*, The Associated Cement Companies Ltd; October, Vol. 78 No. 10, pp. 22–25.

33. Murty C.V.R.; Goel R.K. and Goyal A. (2002), "Reinforced Concrete Structures", Earthquake Spectra, Earthquake Engineering Research Institute, USA July, Supplement A to Vol. 18, pp. 145–185.

34. P.C.Varghese, *Maintenance, Repair and Rehabilitation and Minor Works of Buildings*, PHI Learning Pvt. Ltd, 2014.

35. P.S.Gahlot, Sanja Khanna, *Building Repair and Maintenance Management*, CBS Publishers and Distributors Pvt.Ltd., 2006/2014.

36. Proceedings of National Conference in Civil Engineering Materials and Structures, January 19–21, 1995, Osmania University, AP, India.

37. R. Dodge. Wodson, Concrete Structures, Protection, Repair and Maintenance.

38. R.C. Misra, K. Pathak, *Maintenance Engineering and Management*, PHI Learning Pvt.Ltd., 2009.

39. Rodriguez M and Park. R (1994), "Seismic Load Test on Reinforced Concrete Columns Strengthened by Jacketing," *ACI Structural Journal*, Vol.91, No. 2, April-March, pp. 150–159.

40. Santhakumar, A.R. (2006), *Concrete Technology*, Oxford IBH Publication, New Delhi.

41. Seth, A. (2002), "Seismic Retrofitting by Conventional Methods", *The Indian Concrete Journal,* The Associated Cement Companies Ltd; August, pp. 489–495.

42. Stoppenhagen, D.R.; Jirsa, J.O. and Wyllie, Jr; L.A. (1995), "Seismic Repair and Strengthening of a Severely Damaged Concrete Frame", *ACI Structural Journal,* American Concrete Institute, March-April, Vol. 92, No. 2, pp. 177–187.

43. Suriya Prakash S. and Alagusun Daramoorthy, P. (2007), "Experimental Study on Masonry Wallettes and Shear Triplets Externally Bonded with Glass Fibre Fabric", *Journal of the Institution of Engineers.*

44. Tomlinson, M.J. (1981), *Foundation Design and Construction,* Pitman Books Ltd. London.

45. Ziraba Y.N.; Baluch M.H; Basunbul I.A; Sharif A.M; Azad A.K. and Al-Sulaimani G.J. (1994), "Guidelines Toward the Design of RC Beams with External plates", *ACI Structural Journal,* American Concrete Institute, Nov-Dec; Vol. 91, No. 6, pp. 639–646.

Index

A

Abrasion 13, 42, 148
    resistance 80
Accelerators 137
Accidental loading 17, 39, 45
Acid
    attack 7, 20, 40
    etching 189, 200
Acoustic emission 61
Acrylics 119, 144
Active cracks 183
Additives 153
Adhesive strength 165
Admixtures 140
Advanced
    permeability 73
    resistivity 78
    testing 76
Aggressive water 7, 21
Air permeability 74
Alkali
    reaction 20
    silica reaction 40, 81
Alkalinity 25
Anchors 202
Anticarbonation 189
Aramid fibers 151
Arch bridges 298
Archeological reconstruction 234
Arches 233
ASR detect 81
Autogenous
    healing 187
    shrinkage 10

B

Base isolation 282, 284
Basic principles of corrosion 24
Beam retrofitting 251
Beams 214, 240
Bentonite 128
Biological attack 22
Bituminous
    coatings 118
    materials 115
Blanketing 186
Blast cleaning 201
Blasting 182, 192, 196, 198
Bond strength 165
Bondage 230
Bonding
    agent 132, 202
    materials 131, 175, 186
Butyl sealant 124

C

Calcium sulphate 115
Capo test 69
Carbon fibers 151
Carbonation 19, 293
    shrinkage 8
    test 292
Cavitation 8, 13, 38, 40, 42
Cement based coatings 128
Cementitious materials 112
Chemical
    admixtures 109, 135
    attacks 40
    cleaning 201
    reactions 18, 39

Chloride
    attacks 18, 19
    content 79
Coal tar epoxy 119
Codal provisions 31
Coefficient of thermal expansion 165
Column 239, 240
    retrofit 247
Composites 217
Compression testing 70
Compressive strength 175
Concrete jacketing 247
Condition survey 51
Cooling 11
Core drilling 38
Corrosion 6, 24, 26, 38, 41, 45
    damage 84
    errors 7, 45
    inhibitors 4
    mechanism 61
    process 26
    protection 32
Cover 5, 31
Crack resistance 4
    surveys 37
Cracking 35, 38
Cracks 16, 17
Crazing 12
Creep 13, 176
Crushing 189
Crystallization 21
CSF 139
Curing 5
Cutting 199
Cyclic loading 147

D

Dam safety 220
Damages 30
Deep foundation 258
Deficiencies of foundation 258
Deflection 14
Deformation 14
Design
    errors 7, 16, 45
    mistakes 41
Destructive testing 94, 101
Detailed investigation 90
Diamond cutting 193, 194
Diaphragm 224
Differential thermal analysis 92
Dilatometric test 92
Direct load test 91
Discoloration 14, 37

Disintegration 36
Distortion 7, 36, 38
Domes 233
Drilling 187
Dry
    pack method 221
    packing 186
Drying shrinkage 8, 9
Durability 5
Durable repair 181
Dusting 36
Dynamic method 56

E

Efflorescence 20, 37
Electric current 18
Electrical
    methods 60
    process 26
    resistivity 25
Electrochemical process 26
Electromagnetic method 58
Energy dissipation 246
Epoxide resins 142
Epoxies 119, 125, 145
Epoxy
    coatings 141
    lattices 133
    mortars 141
    resins 120, 140
Erosion 38
Essential parameters of coating 110
Evaluation of cracks 100
External
    post-tensioning 218
    reinforcement 217
Externally bonding technique 215
Exudations 37, 40

F

Fatigue strength 147
Ferrocement 154
Fiber
    reinforced concrete 123, 146
    reinforced grouts 116
    reinforced polymer 152, 251, 274
Fire 8, 40
    effect 14
Flexible sealing 187
Fly ash 138
Formwork movement 15
Foundation 233, 241
    settlement 15

Fracture test 68
Freeze and thaw 10, 19, 40, 43, 45
FRP 153
    wrapping 253
    composites 152

G

General
    classification of concrete removal methods 191
    guidelines for seismic rehabilitation
      of existing buildings 222
    remarks 255
Geology 337
GFRC 149
GGBS 133, 139
Glass fiber 121, 149
Global
    deficiencies 237
    retrofit 243
    stiffness 223
    strength 222, 230
Grinding 189
Grouting 183
Grouts 115
Growing cracks 183
Gunite 206

H

Half-cell potential 75, 292
Heating 11
Heritage structures 232
High alumina cement mortar 123
Hirakud dam 334
Historical data 38
Honeycombing 8, 12, 38
Hydro milling 200
Hydrolysis 20

I

Impacting 191
Implemented case 270
Indirect test 91
Infill walls 243
Investigation
    strategies 85
    of damage 98
Isolation cracks 35

J

Jacketing 215
Jetty 313

L

Lar dam 337
Latex 132
    coatings 145
    emulsions 132
    modified mortars 113
Limpet test 69
Liquid floor hardeners 122
Load path 223
Local
    deficiencies 239
    retrofit 246
Low heat cement 15

M

Magnetic method 58
Major causes for distress in concrete 6
Masonry 251
Mastics 118, 124
Mechanical cleaners 241
Membranes 129
Metallic
    aggregate grouts 116
    dampers 285
Methods of
    execution 266
    repair 182
    underpinning 267
Micropiles 261
Milling 192
Mineral acids 18
Mix design 5
Modulus of elasticity 175
Molten sulfur 115

N

Natural fibers 150
Needle beams 266
Neoprene 119
Nondestructive testing 51
Nonengineered structures 227
Nuclear method 58

O

Observation 98
Organic polymers 142
Overlays 122, 186

P

Patch repair 308
Patching materials 112

Pattern cracking 35
Penetration
    resistance method 66
    techniques 65
Permeability 176
    test 72
Petrographic techniques 71
pH value 164
PIC 144, 146
Plan irregularities 237
Plastic
    cracks 9
    settlement cracks 9
    shrinkage 8
PMCC 143
Polyesters 118
    resins 113, 143
Polymer 142
    concrete 113
    grouts 116
Polymeric
    coatings 144
    materials 141
Polypropylene 150
Polyurethanes 142
Pop outs 12, 38
Porosity 18
Pozzolana cement 15
Precast members 242
Prefabricated sheet membranes 130
Prep work 201
Presplitting 200
Probes 67
Propagation period 26
Protective coatings 117
Pulse
    attenuation method 56
    echo method 57
    velocity 54

Q

Quartz 14
Questioning 98

R

Radar technique 61
Radioactive method 57
Radiography 61
Rebar data scan 60
Rebar locator 59
Rebound hammer 292
Rehabilitation 293, 302, 309, 319, 321
Removal methods 191

Repair
    materials 110
    procedure 180
    techniques 228
Resin 153
Resistivity mapping 78
Resonant frequency 56
Resurfacing 189
Retrofit strategy 243, 291

S

Salt attack 18, 20
Sand
    blasting 189
    mixtures 115
Saturation 6
SBR 132
Scalling 36, 37, 38
Sealants 124
Sealing 183
    materials 124
Section enlargement 219
Seepage 36, 37
Seismic vulnerability 222
Semidestructive testing system 65
Settlement 15, 43
Shake hardeners 122
Shoring 266
Short spanning 218
Shot blasting 201
Shotcrete 228
Shrinkage 8, 40, 43
Sifcon 156
Silanes 145
Silicone 125
Simcon 156, 157
Spalling 37
Splitter 201
Steel
    braces 245
    jacketing 250
Stitch
    cutting 199
    drilling 194
Stitching 184, 229
Strengthening 219
Strip foundation 262
Structural
    concrete strengthening 213
    integrity 85
    preservation 328
    reassessment 325
Subgrade movement 15

Sulfate attack 40, 41
Superplasticizers 135
Surface
    coatings 188, 203, 211
    hardeners 121
    mapping 38
    permeability 74
Swelling 12

T

Temperature
    change 40, 44
    gradient 18
    variation 13
Thermal
    cutting 199
    expansion 178
    movement 14
Transition zone 5
Tuned devices 286

U

UPV test 53
Uranyl acetate 233, 276, 278, 280, 282
Urethanes 118, 144

V

Vacuum impregnation 187
Vertical irregularities 238
Vibration method 56
Viscous dampers 285

W

Water–cement ratio 84
Water permeability 74
Waterproofing materials 128
Weathering 8, 11, 40
Wetting 11
Windsor probe 66
Workability 5